# ANTARCTIC WHALING

# ANTARCTIC WHALING

## A Case Study in Near Extinction

John Sheail, Paul Rodhouse and John Dudeney

5m Books

First published 2024



Published by
5M Books Ltd
Lings, Great Easton
Essex CM6 2HH, UK
Tel: +44 (0)330 1333 580
www.5mbooks.com

Follow us on
Twitter @5m_Books
Instagram 5m_books
Facebook @5mBooks
LinkedIn @5mbooks

A Catalogue record for this book is available from the British Library

ISBN 978-1-78918-241-5
eISBN 978-1-78918-345-0
DOI 10.52517/9781789183450

Book layout by Cheshire Typesetting Ltd, Cuddington, Cheshire
Printed by CPI Antony Rowe Ltd, UK
Photos by the authors unless otherwise indicated
Front cover: Detail from an oil painting by Paul Rodhouse, Albatross and Dias at Grytviken, South Georgia

For Gillian, Laura and Helen

# *Contents*


**Foreword** **x**

**Preface** **xii**

**Acknowledgements** **xv**

**Abbreviations** **xvi**

**PART I THE HISTORICAL CONTEXT**

**1 Introduction** **3**

**2 Pursuit of the rorqual** **10**

**PART II THE BRITISH/NORWEGIAN ASCENDANCY, 1900–25**

**3 Southern whaling** **17**

3.1 Confused beginnings 23
3.2 The first four leases 27
3.3 Licensing the floating factories 30
3.4 A British lease 33
3.5 Legitimising regulation 36
3.6 A second British lease 42

**4 Maximising productivity** **45**

4.1 A resident magistrate 48
4.2 The first Antarctic magistrate 52
4.3 Towards closer regulation 55
4.4 The further ordinances 60
4.5 The political dimension 63

| | | |
|---|---|---|
| **5** | **The safeguarding of whale stocks** | **68** |
| 5.1 | Foreign Office enquiry | 73 |
| 5.2 | The Whaling and Protection of Whales Committee | 76 |
| 5.3 | The wider fisheries context | 80 |
| **6** | **Weapon of war** | **84** |
| 6.1 | The Dannevig Pledge | 86 |
| 6.2 | Salvesens – 'the sturdy beggars' | 88 |
| 6.3 | The wartime catches | 91 |
| **7** | **Research and the development of the Falkland Islands** | **95** |
| 7.1 | Realising the colony's potential | 97 |
| 7.2 | The committee's findings | 100 |
| **8** | **The heyday of Dependencies whaling** | **105** |
| 8.1 | Local direction | 107 |
| 8.2 | Fraudulence | 111 |
| 8.3 | The Ross Sea | 113 |

**PART III PELAGIC WHALING, 1925–50**

| | | |
|---|---|---|
| **9** | **Freedom of the seas** | **119** |
| 9.1 | 'Ice-front whaling' | 122 |
| 9.2 | The diplomatic response | 125 |
| 9.3 | The erosion of authority | 129 |
| **10** | **Market collapse** | **133** |
| 10.1 | Re-appraisal | 137 |
| **11** | **Market recovery** | **140** |
| 11.1 | Quota restrictions | 143 |
| 11.2 | The Norwegian boycott | 146 |
| 11.3 | The Japanese and German whaling fleets | 149 |

**PART IV THE DISCOVERY INVESTIGATIONS**

| | | |
|---|---|---|
| **12** | **Establishing the Discovery Investigations** | **153** |
| 12.1 | The Discovery Committee | 157 |
| 12.2 | The despatch of *Discovery* | 160 |
| 12.3 | The initial Investigations | 164 |
| 12.4 | Challenged by the Governor | 169 |
| 12.5 | Challenged by the Secretary of State | 174 |
| 12.6 | The near-demise of the Investigations | 178 |

**13 International regulation** **182**
13.1 The International Whaling Convention 185
13.2 The Whaling Industry (Regulation) Act 189

**14 The London Conferences** **192**
14.1 The International Whaling Agreement 1937 194
14.2 The International Whaling Protocol 1938 197
14.3 The informal London meeting 1939 201

**15 The further disruption of war** **205**
15.1 Readiness for peace 209
15.2 Recovery amidst crisis 212
15.3 The wartime protocols 213
15.4 Rebuilding the fleet 217

**16 The United States – the superpower** **220**
16.1 The Washington Conference 223
16.2 The International Whaling Commission 227
16.3 A post-war German whaling fleet? 230
16.4 Allied dissension 232
16.5 The resumption of Japanese whaling 236

**PART V A SCIENTIFIC OVERVIEW**

**17 The Discovery Investigations – scientific opportunities, successes and failures** **245**
17.1 Whale science, 1925–42 249
17.2 The investigations post-Second World War 258
17.3 An appraisal of the Applied Whale Research 260
17.4 The wider science of the Investigations 267
17.5 A twenty-first century perspective 271

**18 Epilogue** **279**

**Appendix: Biographical notes on key scientific staff of the Discovery Investigations** **282**

**Sources: Manuscript material in The National Archives** **286**

**Bibliography** **295**

**Brief author biographies** **313**

**Index** **314**

# *Foreword*

This is a compendious examination and interpretation of the complex circumstances of the Southern Ocean, revealing the consequences of the profound human effects of the over-exploitation of its biological resources. The preface notes grave anthropogenic perturbations in the ecology of the far southern regions. Whereas massive over-exploitation of two species of Antarctic seals during the 1800s had major consequences, this was far less than those following the vast reduction in populations of whales caught during much of the twentieth century. Fur seal populations have largely recovered, and elephant seal numbers are now relatively stable after a long period of scientific management. The many species of whales, all relatively slow breeders, are eventually showing some early indications of recovery. In contrast, the balance of the food web of the oceans remains so profoundly affected that the consequences are far-reaching, indeed influencing global systems.

Understanding of the complex dynamics of the oceanic ecosystem has developed slowly over a century of research. The book presents a detailed history of the acquisition of scientific knowledge of the Southern Ocean in general. It places particular emphasis on the science and the powerful influences of economics and politics. These last two factors were demonstrably at variance with what may be called 'safe stewardship', an issue not generally appreciated until it was far too late to limit devastation and its consequences. In addition, the distinct reasons for which many, essentially individual, opinions were formed are elucidated.

The vast diminution of whale populations during the twentieth century was a tragedy not only for the whales but also for the Southern Ocean ecosystem. Ramifications, from primary production of phytoplankton to the biodynamics of the largest of animals, have made a major, but lesser-known, contribution to global climate change. The role of whales in nutrient cycling and sequestration of carbon have become appreciated only recently. These are additional reasons for scientific investigation of the recovery of the ecosystem and the changes in ecosystem dynamics that may be expected to occur. Whale numbers are currently increasing slowly, an indication of slow and ponderous ecosystem recovery.

Many factors were involved: dichotomies between various countries and international relationships, private enterprise and government, scientific research and administration, exploitation and conservation. The book looks at the period from the beginning of modern Antarctic whaling in 1904 to the establishment of the International Whaling Commission in 1946 and the ramifications to the present, relating them to scientific developments and political upheavals at the time. Thus, the historical correlation with world wars, economic depression, and industrial development is emphasised. A historical review of these events is very timely.

The structure of the book is appropriate for comprehending the diverse, but interdependent, factors involved. The five sections, each with subdivisions, efficiently describe the historical origin and rapid evolution of circumstances. The book is opportune; the chronology of the development of the Southern Ocean, with its devastating history of whaling and other exploitation, are major current concerns. The analytical value of hindsight due to modern knowledge is immense; whatever foresight can be applied will lead to difficult predictions of the future for the Southern Ocean and the vast regions affected by, and dependent on, it.

R. K. Headland

# *Preface*

Something remarkable is happening in the Southern Ocean. The great whales are recovering from the near extinction caused by the gross over-exploitation of stocks in the greater part of the twentieth century. Their destruction was a tragedy both for the whales and for the whole ecosystem of the Southern Ocean. It may indeed have been a contributory factor to climate change on a global scale by removing the whales' role in fertilizing the ocean and thereby sequestrating the carbon. It is a further reason for scientists to study the present recovery of stocks for the changes made to the ecosystem, as the individual populations increase.

This book's purpose is to provide an historical context to the present-day recovery process. It does so by looking at how the southern whaling industry developed and was administered, and latterly studied scientifically. How significant were the various attempts to manage stocks more sustainably, and why did they fail? The book breaks new ground by using the archival record to show how the United Kingdom Government sought to administer what, in the first quarter of the twentieth century, was a largely Norwegian industry operating from British colonial waters. The move towards a largely pelagic industry in the second quarter of the century called for more explicitly international forms of regulation and, by 1950, the appointment of an International Whaling Commission. The book, in focusing on the first half-century of southern whaling, has two distinct but interlinked strands: the political and the scientific, drawing on the archival record now preserved in The National Archives at Kew in southwest London.

John Sheail is the lead author for Parts I to IV and Paul Rodhouse the lead author for Part V. The book contains two distinct types of writing, historical narrative (Parts I–IV) and scientific review (Part V) each with their distinct conventions.

Part I summarises the rise and fall of previous whaling industries over some 900 years of activity and the emergence of a rudimentary knowledge of marine life.

Part II focuses upon the British/Norwegian ascendancy of Southern whaling in the first quarter of the twentieth century. It follows more particularly when and where, how and why, the Colonial Government at Stanley in the Falkland Islands and the Imperial Government in London played so large a part in what remained a largely Norwegian industry in the waters off South Georgia and the other archipelagos of the Falkland Islands Dependencies.

Part III of the book covers the second quarter, 1925–1950, during which time pelagic whaling, freed from having to use any particular port or territorial water, ranged across the Southern Ocean. Largely unregulated, there appeared to be no limit to the volume of whale oil which might be produced. Two factors soon intervened, the most immediate being the collapse in whale oil prices as a direct result of over-supply and secondly a marked weakening of market demand for whale-oil products. Part Three accordingly covers both *Det Stille Aret* (the Silent Year of 1931–1932) when the Norwegian fleet never put to sea, and the various endeavours made by the companies to agree a quota system by which oil prices might recover. It was a recovery process marred by both the fear of unemployment among the Norwegian workforce and entry of German and Japanese companies.

Part IV of the book focuses more explicitly upon the surveys and research undertaken. Both the Norwegian and United Kingdom governments had quickly recognised the need for much greater knowledge of the effects of whaling upon stocks. The outcome was the Discovery Investigations. The book continues to make close reference to the archival record in recounting how the Investigations developed – financed as they were from an oil tax levied upon the whaling companies. It was through a dialogue between government officials and scientists, and more particularly those of the Discovery Investigations and the International Council for the Exploration of the Sea (ICES), that a demonstrably international basis began to emerge for regulating what had become a largely pelagic industry, namely the International Whaling Convention of 1934 and its later amendment by the London Agreement and Protocol of the later 1930s.

The Second World War adversely affected the industry, first by its being too dangerous for the whaling industry to operate and then its hurried re-establishment to meet the acute post-war shortages of oils and fats. Whaling historians have highlighted the United States' role in founding what became the International Whaling Commission. Much less attention has been given to the role similarly played by the United States Government in re-establishing a post-war Japanese whaling industry.

The concluding Part V of this book appraises what proved to be the United Kingdom Government's most significant contribution to whaling history – the Discovery Investigations. Although the Second World War had prevented further practical work from being undertaken, the Investigations survived and became, in the later 1940s, wholly government financed. They became the most major and sustained study of cetaceans and their habitats ever to have been attempted. Although the phrase 'the precautionary principle' had not yet been coined, the Investigations had encapsulated, from the start, the foresight and concern increasingly shown by those scientists employed over the course of a half-century or so.

The initial regulations succeeded in raising productivity through limiting the size of the industry, but there was never enough knowledge and understanding needed to prevent the near extinction of the humpback whale and an ever increasing threat to stocks of blue and fin whale. The basic tools for doing so were lacking. There was neither a theory of whale population dynamics nor an effective method by which changes in population size could be assessed. The Investigations never succeeded in reducing the pace of slaughter. The whaling companies took no direct part, their presuming the annual whaling expeditions would have become far too costly to mount well before stocks were so diminished as to imperil the survival of the species involved.

**References to material cited from The National Archives will be found in the Sources section and the published works in the Bibliography. The book, as a diplomatic history, follows the practice of the thousands of individual documents in The National Archives in capitalising the names of particular territories and posts of political and administrative responsibility as the most concise way of highlighting their importance.**

**Citations of material published in Parliamentary Papers are denoted by 'Parliamentary Debates' and to the published records of Foreign Relations of the United States by the initials 'FRUS'.**

**Monetary values, weights and measures are expressed in the form used in the relevant documentation**.

John Sheail, Paul Rodhouse and John Dudeney
New Year's Day 2024

# *Acknowledgements*

The authors are deeply indebted to the many who have helped and encouraged us in the writing of this book. We particularly acknowledge the support of colleagues at the British Antarctic Survey and, very obviously, the staff of the Search Room of The National Archives. Nigel Balmforth, Alessandro Pasini and Jeremy Toynbee at 5m Books encouraged and patiently guided our progress. Bonnie Pickard at British Antarctic Survey MAGIC assisted with preparation of maps. Josh Holder at BAS provided IT support. Ieuan Hopkins, BAS Archives, assisted our search for images of early Antarctic whaling. We are deeply grateful for the copyright consent to publish from this archive. Peter Burkill, retired director of the Sir Alister Hardy Foundation for Ocean Science, and Victor Smetacek, Alfred Wegener Institute, advised on some current and relevant biological oceanographic research. Mark Belchier advised on current CCAMLR activity. Richard Barling read and provided most helpful criticism on a draft of the text. We are particularly grateful to Bob Headland for the foreword which he has written.

John Sheail holds a Research Fellowship of the Centre for Ecology and Hydrology, and Paul Rodhouse an Emeritus Fellowship of the British Antarctic Survey, both of which are component bodies of the UK Natural Environment Research Council. John Dudeney is former Deputy Director of the British Antarctic Survey.

Much of the text was written during successive COVID-19 pandemic lockdowns, and would not have been possible without access to the US Coast Guard copy of the *Discovery Reports* held at the Library of the Marine Biological Laboratory/Woods Hole Oceanographic Institution and made available via the online Biodiversity Heritage Library https://www.biodiversitylibrary.org/.

# *Abbreviations*

| | |
|---|---|
| BANZARE | British Australian New Zealand Antarctic Research Expedition |
| BMNH | British Museum, Natural History |
| BWU | blue whale unit |
| CBE | Commander of the British Empire |
| CCAMLR | Commission for the Conservation of Antarctic Marine Living Resources |
| CPR | continuous plankton recorder |
| DI | Discovery Investigations |
| DSIR | Department of Scientific and Industrial Research |
| EBFM | ecosystem-based fishery management |
| FAO | Food and Agriculture Organisation |
| FRS | Fellow of the Royal Society of London |
| GLG | growth layer groups |
| HMS | His/Her Majesty's Ship |
| ICES | International Council for the Exploration of the Sea |
| IEFO | International Emergency Food Organisation |
| IOS | Institute of Oceanographic Sciences |
| IUCN | International Union for the Conservation of Nature |
| IWC | International Whaling Commission |
| KEP | King Edward Point |
| MAGIC | Mapping and Geographic Information Centre |
| MBA | Marine Biological Association |
| MSY | maximum sustainable yield |
| NIO | National Institute of Oceanography |
| NOC | National Oceanography Centre |
| PD | Parliamentary Debates |
| RN | Royal Navy |
| RNVR | Royal Naval Volunteer Reserve |
| RRS | Royal Research Ship |
| SCAP | Supreme Commander for the Allied Powers |
| SHAEF | Supreme Headquarters Allied Expeditionary Forces |
| SSO | senior scientific officer |

# ANTARCTIC WHALING

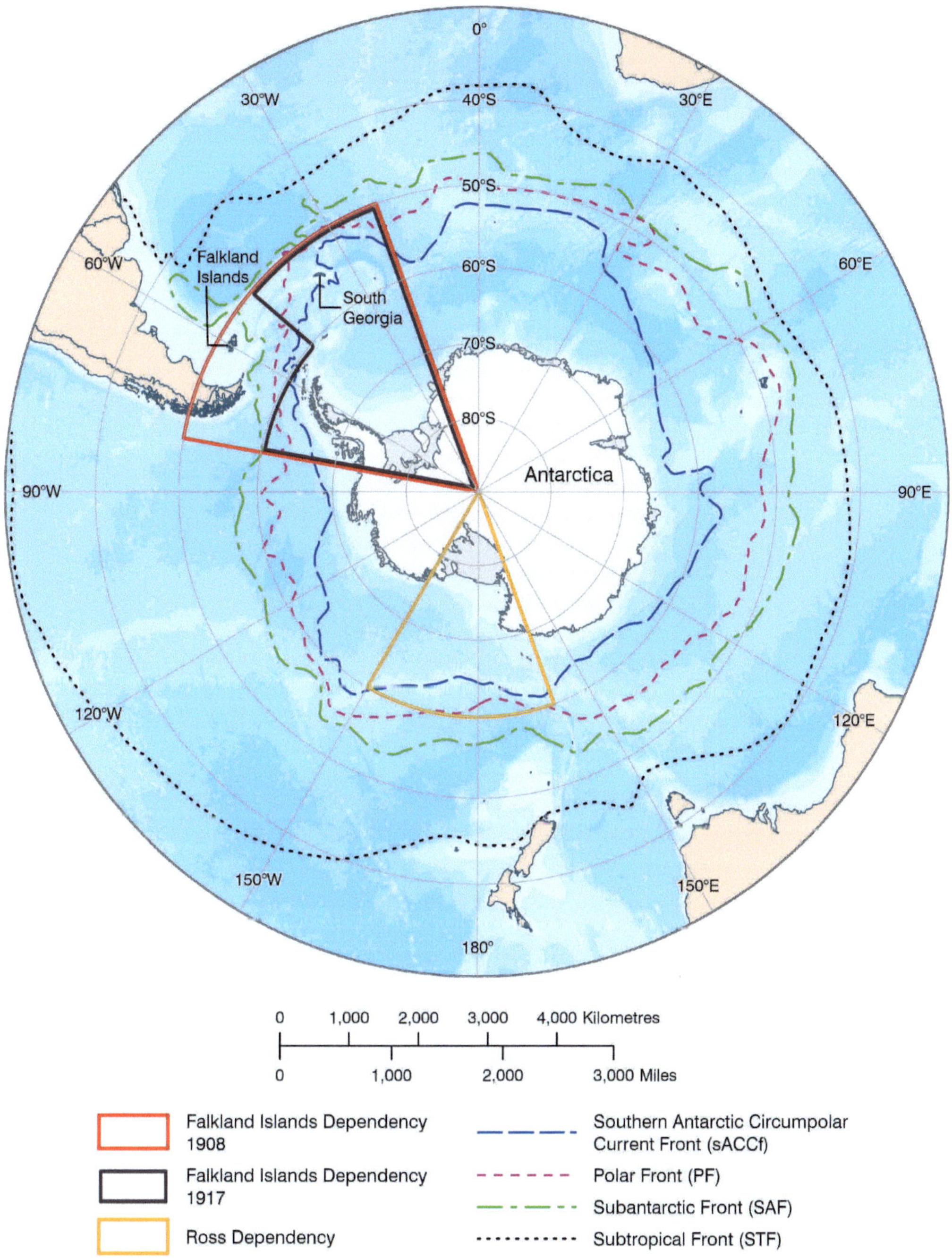


**Figure 1** A map showing the Antarctic continent and Southern Ocean with the main oceanographic boundaries of the Southern Ocean indicated as well as two geopolitical boundaries of relevance to the narrative. Polar fronts taken from South Georgia GIS, accessed 2023. Subtropical front extracted from 'Orsi, AH., Harris, U. (2019) Fronts of the Antarctic Circumpolar Current – GIS data, Ver. 1, *Australian Antarctic Data Centre*. https://data.aad.gov.au/metadata/records/antarctic_circumpolar_current_fronts, Accessed: 2023-02-14'. © British Antarctic Survey, UKRI.

# PART I

## *The historical context*

CHAPTER I

# *Introduction*

'Whaling provides one of the most salutary instances of failure in resource management.' Michael Graham wrote those words in his contributory chapter to a monumental international conspectus, *Man's Role in Changing the Face of the Earth,* published as a symposium volume in Chicago in 1956. Graham was Director of the Fishery Research Laboratory at Lowestoft in England, and the history of whaling was for him a sad one; it would become a tragedy if it ended 'with these last resources being cut off altogether from commercial exploitation because stock levels had become too low for the fishery to be profitable'. There was no other example of 'the world-scale production of the ocean' being so concentrated as 'in the body of a whale'. Graham's words continue to have a resonance in the present-day studies of how stocks are recovering and their significance for environmental wellbeing.

Whales are mammals, but the late twentieth-century, quantitative science that was applied to their conservation and management (Allen 1980) was derived largely from wider studies of fish population dynamics. Some of the key players in fisheries science, including Kenneth Radway Allen, Douglas Chapman and Sidney Holt, achieved importance in the conservation and management of whale stocks. The development of marine biology and fisheries science during the nineteenth century provides the context in which to follow the course of British involvement in Antarctic whaling during the twentieth century.

Early scientific studies of the open sea were made on the voyages of Captain James Cook between 1768 and 1780, with the naturalists Joseph Banks and Daniel Solander, aided by the artist Sydney Parkinson, on the first voyage aboard HMS *Endeavour* and Johann and Georg Forster on the second voyage in HMS *Resolution.* Progress paused with the Napoleonic Wars, but these scientific studies resumed through the practice of the British Admiralty attaching naturalists to its survey vessels. Charles Darwin joined HMS *Beagle* in 1831 with the most far-reaching consequences for biological science. Thomas Henry Huxley sailed on HMS *Rattlesnake* in 1846, both as ship's surgeon and naturalist. Joseph Dalton Hooker was part of the James Clark Ross Antarctic expedition of 1839–43 with HMS *Erebus* and HMS *Terror.* Edward Forbes joined HMS *Beacon* in surveying the Aegean, having already done much biological dredging around the British Isles. Charles Wyville Thomson succeeded, with the Royal Society, in persuading the Admiralty to provide HMS *Lightning* in 1868 and HMS *Porcupine* in 1869–70 for biological cruises off the Faroes, the West Coast of Scotland, the Bay of Biscay and the Mediterranean.

It was again Thomson, this time with William B. Carpenter, who encouraged the Admiralty, again through the Royal Society, to combine a circumnavigation of the world with the first deep-sea research expedition from aboard HMS *Challenger*. The 50 large volumes chronicling the voyage, edited by Thomson, became the foundation of the new, interdisciplinary science of oceanography. There is, however, only one mention of cetaceans, and then only about whale bones dredged from the seabed (Turner 1880). Sir John Murray succeeded Thomson as the leading figure in British marine biology. Among many activities, he collaborated with the Norwegian marine biologist Johan Hjort and others in writing *The Depths of the Ocean* (1912), describing their joint investigations of the North Atlantic aboard the Norwegian vessel *Michael Sars* (Murray and Hjort 1912). Hjort was appointed to the chair of marine biology at the University of Kristiana in 1921, where he became closely involved in whale biology and the whaling industry generally.

The nineteenth century was one of burgeoning public interest in marine biology, stimulated by several British naturalists (Yonge 1949). Most notable among these was Philip Henry Gosse who pioneered the use of aquaria and stocked the first public aquarium at the London Zoo in 1853. Marine laboratories were opened by the French at Concarneau (1859), Arachon (1863) and Roscoff (1871), and the zoological laboratory founded by Anton Dohrn (the German zoologist) was opened at Naples in 1872. The first British laboratories were at Plymouth (1884) and St. Andrews (1896), followed by others at Aberdeen, Cullercoates, Lowestoft, Millport and Port Erin. The United States' Woods Hole Marine Biological Laboratory was opened in 1888. Prince Albert I of Monaco founded the Institute Oceanographique at Monaco and in 1898 began financing four oceanographic expeditions, two of which were voyages to the Arctic which included William Speirs Bruce.

The study of whale biology can be traced back to ancient Greece. Aristotle (384–22 BCE), in his study of the eastern Mediterranean, observed that whales had red blood, breathed air and suckled their young. Describing them as 'blooded animals', he coined the term 'cetacean' (Romero, 2012). The Swede, Karl Linnaeus (1707–78), identified several baleen and toothed whales: the Greenland, blue, fin, killer and sperm whales, as well as the narwhal, harbour porpoise and common dolphin. The study of whale biology throughout the nineteenth century continued to be largely anatomical, based on the skeletons of stranded animals (Flower 1864). As with so much of nineteenth-century zoology, it was purely descriptive.

Johan Hjort remarked, as late as the 1920s, that no other group of mammals was so little known as whales. However much the individual whalers might have learnt of their quarry, there remained a dearth of scientific observation made of whales in their natural element. The Prince of Monaco and William Speirs Bruce had been exceptional in studying specimens they themselves had actually caught on their expeditions to the Arctic.

There had been a scientific interest in fish for centuries. Francis Willoughby's comprehensive treatise on fishes of the Mediterranean and North Atlantic was published in 1686, ten years after the author's death. The first comprehensive work focusing on British fish species was written by Thomas Pennant (1726–98). Then in the mid-nineteenth century, William Yarrell published his *History of British Fishes*. This went far beyond illustration and description to encompass aspects of anatomy, behaviour and the ecology of freshwater and marine fish. Both Frank Buckland (1826–80) in Britain and Spencer Fullerton Baird (1823–87) in America recognised the increasing consequences of fishing on stocks and called for scientific study of its effects. Buckland was a practical, if somewhat eccentric, natural historian. As the Inspector of Salmon Fisheries and a member of numerous fishery commissions he did much to instigate the statutory regulation of freshwater and marine fisheries. Alongside his own experiments in fish culture, he encouraged wider studies of fish ecology for their bearing on commercial fisheries. It was largely through his writings that an international fisheries exhibition was held at South Kensington in 1883, albeit three years after his death. The exhibition was opened by the Prince of Wales (the future Edward VII) (Huxley 1885; Lee 1992).

Thomas Huxley was both 'Darwin's Bulldog' and the leading biologist of his day, making it perhaps all the more surprising that he should have ridiculed the notion that the marine fisheries could ever be over-exploited. A Royal Commission of 1866, with Huxley as chairman, finished with the most extraordinary pronouncement that 'with fish in abundance and a growing fishing industry several old acts of parliament restricting fishing' could be abolished. Huxley, in his inaugural address to the 1883 International Fisheries Exhibition, professed that 'probably all the great sea fisheries, are inexhaustible; that is to say, that nothing we do seriously affects the number of the fish. Any attempt to regulate these fisheries seems consequently, from the nature of the case, to be useless.' Although conceding that oyster and salmon stocks might be over-exploited, he made no mention of the industry's effect on the northern whaling grounds. His only reference to cetaceans was their alleged predation of cod. A fuller and much-needed knowledge of the life histories and ecologies of fish was beginning to emerge, perhaps most famously through the founding of the Marine Biological Association and, once sufficient funds were raised, its laboratory at Plymouth (Garstang 1900). It is curious, given the developing interest in fisheries and marine biological sciences how little attention was paid to the plight of whales, particularly at a time when the previously uncatchable rorquals were under increasing pressure. Old-time whaling had been conducted from sailing ships and rowing boats in pursuit of right and sperm whales, as they broke the surface. It was not only possible to kill them with hand-held harpoons, but their carcases floated long enough for much of the blubber and baleen to be removed. The blubber would be cut into chunks and stored, or cooked immediately for 'train oil' and stored in wooden barrels for use in lighting, lubricants and soft soaps. The most highly prized parts were often the long baleen plates, which the right whale used to filter its planktonic food from the seawater. The strength and elasticity of those fibrous strips meant they were ideally suited for women's skirt hoops and corsetry, brooms and brushes, fishing rods and riding whips. Most of the species amenable to this form of hunting and market use were close to commercial extinction by the late 1800s (Harmer 1928; Sanger 1991, 127–137).

The Basques and others had begun to hunt the North Atlantic right whale *Eubalaena glacialis* around the Bay of Biscay from at least the eleventh century, moving from one whaling ground to another as each was commercially worked out, reaching as far north and west as Iceland and Greenland by the sixteenth century. The Dutch, and less successfully the British (often employing Basque harpoonists), turned to the Greenland whale, *Balaena mysticetus*. The year 1697 was a record year, when as many as 200 were caught, about half of them by the Dutch in Spitsbergen waters. In the Pacific (where *B. mysticetus* is known as the bowhead), whalers had begun from 1841 to exploit the stocks north of the Bering Strait and in the Sea of Okhotsk. The catch had so dwindled by the 1880s that the waters were largely abandoned (Brown 1927, 157–158; Hohman 1928; Gambell 1993).

A whaling industry had developed after 1810 on the North American eastern seaboard, from Maine to South Carolina, mainly for humpbacks (*Megaptera novaeangliae*) and some fin whales. Large catches continued to be reported in the 1870s, but the shore-based industry was largely abandoned by 1895. Inshore whaling opened on the California coast in 1895, mostly for humpbacks and Pacific grey whales *Eschrichtius robustus* (with the occasional North Pacific right whale *Eubalaena japonica* and a few rorquals) but had also ceased by 1900. A southern sperm whale industry encompassed both the South Atlantic and the Indian Oceans, and by the early nineteenth century the waters off Peru, Chile and Galapagos Islands and, later still, the Antipodes. A largely American initiative, the biggest catch was in 1836 – this too had become largely defunct by the 1900s (Hohman 1928; Jackson 2005, xiv).

Exploitation of the southern right whale *Eubalaena australis* had started in the early nineteenth century when the New England sperm whalers had come across individuals, as their vessels ventured into southern seas. The right whale proved so abundant off South Africa, Australia, New Zealand, Kerguelen, Crozet, St Paul, the Amsterdam Islands and off Chile, as to become the whalers' principal quarry. Nearly 200,000 were killed between 1804 and 1817, but thereafter they almost vanished. The North Pacific right whale *Eubalaena japonica* followed a similar pattern, being caught initially off the coast between Vancouver and the Aleutians, and off Kamchatka and in the Okhotsk Sea. It had largely disappeared from the catch records by the late nineteenth century.

There are numerous descriptions of whales being caught and flensed and their oil tried out and stored in wooden barrels of six to the volumetric ton. William Scoresby published his history and description of the northern whale fishery in 1820, being a firsthand account of all aspects of the industry up to that time. Thomas Beale followed with *The Natural History of the Sperm Whale* in 1839. Among literary works, Frederick Debell Bennett published *Narrative of a Whaling Voyage around the World* in 1840. Frank Bullen's *The Cruise of the Cachelot*, an account of his whaling circumnavigation on a Yankee whaler, appeared in 1898 to great acclaim from Rudyard Kipling among others. It remains in print. Herman Melville's novel *Moby Dick* draws on the true story of the wreck of the whaleship *Essex*, but, although published in 1853 and closely detailed in its description of the old whaling practices, it excited little popular interest until well after the author's death.

The whaling industry, with the emerging technologies of the time, had already begun to set its eyes on the Southern Ocean, and most obviously South Georgia, the largest island of the Falkland Islands Dependencies. Harrison Matthews encapsulated South Georgia's importance in the subtitle of his book about the island: *The British Empire's Sub-Antarctic Outpost* (Matthews 1931). The inter-war years are generally perceived as 'the tranquil high noon' of the British Empire – 'a period of confident consolidation and development in which a whole philosophy of colonial rule was perfected' (Judd and Slinn 1982, 41–42). The Falkland Islands, which became so closely associated with the whaling industry, provide a highly relevant example of how those global and diverse possessions continued to evolve. As their previously untapped natural resources came to be developed, there was need for further adaptation in British colonial rule.

The Falkland Islands Dependencies provide a highly relevant example of the development of the polar regions, the protection of whale stocks and imperial experience of those dispersed territories (Howkins and Roberts 2023). Robert Headland has, in separate volumes, provided both a comprehensive overview of South Georgia and an exhaustive chronology of Antarctic exploration, of which South Georgia forms an essential part (Headland 1984 and 2009). Peder Roberts has shown the extent to which the waters beyond the Falklands Islands became part of a global networking of European experience (Roberts 2011). But then, other parts of the world were undergoing a similar process of assimilation – perhaps most famously the North American continent, where the twentieth century had begun with tremendous optimism, if not arrogance, as to what those pioneers might achieve. The historical focus has rightly been on those who actually cleared the forest and ploughed the first furrows; but it was the individuals with wealth and privilege, on the Eastern Seaboard of the United States, who brought order and moderation to what was being accomplished. They saw the need for a new kind of ethic 'Progressive Conservatism', which emphasised productivity through closer, more directed forms of management, based on scientific study (Rothman 2000, 3–10).

For all the rhetoric and self-reliance of the settlers, the lands of the American West were just as much colonies as those territories forming the dependencies of the Falklands. Without the economic investment and governmental direction of the North American East, there would never have been the physical infrastructure to export, let alone market, the produce of the Western farms, ranches, forests and mines. Again, it was not those close at hand but those much further removed who saw the 'superabundance' for what it was, merely a chimera. Whether at the pioneering fringe of North America, or the Southern Ocean, the imperative was not to withdraw from what was being attempted but to develop a more sustainable mastery of nature, where the resources would be of lasting value – if used properly (Albanese 1990, 107). It was elitist to the extent of being based not on practical know-how but on the professional acumen of government (Hays 1959; Rothman 2000, 14–16).

Richard White has written of 'a quiet revolution' in the American West, as government agencies came to play an essential part in the use of the resources, assuming a role and significance previously unthinkable. However petulant and angry the homesteaders may have been, they became remarkably adept in denouncing the constraint and interference imposed on them while accepting the individual benefits. So too were the Norwegian whaling companies in their relationship with the Falkland Islands Government in the early development of the southern whaling industry. There were still instances of unbridled spoliation, but, more positively, a kind of working arrangement emerged that was summed up in the Sierra Club's aphorism, 'Not blind opposition to progress, but opposition to blind progress' (White 1991, 392–393 and 407).

Commercial whaling began in the Southern Ocean (the world's fourth largest ocean) a decade before the First World War and ended two decades after the Second World War. As a 'modern trade, with new sorts of whales, new techniques and a totally different order of magnitude', it represented a complete break with the past. By providing raw material for the soap and margarine industries, whaling became an integral part of modern oil technology. The first substantial histories of the transformation appeared half a century ago. Gordon Jackson's *The British Whaling Trade* described the twentieth century, and more particularly the inter-war years, as the most productive in catch size and the first to give thought to safeguarding whale stocks (Jackson 2005, xiii–xiv). The Norwegian whaling historians, Arne O. Johnsen and Johan N. Tønnessen, from their exhaustive study of the industry, see the post-war years of the 1950s and 1960s as little more than the coup de grace, and Bjørn Basberg dismisses them as 'the final fling' (Johnsen 1959; Tønnessen 1967–70; Tønnessen and Johnsen 1982; Basberg 2004, 35).

Bjørn Basberg once remarked, and has demonstrated so well in his writings, that 'history is not written once and for all' – there are always gaps to be filled, new questions to be asked and fresh material to examine (Basberg et al. 1993). Both Jackson and Tønnessen and Johnsen used the company and government records available at the time of their writing. Tønnessen and Johnsen's magisterial four-volume study *Den Modern Hvalgansts Historie*, was commissioned by nine of the Norwegian whaling companies between 1959 and 1970. A condensed English translation was published some ten years later (Husby 2018, 5). Ian Hart has published five substantial volumes focused more specifically on the Falkland Islands Dependencies, using company records and the archive of the Falkland Islands Government. His first volume, *Pesca* (*Compañia Argentina de Pesca*), shows how this company's history exemplified the evolution of modern whaling through 'initial success, high hopes and vast profits, but culminated in its decline in the 1960s and final collapse' (Hart 2001, 3) His later volume *Whaling in the Falkland Islands Dependencies 1904–1931* looks more generally at the fortunes of both shore- and bay-based whaling (Hart 2006). A third volume, *Antarctic Magistrate*, has through the photography of Edward Binnie, provided a remarkable insight into the life and times of that emerging industry (Hart 2009). A fourth volume, *Whale Factory Ships and Modern Whaling*, focuses on the vessels employed, and a fifth, *Austral Enterprises*, on the distinctive histories of the Southern and the Kerguelen Whaling and Sealing Companies – again drawing largely on company records and those of the Colonial Government (Hart 2016 and 2020).There continues to be no comprehensive history written from the perspective of the Imperial Government in London. Graham Burnett, in his book *The Sounding of the Whale. Science and Cetaceans in the Twentieth Century* (Burnett 2012), is severely critical of the Imperial Government in London, but seems to have entirely ignored what may be found in The National Archives. Kurkpatrick Dorsey has shown a much closer engagement with the 'realities of government', drawing as he does on an impressive range of archives, including The National Archives at Kew (Dorsey 2013). The result is his very comprehensive study *Whales and Nations: Environmental Diplomacy on the High Seas.*

By focusing upon the United Kingdom Government, the historian may more easily discover how intervention by governments, and then international action, was called for in the case of cetaceans. The United Kingdom had an obvious interest in the southern whaling industry. Dudeney and Sheail (2019), in *Claiming the Ice. Britain and the Antarctic 1900–1950*, focused on the factors causing the British government to make and retain a firm territorial claim to that part of the Antarctic continent known as the Falkland Islands Dependencies (see Figure 1) which included the sub-Antarctic islands. As Governor Hodson reminded those who assembled to welcome him on his first arrival at Stanley in June 1927, 'The Colony, with its Dependencies, extends to the South Pole and is in area nearly one quarter of the British Empire'. The precedents for establishing a global whaling industry in so remote a part of the world were however far from good. As David Day (2012, 125) writes of the earlier sealing industry, the impact on the rookeries of South Georgia and the South Shetlands had been devastating. How then did the whaling industry manage to prosper for half a century? Was there a near infinite store of whales, as the industry would have government officials believe? Did those entrepreneurs practice forms of self-control, or was the comparative longevity of whaling a consequence of government regulation?[(1)]

Garrett Hardin highlighted a phenomenon familiar to historians of communal husbandry. However beneficial it might be for individuals possessing rights to common pasture to add further livestock, there came a point when the carrying capacity of the grazing pasture would collapse with devastating consequences for all those dependent on it for a livelihood (Dorsey 2013, vii–ix). Should any individual desist from adding more animals, others stood largely to benefit. William Cronin writes that it became weirdly rational to increase the grazing pressure so that every individual within the farming community had benefitted before the inevitable collapse. It was an approach that fisheries economists have also become familiar with in a more modern context. Arthur McEvoy writes extensively of it in *The Fisherman's Problem. Ecology and Law in the California Fisheries 1850–1980*. Stock preservation, whether on land or water, might be achieved only through the privatisation of the common resource or by international regulation (Hardin 1968; McEvoy 1990).

Cronin writes in Kurkpatrick Dorsey's study, *Whales and Nations*, that the story becomes even more complicated when the prejudices and perceptions of the individuals exploiting the resource are taken into account. Whalers were hardly likely to accept any restraint when they saw the possibility of a further remunerative year. In any negotiation of international treaties, objections were raised once discussion moved from the generic statement to the specific ruling of how many whales might be taken, and on what basis (Dorsey 2013, xxii and 4). This volume explores that interface through the records of the Imperial Government in London, and more particularly, the archives of those scientists appointed to appraise and advise on the husbandry and preservation of whale stocks in the Southern Ocean.

Everything about an empire rests on the efficacy of the 'messengers' and systems by which they communicate. It might, in the early twentieth century, take three months for the Colonial Office and Governor to consult one another (Husby 2018, 80). A question for this book is how the immense geographical challenge was met for a particular industry, operating in the southernmost waters. Because the various civil bureaucracies, whether in the furthest reaches of the Southern Ocean or at the Metropolitan centre, needed 'manageable pieces of knowledge', the British Empire was 'a paper empire' in a very real sense (Jeffries 1956; Richards 1993, 1–4). Technical breakthroughs such as the telegraph and wireless were all the more valuable because, as David Reynolds characterises the British Empire, it was 'an empire on the cheap'. The Imperial Exchequer in London operated on the basis of 'getting little out' and, for its part, investing little by way of Exchequer monies in the individual territories (Reynolds 2019, 42–43).

Rather than use the expression 'the British Government decided', this book endeavours to identify who actually determined the main lines of policy. Who drafted the outgoing telegrams and despatches, instructing or recommending that their recipients take some required action? Those formal missives were frequently accompanied by more private, semi-official exchanges, and often copious minute-keeping within the respective departments, all meriting equal attention as the next steps in policymaking were formulated (Hyam 1979; Woodward 1949). The relevant officials were so well versed in the thinking of their respective Parliamentary Under-Secretary and the Secretary of State that they were often able to draft the policy decisions themselves – and indeed took the policy decision where the point at issue had little political importance (Smith 1990). These senior officials were invariably briefed by more junior officials, who often had greater knowledge and experience of the particular matter in hand. The historian must therefore probe deeply into the official archives to discover how the particular knowledge and understanding were acquired, and how they were conveyed to those who ultimately decided the relevant government policy (Vital 1968, 92–93; Lowe 1997).

CHAPTER 2

# *Pursuit of the rorqual*

The mid-nineteenth century was 'the apogee of American whaling'. The output of whale oil rose by 20 per cent and baleen by 80 per cent, and their sales price by 97 and 133 per cent, respectively. But the Americans also pioneered large-scale petroleum development. Within a couple of years after the first oil well being drilled, Pennsylvania was producing more than 2 million barrels of oil annually. Kerosene was both cheaper and a better illuminant than whale oil. Petroleum-based lubricants soon displaced whale oil for a multiplicity of purposes (Hohman 1928, 289; Davis and Gallman 1993, 61).

The United States' whaling fleet might well have recovered from its heavy destruction during the Civil War and from the loss of almost an entire expedition to the Alaskan ice in 1871 had it not been for what Elmo H. Hohman describes as 'a simultaneous sapping of recuperative power'. The capital which could have brought recovery to the whaling industry was now being invested in manufacturing, mining, farming and commerce. Even those who had previously supported the New England ports now found cotton mills more remunerative than whaling. Manning the whaling fleet was increasingly affected by the California gold rush and the general development of the West, culminating in what Hohman (1928, 300) describes as 'the halcyon days of internal development' following the Civil War. The New Bedford fleet of over 300 whalers in 1860 shrank to 136 in 1880, and a mere 14 by 1905. Whalebone might still fetch a high price but never enough to cover the poor return on whale oil. The late nineteenth century was probably the most peaceful period for whales for some 300 years (Slijper 1962, 31).

United States whaling was therefore in steep decline when in the 1860s the Norwegians began to devise techniques for catching whale species that had previously been too fast and dangerous to hunt. Known collectively by the Norwegian word *rorqual-val* (grooved whale for their characteristic longitudinal pleats along the throat), these were the blue whale (*Balaenoptera musculus*) and fin whale (*B. physalus*), and, much less commonly, the sei (*B. borealis*) and Bryde's whale (*B. edeni*). These four species, along with a rorqual-like smaller whale, the humpback (*Megaptera novaeangliae*), became the mainstay of the twentieth-century catch (Burton 1973; Ellis 1992, x–xi).

A variety of entrepreneurs had begun to identify the various pieces to the puzzle required to catch rorquals. A Scottish company, W and G Young of Leith, had developed a harpoon head with hinged barbs in the 1830s which, upon impact, smashed a glass of prussic acid intended to poison the victim. A French scientist, M. Louis Thiercelin, published a firsthand account of its being used on an Arctic voyage of 1866. The whales died within ten minutes or so (Hornell 1950, 179–180; Smith 1994, 220). The priority was to find a method of slaughter that retained the value of both the carcase and the oil. Thomas Welcome Roys and Gustavus Adolphus Lilliendahl made innovative strides, but, as Gordon Jackson comments, the otherwise booming Scottish economy contributed nothing by way of support. It was rather the Norwegian Svend Foyn who saw 'the revolutionary possibilities'. By what Johan Hjort describes as 'a stroke of genius', Foyn found them in the combination of steamships, guns and explosive shells, which together opened 'immense economic vistas' (Hjort 1932, 29; Jackson 2005, x).

Svend Foyn was also 'in the right place at the right time'. Under pressure from competing sealing interests, he used profits from his highly successful business to embark, from 1864, on a series of trials, the first being to use steam-driven vessels for hunting whales in the Barents Sea, where 30 blue whales were caught in 1869. But something more than steam power was needed, namely, an accumulator which, through a system of pulleys and springs, took up the strain on the rope between the catcher and the whale and drew the whale toward the catcher like a fishing rod. But alongside the gun, accumulator and steam winch, Foyn also provided a projectile that stood a chance of killing the whale outright. Over some nine years, Foyn and Hans Esmark perfected a detonating device that exploded like a grenade on entering the whale's body. So, the basic whaling unit had come to comprise a small but strongly built steamship of 80 to 90 feet in length, with a crow's nest mounted on the single mast and a harpoon gun on its bows. The harpoon, to which a rope was attached, was shot out with a 'bomb' installed within the hollow tip so that it exploded on entering the body, often killing the whale outright. Using a strong rope of some 40 fathoms, the whalers could speedily haul the body alongside the boat where a steel tube was thrust into the body cavity; air was pumped through this to prevent the whale body from sinking (Jenkins 1923, 182; Budker 1958, 103–115; Hardy 1967, 191; Jackson 2005, 143–144).

These three inventions – small steam vessels, exploding harpoons and compressed air inflation – provided the technological breakthrough required to revitalise the industry (Dorsey 2013, 7). But it was also the infrastructure developed along the Finnmark coast that gave the Norwegians the commercial head start – one they retained for over half a century (Ellis 1992, 256–257). The shore stations were so close to the whale migration routes that the carcases could be hauled that same day by powerful winches onto the flensing plan, where the blubber was removed by skilled flensers, making two or three longitudinal incisions down the whale's side to facilitate the separation from the underlying flesh (see Figure 2). The blubber was then cut up into smaller pieces and fed into a large cauldron. A blue whale might yield between 50 and 70 barrels of oil. Whereas most of the previous Arctic expeditions had simply stripped the blubber, Foyn showed how an additional amount of oil (albeit of poorer quality) could also be obtained from the meat and viscera. This, in turn, caused him to develop techniques for cutting up and boiling those body parts for their second-grade oil. The eventual residue could then be dried, ground up and used for cattle feed (called guano). Whalers had never before achieved such productivity (Jenkins 1923, 182 and 189; Ellis 1992, 259–262; Jackson 2005, x).

Both the technical and marketing challenges of whaling had passed to Norway. With their 'less promising hinterland' for development, the Norwegians had a much stronger 'economic pull toward the sea'. The Vestfold region of southeastern Norway provided the economic impetus – more particularly the sealing and global shipping communities of Larvik, Sandefjord and Tønsberg. It was from there that Svend Foyn had made his fortune, but, as the sealing

**Figure 2** A rorqual being flensed on the plan at Grytviken in 1958. © British Antarctic Survey, UKRI.

industry had declined through over-hunting and coincidentally the merchant shipping industry was affected by the Long Depression, there was renewed interest in the whaling grounds off Finnmark (Basberg 2006). Rather than embark on long and difficult voyages in search of the dwindling stocks of right whale, considerable cost savings could be made from this new-found capability to hunt rorquals off the Norwegian coast and to bring them ashore that same day. Foyn built the first shore station at Vadsø and was granted a ten-year monopoly (until 1883) off the northwest coast of Norway, west of Nord Kap (Niemi 1993, 67–71). This proved so profitable that 20 companies joined the industry once Foyn's monopoly had expired, thereby precipitating what Charles Boardman Hawes calls 'wholesale slaughter and early evidence of severe stock-depletion' (Hawes 1924, 63; Ellis 1992, 259–260).

As well as the ten shore stations that continued to function along the Norwegian coast, and an increase in the size and value of the catches made off Bjørnøya (Bear Island) and Spitsbergen, Foyn and others turned increasingly to the waters around Iceland in 1889 and the Faroe Islands in 1892, and off the northern and western coasts of the British Isles by 1900. As many as 63 vessels, owned by 24 companies, operated there at the peak of activity in 1905, catching an aggregate 2,671 whales, from which 83,000 barrels of oil were produced. It was in those remoter waters from the Norwegian coast that the concept of a floating factory was developed, which, by dispensing with the need for a shore station, sowed the seed for a pelagic industry some two decades later in the Southern Ocean (Isachsen 1929, 390; Niemi 1993, 72–74; Basberg 2004, 26 and 2019, 7).

Whaling activity did not however go unchallenged. Fishermen had commonly assumed the whales, in their movement along the coast, had encouraged the cod and herring to stay closer to the fjords, and by doing so had greatly increased the commercial catch. That same movement was also thought to frighten away the predatory Greenland seal. Fisheries scientists both ridiculed and dismissed any threat of whaling to the fishermen's livelihood (Hjort 1932, 29). Matters came nevertheless to a head when 700 fishermen tore down the smokestacks of the Tanen shore station in what became known as the 'Menhavn riot'. The Norwegian Storting (parliament) enacted a Whale Protection Bill in February 1904, banning all whaling from the Norwegian coast for ten years. Beyond making token protest, most whalers welcomed the ban for the compensation which they received for leaving what had effectively become an exhausted whaling ground (Niemi 1993, 74–78). But wherever they moved, there were soon protests, particularly from local health officials, at the flensed carcases and offal they abandoned on the shoreline. Both bylaws and a Whale Fisheries (Scotland) Bill were enacted in respect of the Scottish whaling stations (Johnson 2015, 14–20).

Johan N. Tønnessen published, in 1970, a much-needed corrective to the overly simplistic belief that it was the depletion of these northern waters which precipitated a southern whaling industry. While acknowledging the technical advances made in whaling, Tønnessen contends that the primary reason for the industry's move to the Antarctic was the disastrous slump in market demand for whale oil. There had been a bumper American maize harvest in 1904, producing unprecedented quantities of corn oil. This, and the unusually plentiful supplies of linseed oil and cottonseed oil meant that the price which soap and margarine manufacturers were prepared to pay for whale oil fell to an all-time low. Whaling companies were compelled to seek grounds where whale stocks were still so high they could achieve the levels of productivity required to compete in price with other oils and fats (Tønnessen 1970; Tønnessen and Johnsen 1982, 160–162 and 176–177).

The only place with that potential – where no whaling for rorquals had previously occurred – was the Southern Ocean. The New England whalers had rarely ventured that far south. James Clark Ross claimed to have witnessed an abundance of right whales during his naval expedition of 1839–43. The Enderby Brothers of London (the leading British sealing and whaling company) had obtained a royal charter in 1848 to establish a whaling station and settlement on the Auckland Islands. This was abandoned after only two or three years, largely because no right whales could be found there. The first steamship to enter Antarctic waters, the German *Groenland* in 1874, was assigned specifically to search for new whaling grounds, but it found none (Aagaard 1929, 23; Payton 2013).

An attempt by the long-established Scottish whaling company, David and John Gray of Peterhead, to send an expedition in the 1870s failed through lack of funds. The company had however published a promotional pamphlet *A Report on New Whaling Grounds in the Southern Seas*. A copy was obtained by the Royal Geographical Society of Australia which, with the Royal Society of Victoria, had recently appointed an Antarctic Exploration Committee. This initiative also failed through lack of funds to embark on the intended scientific and whaling expedition (Tønnessen 1970, 284; Tønnessen and Johnsen 1982, 148–149). Its publicity had however attracted the interest of a Hamburg shipping company, Woltereck and Robertson, which in turn financed Christen Christensen (a shipping and shipyard owner as well as part-owner of Norway's largest sealing company) to send a three-masted sealing vessel *Jason* south, fitted with an auxiliary engine. The expedition's object was to bring back as much baleen from the right whale as could be carried. There was never intention of catching rorquals – the steam catchers of that date could never have made so long a journey, even if coaling points could have been provided. The ship, equipped to catch only right whales, left Sandefjord on 1 September 1892 (Aagaard 1929, 23–25; Basberg 2019, 15).

Much has been written of what Basberg (2019, 6) calls 'the exploratory voyages' of the 1890s. Their longer-term significance stemmed most obviously from Christensen's choice of the 35-year-old Carl Anton Larsen as the master of the *Jason.* A third of the nine-month expedition was spent searching fruitlessly for the potentially lucrative right whale, penetrating the Weddell Sea as far as 64° south. A loss of 36,983 kroner was recorded despite *Jason*'s return with the skins and oil from 6,335 seals. Larsen had seen 'countless' herds of blue whale and humpback but none of the anticipated right whale. Two further Norwegian expeditions followed. Larsen persuaded Christensen to mount another expedition – this time explicitly for seals. The very exceptional ice-free conditions meant he was able to venture farther along the eastern coast of the Weddell Sea than had any previous ship – and indeed very few even to the present. But once again, despite an impressive haul of 6,586 seal skins and blubber and a final but fruitless search for right whales around South Georgia, the expedition recorded a financial loss (Tønnessen and Johnsen 1982, 150–152; Hart 2001, 13–18).[(1)]

Svend Foyn thought Larsen had looked in the wrong place, reasoning that James Clark Ross could well have encountered the right whales in what later became known as the Ross Sea. Foyn accordingly financed what he styled as a commercial and exploratory expedition led by Henrik Johan Bull (an Australian businessman and family member) during the austral summer of 1894–95 aboard Foyn's old sealer *Antarctic.* A right whale was caught off Campbell Island, but no more were seen. Bull's intention of making a second voyage was thwarted by Foyn's death and the deficit of 10,000 kroner incurred from the first one (Bull 1896; Headland 1993; Tønnessen and Johnsen 1982, 153–155; Hart 2001, 19).[(2)]

Two factors had made those expeditions little more than a footnote to whaling history. First, the right whales had become so overhunted in their winter breeding grounds off the Isles Crozet in the southern Indian Ocean that there were scarcely any left to be caught. Secondly, those expeditions had become, as Gordon Jackson characterises them, 'an old dream in new waters' – their object was to catch 'the right or Greenland whale'. Sales of seal skins and oil would never have covered the costs of the nine-month expedition (Jackson 1993, 117; Riffenburgh 2016, 33–61). The expeditions of the 1890s are a distraction, in the sense that the historian's priority is to recount how, once whale oil prices had recovered, every species of whale would be affected by what was becoming an increasingly global industry.

# PART II

## *The British/Norwegian ascendancy, 1900–25*

CHAPTER 3

# *Southern whaling*

There is no mention of the Antarctic in Niall Ferguson's history of 'how Britain made the modern world'. Africa was for him the last blank sheet in the imperial atlas. The driving force was 'a combination of financial power and fire power'. Those same elements also drove the 'scrambles' for North Borneo, Sarawak and New Guinea in the Far East. They were similarly present in 'the commodification of nitrates, phosphates, coconuts and fishmeal', by which the near-contemporary exploitation of the Pacific archipelagos was achieved (Ferguson 2003, 222–239; Cushman 2013). Those same principles drove the 'commodification' of the Antarctic continent and its widely scattered archipelagos, but with the essential difference that in these places there was no indigenous population.

Alessandro Antonello recalls that 'the southern oceans in this time' became 'completely entangled in new imperial and capitalist expansion' (Antonello 2018, 301). Roberts and Howkins (2023) have similarly written of how the bodies of knowledge developed to regulate commercial whaling 'emerged in concert with the industry itself' aimed at 'benefitting the whalers more than the whale'. The United Kingdom Government positively encouraged the Norwegian development of a whaling industry that was prepared to buy leases and licences for the privilege of hunting whales. The Norwegian input is a further illustration of the complexity of the Imperial governance of the colonies (Hodge 2011, 16–19). It was essentially 'a marriage of convenience' between the individual whaling companies and officials of the Colonial Government at Stanley, with grumblings on both sides – the Norwegian at the costs and constraints imposed, and British regret at the Norwegian dominance of the industry (Basberg and Hacquebord 2023).

Both explorers and whalers had begun to acquire an intimate knowledge of the islands and coasts of the Southern Ocean. There were as many as 15 exploratory expeditions of seven nationalities between 1897 and 1917. Among them was Adrien De Gerlache aboard *Belgica* (1897–99) and Jean-Baptiste Charcot on *François* (1903–05) and shortly afterward *Pourquoi Pas?* (1908–10). Their charts of the Antarctic Peninsula identified 'the good harbour at Port Lockroy in Wiencke Island' and the shelter afforded by Port Charcot on Booth Island (Boothe 2011). But whereas De Gerlache and Charcot sought the greatest publicity for their ventures, there were good commercial reasons why the whaling companies should be so reticent in revealing where their 'hardy little steam craft' had poked 'their bows, during the course of their hunting, into all sorts of strange bays and bights and corners', which otherwise remained 'totally unsurveyed' (Chatterton 1925, 200–201 and 236).

Carl Larsen was exceptional in being a whaling entrepreneur, an outstanding sea captain and navigator and 'the most scientifically' inclined of whalers. The discoveries he made in the course of his two voyages aboard the whaling schooner *Jason* in the mid-1890s went far beyond what was strictly required for founding an industry (Basberg 2004, 28). William Speirs Bruce described Larsen at the time as having 'a zeal for extending our knowledge of these [polar] regions that would not have been unworthy of the leader of a purely scientific expedition' (Murdoch 1894, 349). The Royal Geographical Society of London published an extract from his journal of those voyages and awarded him the Society's Back Grant in recognition of his 'geographical discourses', meteorological observations and further provision of fossil evidence (Larsen 1894).

Larsen was particularly taken with the development potential of South Georgia, the crescent shaped Dependency of 104 miles in length and up to 22 miles across, lying nearly northwest to southeast. On a similar latitude and almost due east of Cape Horn, the island rose as a mountain range steeply from the sea. Confident he could raise the necessary capital for further sealing and whaling ventures, Larsen wrote to Clements Markham (President of the Royal Geographical Society) in 1896, soliciting his help in obtaining a licence to erect dwellings and shelters in which to extract the whale oil at Cumberland Bay, the largest indentation on the northeast side of the Island. Presently employed as the manager at a whaling station in Finnmark, Larsen asked for the content of his letter to be kept confidential. Further correspondence of 27 April showed him to be much less hopeful of obtaining the funds required 'to undertake fishing in the South' (Headland 1993, 192; Royal Geographical Society archives CB7/1596/23). There is a Larsen family recollection of Queen Victoria granting him a concession for the whole of South Georgia and of his assurance Cumberland Bay would suffice (Hart 2001, 19–21 and 61). No supportive evidence of this has been found in either the British or Norwegian archives.

Larsen did not return to southern waters until he was appointed the captain of Otto Nordenskjőld's Swedish National Antarctic Expedition ship *Antarctic* in 1901. Although a scientific expedition, the ship was also equipped to catch right whales (Basberg 2019, 4). Larsen again saw immense numbers of rorquals and spent a further month in South Georgia waters, where he found all the elements required for a whaling station in Cumberland Bay, which he called Grytviken – the Bay of Pots – after the large metal pots abandoned there by the nineteenth-century elephant sealing parties. Although *Antarctic* was crushed by the pack ice and abandoned on 12 February 1903, Larsen's skill and courage secured his party's survival through the ensuing winter. The entire Nordenskjőld party was rescued by the Argentine corvette *Uruguay* and taken to Buenos Aires on 2 December 1903 (Nordenskjőld and Andersson 1905; Tønnessen and Johnsen 1982, 159–160; Hart 2001, 23–25; Basberg 2004, 31; Rabassa and Borla 2007, vii–xii and 125–126).

The Argentine success in rescuing the Swedish expedition and Larsen's acclaim as something of 'a Polar Hero' meant that those from the Argentine business sector, who had played some part in instigating the rescue mission, were particularly attentive to his boast, made at a celebratory banquet, on 5 December 1903, of seeing whales in their 'houndreds and tousends' (Tønnessen 1970, 284–285). The Norwegian Peder Christophersen was among those present. A leading business figure and president of the local shipping association, he introduced Larsen to German-born Herman H. Schlieper and other principal figures of the Tornquist Bank (the leading immigrant bank). A long and detailed prospectus was drawn up for an Argentine whaling company Compaňia Argentina de Pesca (or Pesca as it soon came to be called). Ernesto Tornquist (the Swedish head of the bank), with his family and close business

**Figure 3** Map of South Georgia showing the location of the whaling stations that were established there in the early part of the twentieth century. The first was at Grytviken in 1904, followed by Godthul in 1908, Leith and Ocean Harbour in 1909, Husvik and Stromness in 1910, and Prince Olav Harbour in 1917. Whaling stations and hill shade are taken from South Georgia GIS, accessed 2023. Land cover is derived from Landsat ETM photography (2003) and is approximate. Coastline has been adjusted to 2021 imagery. Place names are taken from the South Georgia and South Sandwich Islands (SGSSI) gazetteer, maintained by the UK Antarctic Place-names Committee. © British Antarctic Survey, UKRI.

associates, became the chief shareholders, with Schlieper as president and Christophersen as vice-president (Tønnessen and Johnsen 1982, 159–160; Walton 1982, 234; Hart 2001, 29–39).

Peder Roberts writes in his book *The European Antarctic* that Larsen saw whales as 'floating oil barrels', their sheer number promising rich rewards from the sale of the highest quality whale oil (Roberts 2011, 16). Larsen was undoubtedly a skilled entrepreneur, but it would be wrong to overlook the technical progress that had also been made since his earlier ventures on the *Jason*. The steam-powered catchers, employed for rorqual hunting in the Arctic of the 1890s, would have lacked both the bunker space and sturdiness required for operating in the Southern Ocean (Basberg 2019, 8). It was now technically feasible, a decade later, for Larsen to take up Pesca's offer of support. He had returned with Nordenskjöld's party to Gothenburg on 13 January 1904, and arrived a month later at Sandefjord, where he continued to encounter a 'Norwegian negativity' toward both the logistical challenge and continued marketing uncertainties of whaling in the Antarctic (Hart 2001, 38–39).

Pesca's capital and overall direction were Argentine, but its management was Norwegian. Larsen returned to Grytviken (Figure 4), now as manager of a projected shore station, in November 1904. He did so aboard *Fortuna*, a steam catcher large enough to carry 100 tons of coal and powerful enough to tow six blue whales. Whales were so plentiful there was initially no need to pursue them beyond Cumberland Bay. Blubber cooking began on 22 December. As many as 100 whales were caught in the first six months, mostly humpback (Walton 1982, 234). The venture prospered to the extent that the company chartered a considerably larger vessel, the Argentine government's *Guardia Nacional*, to carry coal, machinery and barrel-making materials for Grytviken – reported by William H. D. Haggard (the British Minister at Buenos Aires) to the Foreign Office of 28 June 1905. By a decree of the Argentine Ministry of

**Figure 4** The Grytviken whaling station at full cook in 1949 with Mount Hodges in the background. The track linking Grytviken and King Edward Point on *far right*. © British Antarctic Survey, UKRI.

Finance, the company's barque *Rolf* was permitted to deliver 185 tons of whale oil to Buenos Aires duty free, on the premise that it had been obtained as 'the product of fishing on the high seas'.[(1)]

There were also potentially rich whaling grounds off the archipelagos and the Antarctic Peninsula, or Graham's Land, as it was called in the Colonial Office documentation. On the presumption that it was physically impossible to erect and work shore stations in such climes, the blubber would be tried out aboard what were effectively 'floating factories', anchored in the more sheltered waters close to where the whales were being hunted. To that limited extent, the floating factories had the advantage of being able to follow the whales. Christen Christensen's shipyard at Sandefjord had previously installed a cooker aboard the 450-ton steamer *Telegraf* in 1903. About a hundred whales were processed in the three ensuing summers off Bjørnøya (Bear Island) and Spitsbergen. The blubber, having been flensed from the carcase alongside, was hauled aboard the ship for further rendition on deck. So encouraged, Christensen purchased and refitted a 34-year-old iron steamer, which he renamed *Admiralen*, equipped with four open cookers and a small pressure boiler (Christensen 1935, 115–116; Slijper 1962, 33; Arlov 1993, 81–85).

Larsen's boastfulness of the whales in those southern waters, upon his return to Sandefjord, and more particularly his various purchases for Grytviken, had encouraged Christensen to assemble an expedition of his own, comprising *Admiralen* and the catchers *Ornen* and *Hauken*, with Alexander Lange as manager. The ships became the only ones to have ever operated in both polar regions during a 12-month period. They were also the first truly rorqual expedition ever to be sent to the Southern Ocean, leaving Sandefjord on 21 October 1905 (Isachsen 1929, 391; Ellis 1992, 265; Basberg 2004, 29; 2019, 5).

It could so easily have proved a false dawn. Larsen had been compelled to rely entirely on Argentine capital. So small was the Argentine market for whale oil that most had to be re-exported (at considerable cost and loss of quality) to face the intense competition of a European vegetable oil and animal fats market. As Schlieper made clear to Larsen in a letter, of May 1906, Pesca's directors were 'opposed to any new purchase or investment so long as we are so heavily in debt' (Hart 2001, 106–107). Christensen's decision to send a floating factory was perhaps an even greater gamble (Basberg 2019, 10). Having spent some time off New Island in the Falkland Islands, *Admiralen* moved to what proved to be the considerably more productive waters of Admiralty Bay in the South Shetlands. But despite returning with almost a full cargo of 4,237 barrels of whale oil, the cost of coal and other provisions required for so long and distant an expedition had far outweighed the oil's value (Mørch 1911, 661–662; Christensen 1935, 105–106; Tønnessen and Johnsen, 1982, 172–173; Adie and Basberg 2009).

Christensen and his main shareholders, faced with a deficit of 49,484 kroner, resolved, at an extraordinary meeting at Sandefjord on 6 June 1906, to sell the ship. It was only because of their failure to obtain any offer for the ship, even from Pesca, that they approved a second whaling season. An auxiliary sailing vessel was purchased to economise on bunker and other overheads. Much more critically, the market price for Number 1 whale oil rose to £24 per barrel, enabling them to make a small profit in the second season. So productive were those waters that *Admiralen*'s holds had been filled by 12 February. Christensen was sufficiently encouraged to invest further in larger and more efficient steam catchers to the point that more whales were caught in the 1909–10 summer than in all the previous austral summers since 1904 (Tønnessen and Johnsen 1982, 174; Hart 2006, 18, 35 and 64; Basberg 2019, 11).

Much more was required than simply producing the greatest volume of oil possible: it had to be of the highest quality. Whale oil for the most part comprises glycerides of saturated and unsaturated fatty acids. The premier grades 0 and 1 are derived entirely from the blubber – clear, pale, straw coloured, and with a faint fishy smell. These grades store well, with an acidity of less than 1 per cent. Grade 2 is amber in colour, and contains more than 2 per cent free fatty acids. The proportion might range from 6 to 15 per cent in the darker Grade 3 oil and be as high as 30 per cent in the even darker, 'evil-smelling' Grade 4. Such darkening in the lower grades was due both to the presence of free, or saturated, fatty acids and the time that had elapsed since the whale was killed (Schwitzer 1956, 134).

Whereas the first oil, exported by Pesca, had been of the highest quality, Arthur H. Brandt (the German broker) complained that subsequent deliveries were darker and more acid. The falling-off in quality was attributed to the introduction of a second catcher which, by doubling the supply of whales, had so overtaxed the shore station at Grytviken that by the time the carcases could be processed, decomposition had increased the free fatty acid content considerably. The reduction in quality was all the more serious because of the renewed competition from other animal and vegetable fats and oils, causing the price paid for even Numbers 0/1 whale oil to have fallen to £15 per ton and as little as £8 for Numbers 2 to 4 oils (over half the whale oil sold). Whaling was profitable only where whales were so abundant that only the best blubber need be used and the rest of the carcase abandoned. Even so, the market for whale oil remained so uncertain as to threaten even the nascent industries that Larsen and Christensen were endeavouring to establish on the highly productive, virgin grounds of the Southern Ocean (Tønnessen 1970, 283).

As Gordon Jackson (2005, 161) phrases it, modern whaling owes its establishment less to Svend Foyn than to the crisis of 1906 in the global supply of animal fats, which encouraged soap and margarine makers to experiment with other fats and oils. Whale oil was so greasy and smelly even in the smallest proportions. Whaling historians have therefore tended to dramatise the technical breakthrough represented by the process of hydrogenation (Hunt 1969). The French chemists, Paul Sabatier and Jean-Baptiste Senderens, had discovered the process around 1900, and the German chemist Wilhelm Normann (using a metal catalyst) first began to apply it to fats and oils in 1903. The physical state of a fat is derived from its chemical composition, and more particularly from the ratio of glycerides of solid saturated fatty acids to unsaturated ones. Normann found that by adding further molecules of glyceride (by means of a nickel catalyst), he could produce a mix of acids similar in their plasticity and digestibility to that of butter. The soap makers, Joseph Crosfield and Sons Ltd of Warrington bought Normann's English patent and were the first to manufacture and retail the product in 1906 (Feron 1969, 83 and 91; Hart 2001, 110–111).

Hydrogenation meant that both soap and margarine manufacturers could take advantage of the general rise of living standards, as exemplified by the increasing demand for soaps and the growing concern at the lack of fats in the diet of the rapidly increasing industrial workforce. The story of those industries in the decade or more before the First World War is one of considerable infighting, both to secure the relevant patent rights and to employ the chemists capable of transferring the principles developed in the laboratory, through the pilot stages, to large-scale use (Hunt 1969, 65–69). The prospects for the whaling industry were transformed. A means had at last been found of converting marine oils into a stable and near odourless fat, capable of being stored for long periods without spoilage (Feron 1969, 92–93; Boldingh 1969, 170–171).

Gordon Jackson writes that for technical, commercial and emotional reasons, attention was rarely drawn to the use of whale oil in soap and margarine production, but there were nevertheless a dozen hardening plants throughout Europe by the First World War, causing the market value of whale oil to rise from £15 to £23 per ton. The large-scale development of the Falkland Islands Dependencies' waters was assured (Jackson 2005, 164–166). The longer-term concern for the oil-processing industry, as Kenneth Hunt writes, was the manner in which the whaling industry continued to behave like 'a seaborne version of the "collecting and gathering" economy as practised by our nomadic forefathers'. Once a locality was hunted out, the hunter-gatherers would simply move on – and so did the whalers (Hunt 1969, 64).

## 3.1 Confused beginnings

The Colonial Governor at Stanley had become the key figure in 'this emerging business' of southern whaling, in which regulation promised considerable profit provided a breeding stock of the particular whale species was preserved and British sovereignty of the relevant islands and archipelagos secured (Roberts 2011, 16; Husby 2018, 63). As a Crown Colony, the Falkland Islands and their dependencies were administered by 'a succession of carefully chosen and experienced autocrats, watchful and considerate of local opinion, but not responsible to it'. They were answerable instead to the Secretary of State for the Colonies, who in turn was accountable to the Imperial Parliament. In the same way the Secretary of State was advised by 'an experienced department', the Colonial Office, the Governor was advised by a Colonial Secretary, who acted as administrator when the Governor was away from the colony (Bertram 1930).

William L. Allardyce was appointed Governor in 1904, having spent a quarter of a century with the Fijian government, during which time he had been made Colonial Secretary. He remained the Governor of the Falkland Islands for the unusually long period of ten years (1904–15) and thereafter was appointed to the Bahamas and briefly Tasmania (Tatham 2008, 37–42). Already immensely experienced, Allardyce was keen to exploit every opportunity afforded. His ten years in office as the Governor of the Falkland Islands marked both the birth and coming of age of whaling policy (Husby 2018, 4). Graham Burnett, in his book *The Sounding of the Whale* demonises him for his rapacity in capitalising on the windfall of whaling revenues (Burnett 2012, 34). Ian Heyburn writes more perceptively of his overriding and proper concern for the Falkland Islands' economy (Heyburn 1980). There was no subvention from the Imperial Exchequer in London for sustaining the colony's finances. As Stephen Constantine writes generally of the colonial territories, every colonial government was expected to cover its outgoings through fiscal revenues from the colony (Constantine 1984, 22 and 25).[2]

Allardyce pressed the Colonial Office for the authority to realise whaling's potential. John S. Risley (the Legal Assistant to the Colonial Office) minuted on 26 June 1905 that there was no legal doubt that South Georgia was a Falkland Islands' dependency. Although 'we have never formally annexed or hoisted the flag there', a previous Governor had invited tenders for the leasehold of the whole island in 1900. No foreign government had challenged his right to do so – nor had anyone tendered for the lease (Hart 2001, 66). The more important question was not so much legality but 'whether to stick to our claim or not'? No one lived there permanently – the island was of 'no use for agricultural or pastural purposes'. There was no report of workable mineral deposits. But then, as Charles Prestwood Lucas (an Assistant Under-secretary) noted, 'The same might have been said of the Yukon Territory a few years ago'. An Argentine company had already shown South Georgia to possess a valuable fishery, and 'if foreigners resort to it, there is no reason why British subjects should not also resort to it'.(3)

A three-way pattern of communication developed between Stanley, London and Buenos Aires. A consul at the British Legation in Buenos Aires reported a conversation with 'Señor Enrique Schlieper, the Secretary of the Fishing Company' in June 1905:

> *So far the Company has not been a success. This is due partly to the fall in the price of whale oil from £20 to £10, to the expenses of freight of the oil from here to Europe and the small prices realized for the whalebone that has been obtained, owing to the fact that only small whales have been caught.*(4)

There was, however, a fair market in Buenos Aires for the oil, which was used in soap making, tanning and as a lubricant. Pesca operated three vessels – a small whaling steamer, an anchored hulk that was used as a store ship and headquarters, and 'a small sailer' which brings stores 'to the hulk from Buenos Aires and brings back the results of the fishing'. The round journey took 65 days. Schlieper had emphasised that 'all the fishing was done on the high seas'. South Georgia was otherwise 'uninhabited save by penguins and seals etc.'

With no government officer stationed on the island, it was only from what Pesca chose to disclose that the Colonial Government became aware of what might be happening in its dependency. Frederic D. Harford (the Chargé d'affaires) forwarded to the Foreign Office and Governor the substance of a letter, published by the Buenos Aires newspaper *La Nacion*, written by an officer aboard the transport ship *Guardia Nacional*, which had been chartered to ferry supplies to Grytviken. 'The small steam whaling vessel *Fortuna* had piloted them into Cumberland Bay, where Carl Larsen, and his wife and child were found to be living with 'about thirty men under his orders', all Swedes and Norwegians, in houses built of wood with zinc roofs. There were poultry and pigs, and plans to import some cattle and sheep. The officers of *Guardia Nacional* had made a survey of the bay during the fortnight spent there. Whales had proved so abundant that a steamer of 2,000 tons and additional machinery for a new factory were being acquired, as well as 'electrical appliances to be worked by water from a waterfall'. The climate was said to be very cold but healthy, with the island largely buried in deep snow all winter. Harford wrote, 'I have thought it necessary to give all these particulars in order to show the effectiveness of the recent occupation of South Georgia by this Argentine Company.'(5)

Alfred Lyttleton (the Secretary of State for the Colonies) agreed with Allardyce that 'the question of charging a licence to the Company should be seriously considered'. Lord Lansdowne (the Foreign Secretary) agreed that the Admiralty should extend its normal summer visit by a naval squadron to Stanley, so as to include South Georgia 'to ascertain the facts' and require the company, if necessary, to take out a licence as 'a vindication of British Sovereignty over the Island'.(6)

The Governor had meanwhile taken what proved to be the truly historic step of asserting 'British legal control over all activities taking place in the dependencies'. He had awarded the long-advertised lease for South Georgia to another company, the South Georgia Exploration Company, which a group of landowners and farm managers had formed at Punta Arenas in 1905. A 21-year lease had been agreed for the grazing, mineral and sealing rights (two years in the first instance) upon payment of £10 a year for the sealing rights and the deposit of

£200 as security. A minimal sum of £5,000 was to be invested in works or stock within the first five years, and a small percentage of the annual profits paid to the Colonial Government. Allardyce's concern that the company chairman and a shareholder (both British subjects) were a front for 'less desirable Chilean nationals' was allayed by the appointment of Ernest J. Swinhoe (a Yorkshireman and well-established Patagonian farm manager) as the company manager (Christie 1951, 243; Tatham 2008, 527–528).

Allardyce boasted, in private correspondence with Hugh Bertram Cox (an Assistant Under-Secretary in the Colonial Office), that he had leased 'the most southern portion of His Majesty's Dominions, to wit South Georgia, to an Exploring Company', on what he described as 'the best terms I could'. Angry at how Pesca had established its shore station 'without our authority', he was 'particularly anxious to shew to the Argentine authorities that it was our Dependency, and that we had leased it, and that they were there without our authority', further remarking that 'all these S. American Republics are mighty off-hand to us' (Tatham 2008, 527–528; Hart 2020, 360). Risley protested Allardyce should have first obtained the Secretary of State's consent for the licencing. Charles T. Davis (a newly promoted First-Class Clerk) warned that the Governor had considerably complicated matters. Harry F. Batterbee (a Second-Class Clerk) expected there to be 'lively proceedings' when a party from the Exploration Company met the Argentine party. Allardyce had however begun to put in place a two-path system by which the industry developed. Alongside the leases awarded for the individual shore stations, annual licences might be granted for additional vessels through individual negotiation. On the premise that it would be physically impossible to operate shore stations from the rest of the Dependencies, whaling would function entirely on the basis of annual licences which, through their intended renewal from year to year, would encourage the companies to invest further in the industry.(7)

The immediate priority for the Colonial Office was to preserve the claim to South Georgia, which, Charles Davis adjudged, was most effectively secured through agreeing to a lease with the already established Argentine Company – 'we should have no hesitation in repudiating the Governor's lease'. Lucas recommended that the Secretary of State should therefore neither 'confirm nor disallow the lease agreed by Allardyce, until one of H M Ships has visited the island'. Having obtained the Foreign Secretary's agreement to a naval visit, Lyttleton wrote to Allardyce on 27 September 1905, informing him that there could be 'no disposition of Crown lands in the Colony' without the Secretary of State's 'special sanction'. The fact that an Argentine party was already known to be on the island was even more reason that the grant of any lease should be deferred until the matter had been considered further. The Secretary of State proposed therefore 'neither to confirm nor to disallow the lease' until a naval visit had been made and a report received 'as to the conditions which prevail there'.(8)

The Colonial Office accordingly asked the Admiralty to instruct the commodore that 'the Argentine Company should not be disturbed simply on the ground of the grant of a lease to the other Company, since that lease has still to be approved by His Majesty's Government before it can be regarded as valid'. But with hindsight, whatever the legitimacy of the lease already granted, Allardyce had begun the process of apportioning whatever assets the island of South Georgia, and indeed the other dependencies, might be found to possess. An 'exploring party', led by Swinhoe, complete with livestock and other requirements for farming, arrived at Stanley on 19 July 1905, where both parties signed the initial two-year lease after paying a fee of £10. Despite its being midwinter, the party had left straightway, aboard a sailing vessel for South Georgia, where it remained for three months (Headland 1984, 66; Tatham 2008, 527–528).[9]

Swinhoe formally presented a long letter, upon his party's return in mid-November 1905, recounting that his first intimation of the presence of others on South Georgia was the sight of a small whaling steamer flying the Argentine flag, which

> *was owned by an Argentine Whaling Company who have occupied these Islands, and have erected houses and a boiling-down works, also a jetty for landing and shipping cargo with a potable water arrangement for the steamer, in the southern branch of the sea which forms Cumberland Bay.*[10]

The crew of Swinhoe's vessel had already left for the main purpose of hunting seals while the remainder of the party had camped within sight of the whaling factory, had released the livestock brought with them and had 'hoisted the Flag several times'. It was soon obvious that South Georgia was too frigid for stock raising, even by Patagonian standards, and, with the return of the ship's crew empty-handed, Swinhoe had come to realise from the Argentine example that only whaling was likely to pay. He had finally visited the whaling station on 22 August, where he found 32 men from Norway and Sweden, each having 'signed a contract for three years before the Argentine Consul in Norway'. It was not, however, until 27 September that he had actually challenged 'their rights to occupy the place,' telling them 'they would have to get out'. He had found Lauritz E. Larsen (Larsen's older brother and under-manager) in charge, who had told him he must consult 'his Company in Buenos Aires' (Walton 1982, 235–237; Hart 2001, 66–68; Tatham 2008, 527–528).[11]

The Admiralty agreed to detach HMS *Sappho* from the naval squadron, which was making its annual visit to the Falkland Islands, expressly to visit South Georgia. This in turn had required the Colonial Office to consult the Foreign Office, on 23 October 1905, for instructions on such matters as hoisting the British flag over the whaling station and the terms by which the Argentine company should be permitted to remain on the island. Since the visit had necessarily to be brief, the Argentine party could hardly be expected to agree definitively to the conditions of a lease or to provide 'the deposit of a sum sufficiently large to penalize a failure' to honour them.[12]

Francis Hyde Villiers (an Assistant Secretary of State) advised the Foreign Secretary, Lord Lansdowne, that His Majesty's Government 'have always maintained that the islands are British as a dependency of the Falkland Islands', and that the Governor of the Colony had recently granted 'a British exploring party' a lease for the whole island, which the Colonial Office was presently 'holding in suspense'. Lord Lansdowne had also approved the Colonial Office suggestion of a naval ship appraising the position at first hand. But what action should the naval commander take, if the Argentine party rejected the terms of any lease? Villiers made note that the Colonial Office had, upon reconsideration, agreed that the naval proceedings should extend no further than to offer 'the Argentine people' a lease, and if they rejected it, they should be told that the legitimacy of their presence would be taken up with the Argentine government. This contingency proved unnecessary.[13]

Lauritz Larsen's report of Swinhoe's visit to the shore station and a copy of the 'statement of claim' made by Swinhoe reached Pesca at Buenos Aires in late October 1905. Christophersen and Captain Guillermo Nuñes (of the Argentine Marine Ministry and Pesca's technical adviser) called immediately on Harford at the British Legation, showing him Swinhoe's note 'giving

Captain Larsen notice to quit in peremptory terms'. Harford wrote to the Foreign Office that Pesca was clearly worried the British government might accede to Swinhoe's demand. Christophersen and Nuñes had impressed upon Harford the company's willingness to pay 'an acknowledgement rent', explaining that Carl Larsen had thought it was not necessary to obtain permission for his station. Whether true or not, Harford ventured his 'personal opinion that a refusal to grant the Fishery Company the license they ask for, would have a very bad effect in this country'. A licence payment would have the undoubted advantage of implying the Argentine government's recognition of British sovereignty over South Georgia (Hart 2001, 69).(14)

Allardyce had received a copy of Harford's report of the Pesca meeting in Buenos Aires, shortly before the South Georgia Exploration Company's party returned to Stanley. Swinhoe had vigorously protested, upon landing, at having been so unjustly treated as a British subject, particularly as his Exploration Company had gone to so much expense in obtaining the necessary permissions. Allardyce telegraphed the Colonial Office, on 23 November 1905, that Swinhoe's party had found South Georgia unsuited to sheep and hardly encouraging for mineral development. Although greatly disappointed by the absence of large seal colonies on the part of the island visited, Swinhoe was 'anxious to return as soon as possible' for the whaling. Allardyce himself had 'suspended action until further instruction' on awarding licences (Hart 2001, 68–70).(15)

Pesca had meanwhile met with sufficient success to be reportedly willing to apply for a lease. Schlieper had immediately reassured Larsen of the Fishery Company's intention to secure retention of Grytviken. Harford forwarded for his part, on 2 November 1905, 'a request on behalf of the Argentine Fishery Company that a licence may be granted them to maintain a depot at South Georgia for keeping coal and stores, and for the purpose of boiling whale oil'. Harford had assumed, in his discussions with the company, that 'His Majesty's Government were desirous rather of asserting their rights to South Georgia, than of exacting any onerous terms from the Company'.(16)

## 3.2 The first four leases

Both the Colonial Office and the Falkland Islands Government had played an essentially passive role in developing any kind of whaling industry. The initiative had been almost entirely Norwegian. Every endeavour was however made to meet the companies' requirements, with the Norwegian Legation in London often playing an intermediary role. As a consequence, the Dependencies became a busy whaling ground, bringing prosperity to the whaling communities of southern Norway and some support of the Falkland Islands' economy. Allardyce had been given no choice but to grant Pesca a lease. A meeting of the company's representatives with the Governor's Executive Council had quickly resulted, on 8 March 1906, in a 21-year lease for Pesca dated from 1 January for a yearly rental of £250, requiring, among other things, an annual report of the number and type of whales caught, details of the whale products produced and a record of the weather (Hart 2001, 72–73 and 82–91).(17)

Now with the foreknowledge of the Colonial Office, the South Georgia Exploration Company accepted a 21-year grazing and mining lease (excluding the area leased to Pesca) for a rental of £250 a year and the payment of a 10 per cent royalty on all minerals extracted, subject to no less than an initial investment of £800 in 'a suitable whale oil factory'. Allardyce wrote of his 'considerable gratification' at securing what was in effect the sovereignty of 'the most southern Dependency of His Majesty's Dominions' through revenues amounting to one 'thirteenth of the present total revenue to this Colony'. The Secretary of State, Lord Elgin, gave his 'special sanction' to the arrangements on 9 August 1906 (Hart 2001, 91–93).[(18)]

A series of whaling ordinances followed, as the Falkland Islands' Legislative Council enacted, on 5 October 1906, *An Ordinance to regulate the Whale Fishery of the Colony of the Falkland Islands – The Whale Fishery Order, Number 3 of 1906.* Approved by the Secretary of State, Lord Elgin, the requirements first imposed on Alexander Lange were extended to every expedition, that is, the payment of £25 for a licence to operate a floating factory and two catchers, the embarkation of a customs officer on each boat and payment of a royalty on every whale caught and processed. The Whale Fishery Order had arisen, as Emil Husby writes, from a concern to prevent over-fishing rather than from any ethical consideration of animal welfare. Continued and optimally increased revenue was sought particularly following a decline in the colony's revenues over the previous five years (Husby 2018, 63 and 68; Roberts and Howkins 2023, 17–18). Each licensee undertook, as an acknowledgement of the free use of the colony's harbours and waters, to 'render to the Colonial Treasurer an accurate account of the number and description of whales taken', as well as 5 shillings towards the daily costs of having a customs officer aboard – that person being taken aboard on the outward journey and brought back from the grounds at the close of the season. A second major provision of the ordinance specified those dependencies in which the expedition might operate. A third imposed a limit on how many whales might be taken – this was never acted on. Rather than impose any limit to the catch, a limit was imposed upon both the number of vessels and the area over which they might operate. As well as a licence fee of £250, a royalty of £10 was imposed on each right whale caught, £5 for a sperm whale and 10 shillings for other species. The Governor-in-Council was empowered to impose fines of up to £300 for any breach of the regulations (Tønnessen and Johnsen 1982, 180–181; Hart 2001, 93–94 and 2006, 27–278; Dudeney and Sheail 2019, 43).[(19)]

Although a naval visit was no longer strictly required, HMS *Sappho* left its squadron in Newfoundland to proceed directly to Montevideo, from where the 1,460-mile voyage was made to South Georgia at 'an economical speed to husband her coal for unforeseen contingencies'. South Georgia had an 'evil reputation' for its 'inclement weather, with fogs and drifting icebergs'. The British Minister at Buenos Aires was therefore asked to 'ascertain at all costs the exact site of the Argentine fishing station, as many days might be lost in searching the coast'. Bertram Cox remembered reading an account of the whaling station published in the *Scottish Geographical Magazine.* The editor forwarded the Colonial Office enquiry to its author, Robert C. Mossman. As the meteorologist on William Speirs Bruce's Scottish National Antarctic Expedition of 1901–04, he had agreed to stay on for a further season, following the transfer of the observatory on Laurie Island in the South Orkneys to the Argentine Meteorological Service. He responded by informing the Colonial Office of how little was known of the actual shore station, but there was no doubting that the company Pesca, 'floated largely by Senor Tornguist, a leading financier in Buenos Aires', was 'actively prosecuting whaling in the southern seas'. Mossman had seen for himself, when passing through Buenos Aires the previous March, a brig come 'into port with 60 tons of whale oil which had been shipped at the boiling station'.[(20)]

Commander Hodges completed his prescribed report to the Admiralty on his return to Montevideo on 11 February 1906, by recounting how HMS *Sappho* had first seen 'the snow-capped mountains of South Georgia' through the mist seven days after leaving Montevideo at a point which they surmised to be Cape North. They had turned to run 'parallel to the land', but 'the weather was so thick that it was impossible to determine our exact whereabouts'. Taking soundings at very frequent intervals, they had proceeded cautiously until recognising 'the island and small rocks lying a little to the westward of Cape Buller', after which they proceeded with greater confidence up the coast until they recognised the entrance to Cumberland Bay that evening where, after some two-and-a-half hours of 'sounding continuously', they entered 'the small bay at the head of which the whaling station of Grytviken is situated and a boat having come off to pilot us in we entered the bay and anchored'.(21)

As the ship's officers began to survey what they had already named King Edward Cove, Carl Larsen showed Hodges over the station, comprising his own house, brought from Norway in sections, 'some huts for the men, the boiling factory, a slip fitted with powerful winches for heaving the whales up on shore, blacksmith's shop, etc'. Everything was 'most complete', including a wooden pier where ships could take on and unload cargoes. Electric lighting was shortly to be installed. As well as the dismantled barque *Louise*, which acted as a general store (containing 'a thousand tons of welsh coal'), there were two whaling steamers *Rossita* and *Fortuna*, both of 200 tons, and the *Cachalote*, a steamer of 900 tons commanded by Carl Larsen's older brother, which had been purchased to bring stores to the station and return loaded with whale oil (Headland 1984, 67 and 121).

Hodges went on to describe how the company had begun whaling in mid-December 1904 and boiling from the following month, since when 189 humpback, 22 fin, 18 blue and seven right whales had been caught, yielding 7,434 barrels of oil. The Argentine market at that time could take no more than 6,000 barrels per annum, with the remainder being exported at great cost, and little profit, to Europe. There was 'absolutely no lack of whales' at least in summer, their distance from the shore depending on where the best feeding grounds happened to be. The milder the weather, the closer they came to land, with some even entering the bays. The right whales were the most eagerly sought for their whalebone, but they passed through the area only from March to May. Larsen intended to take the *Cachalote* northward the next winter 'to try to discover where they go and to kill them there'.

With the survey of King Edward Cove completed, HMS *Sappho* left on 5 February. Together with accounts of the birdlife and domestic animals brought to the settlement, Hodges described the island as very mountainous and heavily glaciated, with permanent snow covering all the high peaks and sheltered lower ground even in summer. The vegetation was of the slightest, principally moss and bunches of tussock grass, with little flat land. The damp climate prevented the extensive areas of peat from being used for fuel. There was little whaling after late May, or indeed any outdoor work in the 'continuous succession of gales' of June, July and August. The winds were 'extremely changeable' whatever the time of year, shifting 'very suddenly from one direction to another and increasing and diminishing with equal rapidity'. Hodges had experienced them at first hand, having joined Larsen on a short coastal voyage. 'The squalls come off the mountains with [such] extraordinary violence' that 'big and well found whaling vessels' had to be used. Larsen's intention had been to impress upon Hodges the need for accurate surveys to be made of Fortuna Bay and Antarctic Bay, both of which had good anchorages. Hodges and Larsen agreed on the best vantage points for two Whigham's Automatic lights to be installed – as a Foreign Office official recorded, 'an interesting report of a most inhospitable spot'.(22)

A further threshold was crossed some two years later, when a third and fourth lease was awarded for shore stations, both dated from 1 January 1908, with floating factories being used until the stations were built. Sandefjord Hvalfangerselskab was founded specifically to purchase and refit a second-hand British steamer as the floating factory *Fridtjof Nansen.* With the season so imminent and still no licence obtained, the Norwegian Legation arranged for the expedition manager, Peder Bogen, to call at the Colonial Office, where the Norwegian Minister had secured a meeting with Bertram Cox, who had discovered that Winston Churchill (the Under-Secretary of State) had taken the relevant papers home with him to Blenheim. Cox hoped the licence handed to Bogen was identical with the one previously given to Alexander Lange. Bogen made directly for South Georgia, where his 'Boiling and Manufacturing Establishment' was wrecked on rocks with the loss of nine lives. There was no alternative but for the catchers to return to Norway. Bogen returned the next season fully licensed aboard another refitted secondhand vessel *Fridtjof Nansen II.* The company's shore station at Stromness was opened in March 1913. Dividends of 60, 120 and 100 per cent were paid in the following three seasons. The fourth leaseholder, Tønsbergs Hvalfangeri, similarly used a floating factory between 1907 and 1913, when sufficient capital had been raised for a shore factory at Husvik Harbour (Tønnessen and Johnsen 1982, 174–517 and 183–184; Headland 1984, 258–259).[23]

The arrival of the next Norwegian company was more complicated. Swinhoe's South Georgia Exploration Company was never able to raise the capital required for a whaling fleet, let alone a shore station. The lease was eventually purchased in 1909 by another Sandefjord company, Bryde and Dahls Hvalfangerselskab, which was owned by Thor Dahl and managed by Ingvald Bryde. The Colonial Office drafted a fresh lease from 1 January 1908, which Nansen accepted on the company's behalf, subject to a third catcher being allowed – which the Colonial Office conceded for an additional rent of £100 (Headland 1984, 113; Tatham 2008).

The length and small number of leases granted provided both the structure and permanence required by the companies. However, the Colonial Government soon regretted allowing leases of 21 years which, following the precedent of the Pesca lease, the Norwegian companies had demanded. A fine judgement was accordingly required by the Colonial Government in determining how many leases should be granted. Pesca had responded to the award of the additional leases by seeking a further lease of its own at Jason Harbour, which its catchers had already begun to use for holding 'their catch or to lie during heavy weather' (Figure 5). An 18-year lease was granted from August 1909 (thereby aligning it to the Grytviken lease), whereby the company might operate as many as four catchers off South Georgia, using 'either Harbour as the lessees saw fit' (Headland 1984, 110). It was through such adaptation to operating conditions that the Southern Ocean quickly became the centre of a global whaling industry.[24]

## 3.3 Licensing the floating factories

As with the shore station leases, there was no clarity on how many licences for floating factories should be granted, particularly as there was no way of policing those waters. The Falkland Islands Government had no vessel of its own. The Admiralty saw the retention of the Falkland Islands as essential but had neither the need nor the ability to range farther than Stanley. It had only very exceptionally agreed to allow HMS *Sappho* to visit South Georgia. The Norwegian

**Figure 5** The whaling station at Jason Harbour erected by Carl Larsen in 1909 to gain the right to use two more whale catchers. It was never developed further once the lease was granted. © John Dudeney.

whalers knew well enough the physical constraints under which the Colonial Government operated, but their own Government was in no better position to assert its authority. Norway's Legation in London never went further than to seek clarification of the British claim to the island groups in the Scotia Sea (Dudeney and Sheail 2019, 48–50).

The determining factor for both the whalers and the Colonial Government was their mutual need for one another. The floating factories required harbours, or at least sheltered waters, in which to flense the whales of their blubber and for operating the on-deck plant for 'trying-out' the oil. Although the ship's boilers had been refitted to heat the onboard cookers, they were never sufficient to provide power for the ship's engines and simultaneously cook the blubber. Cooking had to be done either in harbour or anchored in sheltered waters. Nor was there room for all the fresh water required for boiling the blubber. Water-boats were often used for fetching the water from reservoirs built for the purpose on close-by shores. The Governor used this dependency as the obvious pretext for access to those coastal waters and therefore recognition of British sovereignty.

Both the Governor and Alexander Lange were aware of the precedents they were setting. Lange had, upon reaching Stanley on 18 December 1905, asked for a licence to catch whales off the South Shetlands or, if prevented by ice or snow from doing so, from around South Georgia. The Colonial Secretary at Stanley gave him the licence the day after Lange requested it permitting *Admiralen* and its two catchers to operate off South Georgia, and 'to use the harbours of Our Dependency'. Allardyce emphasised, in a further conversation that afternoon, that the licence did not extend to the South Shetlands, which 'did not belong to England'. The licence required that Lange should take aboard a customs officer and deposit him at Stanley at the end of the season, at which point a one-half per cent tax would be paid on the oil and whalebone obtained (Bush 1982, 238, Elliot 1998, 13; Hart 2006 13–14 and 35–39).[(25)]

Lawrence Newing, a Stanley police constable and gaoler, joined the expedition as the customs officer on 21 December 1905. Lange, believing he was too late for the larger whales off the Falkland Islands, headed straight for the South Shetlands – having first put Newing ashore as no longer required. Two months were spent off the sheltered waters of Admiralty Bay (King George Island). Allardyce wrote much later that Lange had returned, four months later, 'more successful than he had reason to anticipate'. In addition to the 97 sei whales caught within the territorial waters of the Falkland Islands and 25 outside (where three sperm whales were also taken), the expedition had caught a further 58 whales off the South Shetlands. The one-half per cent levy imposed on Lange had raised £17 13s 9d. Although an extremely modest sum, Allardyce wrote of how 'my principal object' had been 'to develop a fresh industry, and in this I venture to think that I have been in a measure successful' (Hart 2006, 17–18).[26]

Lange was so well satisfied that before leaving Stanley for Norway, he had purchased a further season's licence. Instead of a one-half per cent duty, Lange had agreed to 'a fixed royalty' on 'every whale taken within or brought within territorial waters'. Ten pounds would be paid for a right whale. 10 shillings for a sperm and 5 shillings for other species. Allardyce anticipated many more applications for a whaling licence once Lange's success was known. Should that happen, he wrote, 'my only regret will be that as in the case of South Georgia this development should mainly be brought about by foreign instead of English enterprise' (Hart 2006, 13–23, 29–31, and 35–48; Dudeney and Sheail 2019, 48–49).[27]

The British flag had effectively followed Norwegian trade to mutual advantage. Through the Norwegian companies' recognition of British sovereignty, Allardyce had achieved his ambition of both increasing trade and obtaining the regulation of the fisheries. The Colony would become more self-sufficient economically; and, by granting the presently-licensed companies a virtual monopoly of the grounds they would be encouraged to invest further. No one appreciated the position better than Fridtjof Nansen, the Arctic explorer and now the Norwegian Minister in London. There were however flaws in the agreements that Allardyce had negotiated. The first and most worrying diplomatically, as he confessed in his despatch of 29 April 1906, was the impossibility, 'in the excitement naturally consequent on whale fishing', of distinguishing the whales caught within territorial waters from those on the high seas. Secondly, while Lange might be trusted to comply with the regulations, it was soon clear, as more licences came to be granted, that 'if fishing was to be properly supervised and controlled it was imperative that a reliable customs officer should accompany each expedition'. There was no other way of verifying 'the number or class of whales caught,' or to protect the colony's revenue.[28]

Harry Batterbee recorded in the Colonial Office, in February 1907, that the presence of a customs officer was 'certainly valuable to prevent poaching', and was therefore likely to be welcomed by licence holders. It was the customs officer aboard Lange's whaling steamer *Admiralen*, who had first alerted the Colonial Government to the Chilean-registered Magellan Whaling Company hunting whales off the South Shetlands without a licence. Adolf Amandus Andresen (the company owner) originated from Sandefjord and had previously been a tugboat captain trading from Punta Arenas, during which time he had encountered both right whales and rorquals. After training in Norway as a gunner, he had mounted a harpoon cannon in the bow of one of his tugboats, harpooning his first humpback in the Straits of Magellan in 1903. Now operating as the Chilean-registered company, Sociedad Ballenera de Magallanes, from a whaling station at Bahia Aquila in the Magellan Straits, Andresen had, from aboard the

*Gobernador Bories*, with its two catchers, become the first to operate from Deception Island in the South Shetlands (Tønnessen and Johnsen 1982, 157 and 175; Headland 1993, 193; Hart 2006, 12 and 41–42; Dibbern 2010, 212).[29]

The Colonial Secretary at Stanley sent a copy of the Whaling Ordinance to the Chilean Company on 9 April 1907, with a demand for a licence fee of £25, emphasising how, for such purpose, 'all whales taken are to be deemed to be taken within the limits of Colonial Waters' (Hart 2006, 44). The company's managing director, Pedro A. de Bruyne 'denied any attempt at evasion – it had been assumed that the South Shetlands were 'neutral like Spitsbergen in the Arctic seas'. He enclosed payment of £75 15 shillings for a licence and the royalties for the 203 whales taken. Now licenced, the *Gobernador Bories* and three catchers joined others operating in the territorial waters of the South Shetlands, together with 'the use of harbours of the colony free from rates and dues, the harbour of Stanley excepted'. Edward B. Binnie joined the vessel as customs officer, arriving at Deception Island in mid-December. Alongside his official report, his diary of the season's activity became a model of what was required of a customs officer. His natural history and photographic collections have come to be cherished by historians (Hart 2009).[30]

That is not to say that every ship's master adhered so willingly to the Whaling Ordinance (Husby 2018, 60–61). It was at Deception Island that Binnie encountered Captain T. Andersen Lystad (the master of Christensen's floating factory *Nor*), who claimed, by his use of a sailing ship, that he did not need a licence. He had sailed directly to the South Shetlands. Binnie visited the *Nor* every other day, inspecting the log kept of the whales caught (Hart 2009, 49). As Allardyce informed the Colonial Office in January 1908, Lystad's purpose was undoubtedly to 'free himself from Government supervision'. Lystad had fortunately been replaced as master by the next summer, but he was only the first of many (Hart 2006, 39–40).[31]

With no further leases available for South Georgia, the impetus had shifted to the South Shetland Islands, where the number of catchers rose from ten in the summer of 1908–09 to 22 in the following summer, with, as Tønnessen and Johnsen (1982, 194) phrase it, 'the great leap forward' to as many as 30 in 1910–11. Output had risen from 41,000 barrels to 93,600 barrels, and to 148,000 barrels in the third summer.[32]

## 3.4 A British lease

There was no stopping the Norwegian companies, even if the Governor and Colonial Office had wanted to do so. It was a collaborative relationship from which both parties benefitted, but fragile nonetheless. The Norwegian companies evaded the regulations wherever possible, and both the Governor and Colonial Office regretted that there was no British whaling company (Dudeney and Sheail 2019, 55–57). Allardyce wrote in January 1908 of his keenness 'to afford a purely British industry all possible facilities'. None had emerged until the Newfoundland Steam Whaling Company applied for a licence to pioneer the South Orkneys whaling ground in December 1907. Captain Nokard Davidsen, aboard the floating factory *Sabroan* (the largest of its class at the time and accompanied by two catchers) had called at Stanley seven days previously, both to collect his licence and to take a customs officer aboard.[33]

Some historical context is needed. The Gray Brothers of Peterhead had proposed 'an exploratory expedition' to catch the southern right whale in the early 1890s. Baron Adolf Erik Nordenskjöld (the Finnish/Swedish geologist and Arctic explorer) was nominated as naturalist, but sufficient funds could never be acquired for the expedition to proceed. Robert Kinnes (the owner of the Tay Whale Fishing Company of Dundee) was considerably more successful in raising sufficient capital for four auxiliary sailing ships. As this was the first British venture to the Southern Ocean for half a century, it was suggested that there should be 'one or more scientific gentlemen' to study the magnetic variation and other such phenomena at so high a latitude. Benjamin Leigh Smith (the Arctic explorer) secured both nautical instruments and the appointment of two young naturalists, ostensibly as ships' surgeons. One was William Speirs Bruce, a medical student and 'enthusiastic naturalist and experienced physical observer'. The other was William Burn Murdoch, who acted as Bruce's assistant. Murdoch published an account of the expedition, which included both his own artwork and a chapter by Bruce (Anonymous 1892 and 1893; Murdoch 1894; Mill 1951, 137–138; Williams and Dudeney 2018, 63–98).

The expedition was as great a financial disaster as the near-contemporary Norwegian expeditions of the mid-1890s. At a chance meeting in the Weddell Sea with William Speirs Bruce, Carl Larsen had confessed that there had been 'countless shoals of blue whale and humpback', but no right whales. Speirs Bruce recalled many years later how one of the Scottish boats had succeeded in harpooning a blue whale. The dorsal fin was so worn down as to suggest that James Clark Ross might have misidentified it and similar others for right whales. A more likely explanation is that Ross had been correct in identifying schools of right whales, but over-hunting on their summer breeding grounds had since driven them to near extinction.[(34)]

Sir John Murray had convened a meeting at the Royal Scottish Geographical Society in Edinburgh of 'representative commercial citizens of Edinburgh and Leith' to learn more about what the expedition had found. The oceanographer William John Dakin also refers to the meeting, at which Bruce and Burn Murdoch were joined by Henrik Bull, their purpose being to press the feasibility of 'a modern whaling station in South Georgia'. Bruce grumbled many years later, 'What did a scientist and an artist (Burn Murdoch), know about whaling, or the Norwegian expert associated with them!' (Parliamentary Papers 1920, 34). The historian of Christian Salvesen and Company also refers to the meeting and how such figures as Theodore Salvesen saw little prospect of a shore station thriving on so bleak an uninhabited island at the other end of the world – even if a workforce willing to go there could be assembled (Dakin 1934, 167–168; Vamplew 1975, 151; Elliot 1998, 12).

Christian Salvesen and Company of Leith were a major shipping broker and agent. As the leading Scottish agent for Norwegian whale oil, it supplied the northern shore stations with coal. Salvesen had, by the turn of the century begun to delegate. His fourth son, Theodore, was given charge of the timber and whale oil trade in 1899, and shortly after formed with his brothers a separate Olna Whaling Company to develop further Norwegian-style whaling stations, first at Olna in the Shetlands and then Thorsvig in the Faroes and Hellesfjord in Iceland. Their more comprehensive use of the whale carcase had increased their profit, allowing the company, within a decade or so, to become the largest whaling company in the world, operating as many as 20 catchers from five whaling stations in both hemispheres, catching 2,350 whales and producing 66,500 barrels of whale oil by 1910. Even so, ship brokering remained the company's most profitable activity until the First World War (Vamplew 1975, 147–149; Somner 1984, 3–6; Vamplew 1990; Elliot 1998, 9–10 and 17–18).

Although refused a lease for a shore station at South Georgia, Salvesen opened one at New Island on West Falkland in 1908; the station buildings of its Icelandic station were dismantled, shipped to New Island and reassembled for the new station, along with 60 men, thousands of oil barrels and a stock of coal in 1908. Whaling began in January 1909 but was never profitable; only 226 whales were caught that season, producing a mere 3,850 barrels of oil and 3,081 sacks of meat and bone meal. Although a fourth catcher enabled as many as 475 whales to be caught in the 1909–10 season, the numbers thereafter fell to 103 and as few as 87 in the 1912–13 summer. The station was finally dismantled in 1915 (Vamplew 1975, 150–155; Tønnessen and Johnsen 1982, 188; Elliot 1998, 14).

The Falkland Islands might have been off the main migration routes, but South Georgia was increasingly seen as 'a lucrative field for the investment of whaling capital'. But with no 'civil servant' able to visit the dependency, and 'no immediate prospect of one doing so', Allardyce had to rely entirely on what those with an obvious vested interest chose to tell him. Lauritz Larsen, whose 'experience and residence in South Georgia' made him, in Allardyce's words, 'probably unique', claimed that over £100,000 worth of whale oil and bone had been obtained during the 1907–08 summer. The sum that Pesca had paid for its annual lease was less than one-quarter of one per cent of the value of the season's catch. Larsen's purpose in visiting Allardyce had been to lobby for a lease of his own. As Allardyce wrote shortly after to the Colonial Office, 'The waters around the Dependency were literally teeming with whales, and . . . there were many more good harbours than those shewn on the Admiralty chart.'[35]

Allardyce proposed an increase in the number of whaling leases at South Georgia both to accommodate 'the only British firm in the business' and mollify the Norwegian minister by also awarding leases to 'two of the earlier "Norwegian applicants"'. Erling Lund was awarded the fifth lease on 1 October 1909, and Ocean Hvalfangerselskabet of Larvik the sixth 10 days later. Formed for the purpose, with a capital of a million kroner and Lauritz Larsen as manager, the station was eventually built at New Fortune Bay. Henrik N. Henriksen (the manager of Salvesen's New Island shore station) had chosen Stromness Bay, where the transport ship *Starlight* arrived in November 1909 to begin erecting the station at Leith Harbour, taking its name from Salvesen's home port in Scotland.[36]

Every shore station was positioned in one of the deep fjords on the northeastern side of the island, with Grytviken on the larger arm of Cumberland Bay and Leith Harbour, Stromness and Husvik on three adjacent arms of Stromness Bay (Vamplew 1975, 154; Headland 1984, 113; Hart 2001, 122–126). The Salvesen factory at Leith Harbour had cost £117,656 when completed – twice the cost of the other stations. The most modern pressure boilers had been installed to process the blubber, along with 36 meat and bone boilers and a guano factory with four driers to make the greatest use of the rest of the carcase. As with all companies, the labour force was predominantly Norwegian, both at sea and on the shore stations, totalling almost 500 people in the austral summers (Vamplew 1975 and 1990; Tønnessen and Johnsen 1982, 187–190; Jackson 2004).

Theodore Salvesen was anxious to expand further and had tried to purchase the South Georgia Exploration Company's derelict lease. Colonial Office officials had welcomed the move but also emphasised that the Secretary of State must necessarily stand apart from any purchase negotiations so as to avoid any possible appearance of favouritism. As officials recalled some five years later, Salvesen never stopped asserting that the Colonial Office had connived to ensure that Bryde and Dahl won the lease – Salvesen's letter of acceptance of the terms had arrived a day too late. Salvesen had thereupon responded by seeking to purchase the lease that had been offered to Erling Lund. As Gilbert E. A. Grindle (a newly appointed Principal

Clerk) confessed, no one in the Colonial Office had thought of the ruse of purchasing another person's lease to secure additional entry to the industry. The Salvesen company historian wrote that Salvesen persuaded Lund to sell his promised lease, subject to the Colonial Office approving its transfer. Theodore Salvesen contended at the time that Lund had made all the running. He also noted that there was nothing in the Whaling Ordinance of 1908 to prevent one company from holding two or more leases. Grindle readily acknowledged that this was true, and Salvesen was a British firm anxious to 'use the whole whale without waste'. But it was also 'rather a grasping firm' seeking to eliminate every possible competitor. The Colonial Office refused to transfer Lund's promised lease. Theodore Salvesen thereupon enquired whether there would be objection to a new British company being formed, registered at Leith, specifically to purchase Lund's promised 'whaling concession', together with 'the pasture and mineral rights both at South Georgia and the South Shetlands, on the same terms as those previously granted to the South Georgia Exploration Company'. Should the Colonial Office agree, 'some of our friends, who are interested in mining ventures and geology', would be invited 'to associate themselves in the new company'.[(37)]

Salvesen's proposal was again an obvious ruse to circumvent the Colonial Office's objection to his company acquiring a further lease at South Georgia, but, as officials reasoned, Lund could himself have turned to Salvesen for the capital required to build and operate his shore station. As Grindle recorded, the Colonial Office had rightly thought it best not to suggest the ruse because 'we do not altogether trust Messrs Salvesen and we should have put ourselves too much in their power by suggesting to them a means of evading the rules'. But now that Theodore Salvesen had thought of it himself, Grindle noted, 'I think we may accept it and wink at the evasion'. It would also 'get rid of Lund who is a man of straw and wants a cession merely to hawk it about'.[(38)]

On the purchase of Lund's interest for £1,200, the South Georgia Company of Leith was formed as a British limited liability company, with a capitalisation of £20,000. As Grindle acknowledged, with so large an investment and appointment of one of Salvesen's partners as managing director, there was every indication of its being 'a genuine concern'. A geologist, David Ferguson, was engaged in 1912 but found nothing of mineral value on the South Georgia lease. The South Georgia Company had meanwhile, through the purchase of Lund's lease, obtained the use of Allardyce Harbour (later called Rosita Harbour) and 'any other harbour not occupied by another company'. No shore station was ever built (the site reverted to the Falkland Islands Government in 1924) as Salvesen had found it considerably more profitable to operate the company's entire fleet from Leith Harbour. The 26-year old steamship *Neko* and four purpose-built catchers were purchased in 1911, together with a further secondhand ship *Horatio* to cook out the carcases, which neither the *Neko* nor Leith Harbour could handle. *Horatio* caught fire in Leith Harbour and was lost, together with its full cargo of oil in March 1916 (Vamplew 1975, 152–156 and 160; Somner 1984, 49–50; Elliot 1998, 20–1).[(39)]

## 3.5 Legitimising regulation

The whaling companies had welcomed the Whale Fishery Ordinance for the access it provided them and the protection it offered from what could otherwise have been unfettered competition, both in the marketing of the oil and the impact on local whale stocks. They might nevertheless voice misgivings, either when calling at Stanley or through the Norwegian Legation in London (Husby 2018, 100–101). The Chargé d'affaires, Johannes Irgens (a Christiana lawyer), sought through the Foreign Office the evidence by which the British claim was made to the respective territories. The Colonial Office responded, as explained in a letter to the Foreign Office of 7 May 1906, that 'no Expedition of any foreign power appears to have taken possession' of those particular territories (Dudeney and Sheail 2019, 49–50).

Far from brashly asserting British sovereignty, Colonial Office officials reasoned that there was every need to avoid rancour and any sense of grievance. The fact that Captain Nuñes and the Guardia Nacionale had been so closely involved in securing the legitimacy of Pesca's development at Grytviken caused Bertram Cox to write in the Colonial Office, in March 1906, that it might be assumed 'the Argentine Government does not question the British title to the Island'. An obvious reason for rejecting Allardyce's suggestion that he should visit the South Shetlands and South Orkneys, to hoist 'the British flag' on them, was that this might imply doubt of 'the sufficiency and validity of the grounds on which that claim was based'. It was much more important 'to preserve the right of British subjects to resort to 'those territories, whether for whaling or other purpose, than 'to impose onerous conditions on foreigners resorting to them for whaling and sealing'.(40)

Irgens continued to probe the Foreign Office, reporting, in February 1906, that 'a company of whale fishers in Sandefjord, the well-known centre for this business in Norway', intended to send 'a whale fishing expedition to the Southern Ocean'. He enquired therefore

1. Which clusters of islands and which countries are belonging to the British empire in the territory between 40°–65° southern latitude and 35°–80° western longitude?
2. Are the South Shetland Islands and South Orkney Islands considered to be international territory?
3. Could the said Norwegian company obtain permission from the proper British authorities to land at some harbour on the Falkland Islands or on South Georgia with a floating whale fishing establishment and to carry out whale fishing from the same harbour, if this should be considered necessary?

The purpose of the proposed expedition was 'merely ordinary whale fishing like similar expeditions to the Shetlands and the Faroe Islands, only with this difference that the present expedition is supplied with a floating boiling and manufacturing establishment instead of the usual fixed establishment ashore'.(41)

Allardyce wrote, in late February 1906, of how 'our Sovereignty over South Georgia', and indeed the South Shetlands and South Orkneys, should be placed on 'a satisfactory basis'. Still seen by many as a 'no man's land', the valuable seal and whale resources made it timely 'to consider the question before the adjacent South American Republics move, as they are pretty certain to do, in the same direction'. The archipelagos were not only attracting an increasing commercial interest, but Alexander Lange had assumed 'they were already a protectorate of this Colony'. Their formal incorporation would permit trade to be 'increased and the fisheries regulated'. The Colonial Office had already consulted the Admiralty as to 'the national status' of the South Shetlands, now they had become 'a matter of any material consequence', and more specifically what further steps should be taken

> *to uphold the British claims to these Islands, as, if they should be allowed to fall into the possession of a foreign state, British subjects would be debarred from fishing or sealing within the territorial waters.*[(42)]

The optimal course was to treat 'the South Shetlands, South Orkneys and Graham's Land . . . in the same manner as South Georgia, as dependencies of the Falkland Islands', thereby requiring the Norwegian companies to apply for a licence in the same way that Alexander Lange had done. Irgens's omission of the South Sandwich Islands in his original enquiry had similarly caused the Colonial Office to overlook them. As Bertram Cox warned the Foreign Office in August 1906, 'Captain Cook seems only to have discovered, and not landed upon, this group of islands'. British claim might consequently not be so strong as that over other islands in the South Atlantic. By the Whale Fishery Ordinance of October 1906, every expedition was made liable to the requirements first imposed on Lange: the payment of £25 for a licence, embarkation of a customs officer and royalty on every whale caught and processed.[(43)]

It was not until 2 September 1907 that the Norwegian government wrote again, this time taking up Sir Edward Grey's offer, as Foreign Secretary, to consider any particular grievances. The most important were the royalties levied on the whales by the Colonial Government, whether they had been caught within territorial waters or on the high seas. The second was the requirement for a customs officer to accompany every expedition, and therefore the need to call at Stanley. The third was to require separate licences for each of the whaling locations off the Falkland Islands, South Georgia, the South Shetlands and South Orkneys. Each expedition should have the flexibility to go where, say, ice conditions were optimal (Dudeney and Sheail 2019, 51).[(44)]

The Colonial Office had already discovered that Allardyce had interpreted the Whaling Ordinance so widely that a royalty was being charged on every whale taken, regardless of whether it had been caught in territorial waters or beyond. Officials were appalled that he should have done so. As Arthur E. Collins (the Principal Clerk) wrote on 18 March 1907, there was very great objection to charging for whales caught on the high seas, particularly where 'the whalers concerned are entirely of foreign nationality'. The so-called royalty should be commuted to an annual licence, as acknowledgement of the facilities provided within the territorial waters, albeit it should be set as low as £100 since nothing was provided other than anchorages to which 'our title is slight'.[(45)]

Batterbee (now a Second-Class Clerk in the restyled Crown Colonies West Indian Section) thought the Governor had made a strong case for imposing a licensing system. The proposed charge was hardly 'excessive considering the large profits which they have made'; the term 'royalty' must however be dropped. Whatever term was adopted, Hubert R. Cowell (a Second-Class Clerk) acknowledged, in September 1907, that 'the Department had not realised a charge was being levied 'on whales caught outside territorial waters'. As Collins further noted, on 2 October 1907,

> *some of those Dependencies are only ours by a questionable title, and to take the opportunity of levying royalties on whales chiefly caught in the high seas because whalers have to boil them down on or in the neighbourhood of these Dependencies, where the Govt provides nothing in the way of civilization, and is only represented by Customs Officers employed to count the whales and maintained at the expense of the whalers, seems to me a predatory proceeding, not usually associated with a British Govt.*[(46)]

Bertram Cox disagreed, minuting that the Norwegians knew perfectly well that

> *we have never set up a claim to whales caught on the high seas. What we do is to claim payment on whales caught in or brought within territorial limits. All the boiling down of the whales and extraction of the oil is done within British territory the vessels putting in to our waters for that purpose whether the actual work is done on board or on shore.*[47]

The Secretary of State should both disclaim 'any desire to tax whales caught on the high seas' and warn the Foreign Office that any removal of the customs officer posted on the whaling fleets would mean that 'no one will know what is going on in these remote dependencies, which is probably what the Norwegians want' (Dudeney and Sheail 2019, 51).

Allardyce acknowledged, in a despatch of 28 December 1907, that the term 'royalty' was 'perhaps somewhat of a misnomer', but there was no better way of checking unlicensed whaling and obtaining some form of payment. It was considerably cheaper than 'sending a man-o-war to patrol more or less un-surveyed waters, seldom free from ice'. The whalers, in making the present form of payment, knew perfectly well how impossible it would be to operate in 'this tempestuous latitude unless the commodious and magnificent harbours of this Colony and its Dependencies were thrown open to them'. The industry would otherwise collapse. The income had also begun to provide the Colonial Government with 'a small source of profit' as 'a consequence of being a source of great wealth to a few foreign Companies'. Grindle (the First-Class Clerk) effectively closed the discussion by minuting that any system of charging for whales that had been caught on the high seas was most objectionable whether called 'a royalty or anything else'. Even if the Norwegian companies were prepared to continue paying the charge, there was a real possibility of the United States' whalers being attracted to Antarctic waters. From the evidence of their nineteenth-century sealing parties, they would hardly 'submit to an imposition which could not be defended before any international court'. The wiser course was to levy 'a licence fee of £250 for whaling in all Colonial waters'.[48]

As the Colonial Office informed the Foreign Office, on 3 April 1908, the Secretary of State recommended that the present royalty should be replaced by 'a fixed fee of moderate amount' – £100 per whaling licence considering that the Colonial Government maintained 'no establishment and does nothing to assist navigation in the dependencies'. Foreign Office officials both agreed and informed the Norwegian Minister, emphasising that the Governor had never sought 'a royalty in the strict sense of the word'. He had merely claimed the right to make charges for 'the privileges of using Colonial waters and harbours'. The royalty system had seemed 'a convenient method' of estimating to what extent the individual whaler made use of those colonial waters and of charging him in proportion. Having put the legitimacy of the previous policy on formal record, Sir Louis Mallet (as Assistant Secretary) indicated that to avoid further misunderstanding, it was being replaced by a fixed charge of £100 per licensed floating factory and two catchers. A further Whaling Ordinance was enacted to that effect on 8 August 1908 (Dudeney and Sheail 2019, 48–57).[49]

In seeking to legitimise the Governor's actions, Colonial Office officials had come to realise how vaguely the dependencies were defined. The Letters Patent of 1843 had simply referred to them as the 'Falkland Islands and their Dependencies'. Allardyce was familiar with the process of defining territories from his quarter-century's residence on the Pacific island of Fiji, during which time the Cook Group and other 'such other islands in the Pacific' had been incorporated within 'the boundaries of the self-governing Colony of New Zealand'. Their limits had been defined by Letters Patent, of 13 November 1901, as 'the intersection of lines of longitude and latitude' (Heyburn 1980; Tatham 2008).[50]

The same interdepartmental meeting, whose recommendations had led to the abolition of royalties, agreed, on 5 March 1908, that 'the common law of England and statutes of general application, and any appropriate Falkland Islands Ordinances', should be extended to encompass the dependencies. The Whale Fishery Ordinances had made their incorporation especially urgent. Whereas the shore station leases constituted 'grants of the Crown', the Governor appeared to have no legal authority to enact the Whale Fishery Ordinance. He might therefore be sued for the recovery of those licence fees and payment of indemnities. But in doing so, Grindle emphasised that the Colonial Government must avoid any 'step which would bear the appearance of annexing the dependencies to the Colony'. That had been accomplished decades ago.[51]

By the Letters Patent of 21 July 1908, as passed under the Great Seal of the United Kingdom and published in the *Falkland Islands Gazette* of 1 September 1908, the Governor of the colony of the Falkland Islands was appointed Governor of South Georgia, the South Orkneys, the South Shetlands, the South Sandwich Islands and Graham's Land, and provided a government for them. The preamble to the Letters Patent identified the groups of islands, and the territory known as Graham's Land, as 'situated in the South Atlantic Ocean to the south of the fiftieth parallel of south latitude, and lying between the twentieth and the eightieth degrees of west longitude' as being also part of His Majesty's Dominions. As a Colonial Office official recalled, a decade later, since 'it was not known exactly what the limits of the territory were, the definition was therefore made so wide that it would be sure to include anything that might be called Graham Land' (see Figure 1 on page xx above; Dudeney and Sheail 2019, 51–55).[52]

The same issue of the *Falkland Islands Gazette* that published the new Letters Patent also laid out the Falkland Islands Whale Fishery Ordinance of 8 August 1908, whereby it was made unlawful 'to kill, take, or hunt, or attempt to kill or take any whale in Colonial waters', without purchase of a licence. The Governor-in-Council might at any time extend the ordinances to regulate the number of licences granted and to define the limits within which the whales might be taken. Penalties could be imposed on conviction before 'a Stipendiary Magistrate or any two Justices of the Peace' of not more than £100, plus forfeiture of any whales, whale oil or whale bone found in that person's possession. A 'special exception' might be made where the Governor or duly authorised person consented to a whale being killed or taken for scientific or any other special purpose (Headland 1992).[53]

There continued to be enquiries as to the legitimacy of British claims to particular waters. Thor Dahl wrote from Sandefjord in September 1911, asking whether Joinville Island was a British possession – there was no mention of it in the Letters Patent. The Foreign Office library found no evidence that it had ever been specifically claimed. The Admiralty adjudged Joinville Island to be 'sufficiently continuous to Graham's Land to come within the jurisdiction of the Governor of the Falkland Islands as a part of Graham's Land'. Adolph A Andresen wrote to the Colonial Office, in April 1913, that he was fitting out 'an Expedition to catch and boil down whales at sea' on the west side of Graham's Land. He accordingly asked how far south the British territory extended so that he might 'try to get in south of that'. He was told the

British claim encompassed the whole of the area 'usually known as Graham Land, together with the islands contiguous thereto, but that in the present state of geographical knowledge Mr Harcourt [the Secretary of State] is not prepared to give the precise latitude and longitude of the southern boundary'(54) (Figures 6a and 6b).

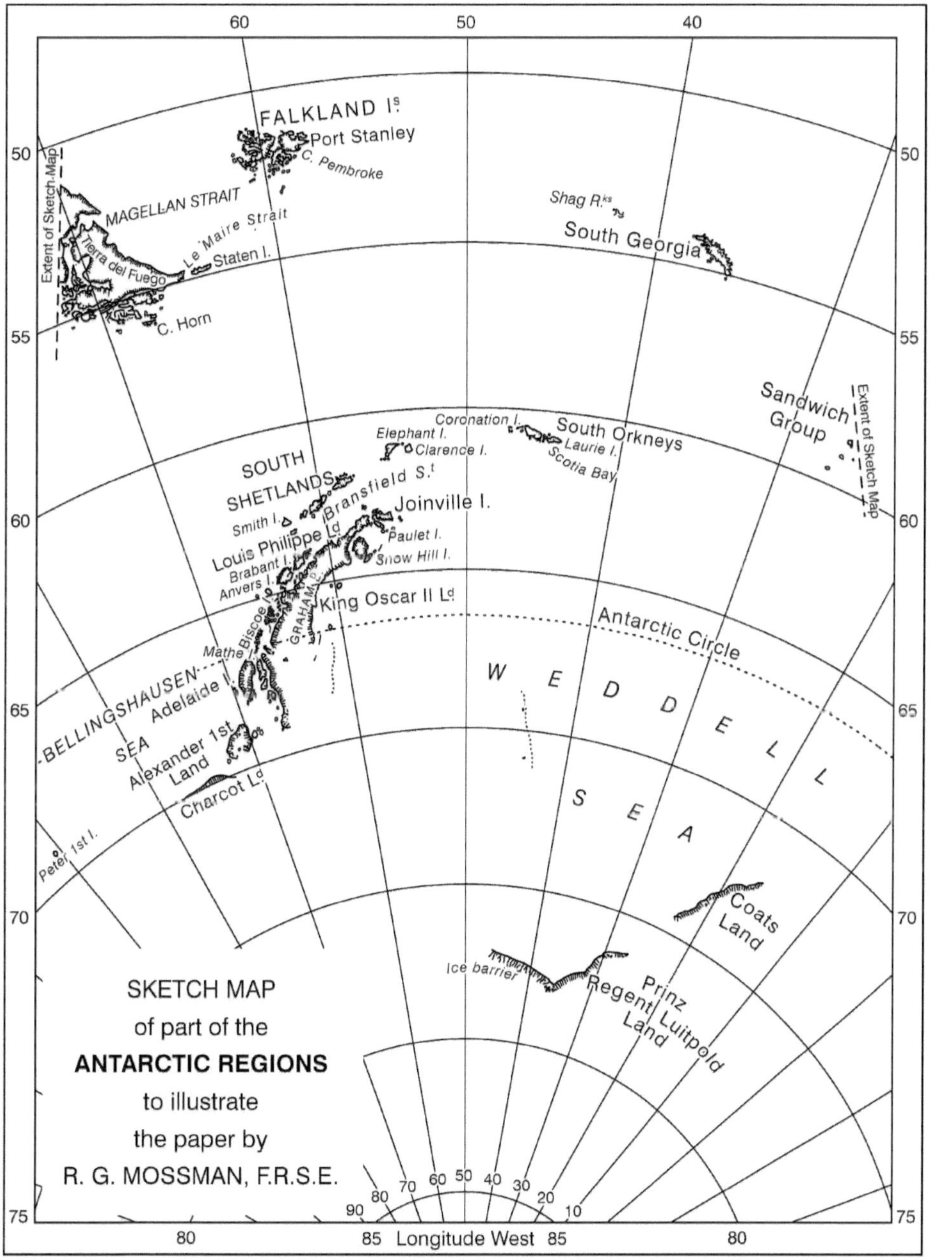


**Figure 6a** Sketch map of the extent of knowledge of the Antarctic region when the area covered by Letters Patent of 1908 and 1917 was defined. Reproduced from a diagram published by Robert Mossman in 1916. Mossman (1916) and reproduction © British Antarctic Survey, UKRI.

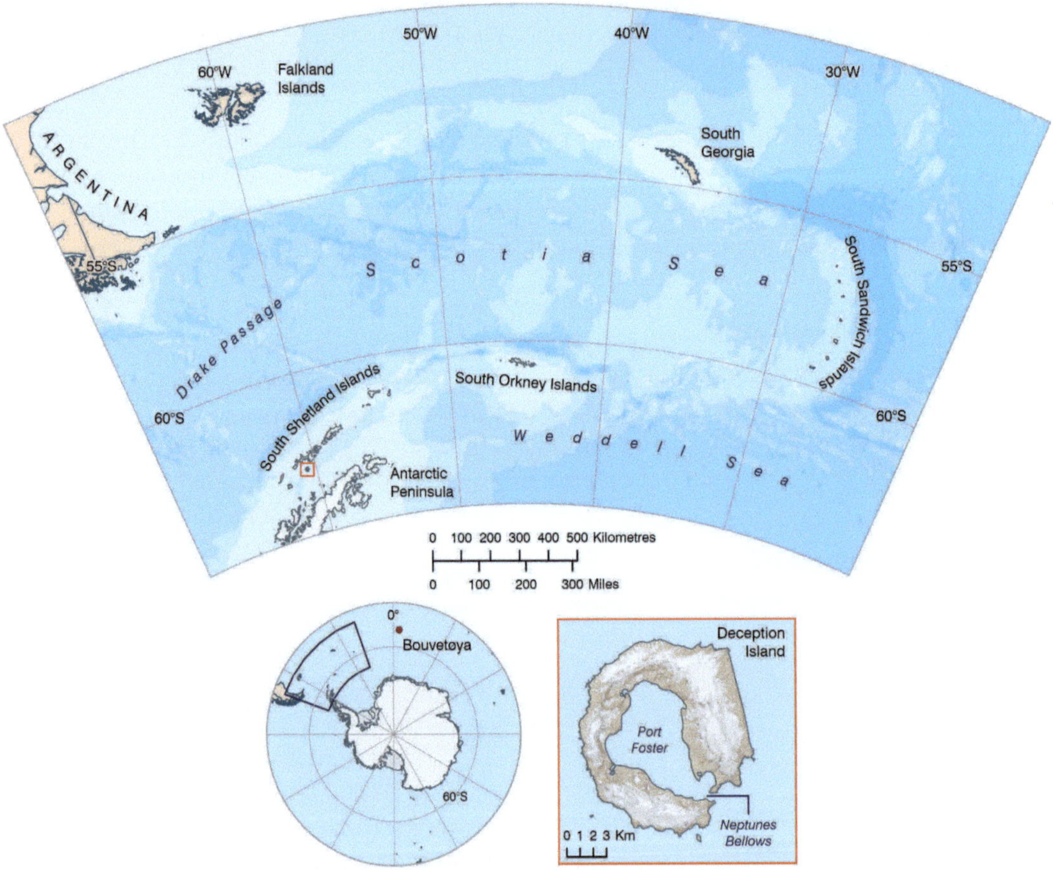


**Figure 6b** A modern map projection of the part of the Falkland Island Dependencies of relevance to the whaling industry for comparison with Mossman's 1916 map. Also shown inset is a map of Deception Island. © British Antarctic Survey, UKRI.

## 3.6 A second British lease

The Colonial Office was never sure how many leases and licences to grant or to whom they should be allotted. Allardyce reckoned, in May 1911, an eighth and even a ninth lease might be justified. Richard Irvin and Sons of North Shields (one of the three largest British trawling businesses) had applied for a licence, but however reasonable it would be to award the next lease to a British company, Alexander Lange had long sought one. While he had failed to use his third annual licence for the South Shetlands, which had accordingly been confiscated, he 'had rendered yeoman service to this Outpost, and done much to assist to bring about its financial prosperity'. An income of over £2,500 might be expected from 'those same Archipelagos' in 1911–12. Without wishing to be hypercritical, Allardyce wrote,

> *There was no getting away from the fact Messrs Irvin had come forward at the eleventh hour, to reap the pecuniary advantage of a lucrative enterprise which has been developed by others, and which it has in no way*

> *helped to build up. While distinctly anxious to give a preference to a British firm as against a non-British one, all things being equal, I am unable to see my way, in the special circumstances of this case, to favour the claims even of my own townsmen.*(55)

However, Richard Irvin and Sons had recently formed a partnership with the Swedish entrepreneur Charles O. Johnson in establishing a South African company, Irvin and Johnson. As Ian Hart wrote, in his volume *Austral Enterprises*, there was already a highly successful sealing and whaling business operating in the guise of the Southern Sealing Company (Hart 2020, 3–20).

Allardyce endeavoured to compromise by suggesting that Irvin and Johnson should be offered a whaling licence for the South Orkneys, causing John H. Irvin to protest how earlier discussions with the Colonial Office had encouraged the Southern Whaling and Sealing Company to purchase and refit 'a large tank steamer' as a floating factory prior to a shore station being built at South Georgia. The company had been formed for the purpose, with its entire share capital of £60,000 being provided by 'my Father, brother, myself and Mr Johnson who is in partnership with us in South Africa'. A further letter described how the floating factory was being equipped with 'the most modern machinery' and that its three catchers were already under construction at the Smith's Dock in Middlesburgh (Hart 2020, 27–28).(56)

The Colonial Office again emphasised, in late May 1911, that there were no further leases or licences available. Richard Irvin left straightway for London, telegraphing his wish for 'an interview the following forenoon'. A telegram, sent to the National Liberal Club, assured him 'a Member of the Department would be pleased' to receive him. As Irvin recounted, in a letter of 10 June, he had 'explained to the Gentleman I saw . . . the serious position in which we were now placed'. A licence had become 'a matter of almost life and death to us financially'. Both he and his father, three brothers, and Johnson had provided £50,000 of the total capital of £80,000 so far invested, with the balance 'borrowed upon their personal security and mortgages over the property'. They had done this on the earlier assurance that their application would be looked upon very favourably. They had been given high hope of licences for South Georgia or the South Shetlands, or both, and had 'felt justified in going ahead'. It was intended that the Southern Whaling and Sealing Company should be registered and insured within the United Kingdom. Although Norwegians would have to be employed, Irvin wrote that 'we are introducing as many Britishers as possible into the crews so that they might become acquainted with the work'.(57)

Darnley remarked that the industry had become so remunerative, every individual and company was leaving 'no stone unturned to try and get a footing there'. Even so, Grindle saw no reason to throw a further lease 'at the head of a Norwegian', however encouraging the position might be. There might not only be more aspiring British companies, but 'we are being pressed by the British Museum "to protect the poor whales"'. Edward J. Harding (a Second-Class Clerk) summarised the position. Seven leases had been awarded to date, to Pesca, Sandefjord, Tønsberg, Bryde and Dahl, Ocean, Salvesen and the South Georgia Company. The last two were 'really the same firm'. There were two points for decision: how many more leases should be awarded and to whom should they be given? Allardyce had recommended eight and possibly nine. Harding identified four credible applicants: Irvin and Sons, Alexander Lange, Johan Larsen and the Olna Whaling Company. Larsen might be excluded as 'a foreigner' and the Olna Whaling Company as essentially Salvesen's. Lange was a pioneer figure, but his promise of 'a considerable sum of money . . . locally for shares in his Company' might suggest the Governor was not wholly disinterested in supporting his application. However, Allardyce had rightly protested at the way Messrs Irvin were essentially taking 'advantage of the experience of others' before making any 'attempt to come in'. The question was whether this tardiness should 'outweigh the value of having another substantial British firm in the business'. Harding thought that if only one further lease was to be awarded, Irvin should have it. If two, then Lange should also have one.(58)

Grindle saw four reasons for holding back: first, to protect the whales, and secondly, out of consideration for the existing lessees. A third might be to keep the workforce at South Georgia within limits – there was only a magistrate and two constables to maintain order. A fourth reason was the length of each lease and the need therefore to go slowly about increasing the number. While none of these reasons should determine in itself the particular number of leases, Grindle thought it prudent to award only one at a time. Should that be so, the Secretary of State would favour 'a British firm', particularly as a refusal would very seriously affect Messrs Irvin. With 'some hesitation', George V. Fiddes (an Assistant Under-Secretary) agreed that the Governor might be overruled and the lease offered to Irvin. Lord Lucas (the Parliamentary Under-Secretary) thought it did 'not seem quite right to refuse him (Lange) everything'. He had the stronger claim and

> *we ought to find something for him if we can. The scientific opinion on the killing of the whales is very vague and there is no reason to think that too many whales are being killed at present or that the grant of a 9th lease in South Georgia will lead to this. I am not sure that we are not being rather over cautious in holding back over the grant of the 9th lease.*(59)

It was at this point that Darnley recalled the Governor's keenness to establish a shore station at Deception Island in the South Shetlands. On the premise that Lange knew the South Shetlands better than anyone else, he should be offered a lease for its development. As Lord Lucas minuted, 'Mr Darnley has found a very good solution' and, without waiting for any response on the part of Lange, the eighth lease was awarded to Irvin for the erection of a shore station on South Georgia. As Irvin acknowledged, on 28 June 1911, it had 'lifted a very great weight off our minds for which we are extremely grateful'.(60)

Seven companies operated from nine locations in South Georgia by 1914: Pesca at Grytviken and Jason Harbour; Sandefjord (Stromness); Tønsberg (Husvik); Bryde and Dahl (Godthul); Ocean (New Fortune); Salvesen and the South Georgia Company (Leith Harbour and Rosita Harbour); and the Southern Whaling and Sealing Company from Prince Olav Harbour. Charles Johnson had typically secured the most advanced floating factory; it was a British cargo vessel which, once refitted at Sandefjord as the *Restitution*, achieved a productivity of 39.1 barrels of oil per whale, and 14.7 bags of guano. The building of the shore station at Prince Olav Harbour was severely delayed by Salvesen's calculated indecision of whether to exercise the company's right to erect a shore station there or at some other point, in accordance with the terms of the South Georgia Company's lease. The 21-year lease was not finalised until September 1913 and the erection of a modern factory and employment of two catchers authorised. The shore station was opened in 1917 (Headland 1984, 15; Hart 2020, 33–48).

CHAPTER 4

# *Maximising productivity*

There was a remarkable turnaround. The Southern Ocean had begun, after only six years, to overshadow Arctic waters in commercial importance. Operating costs were much higher, but so too was the productivity obtained per catcher. Tønnessen and Johnsen (1982, 176) emphasise that it was productivity rather than their production which made the southern grounds so remarkable. The Arctic, far from being abandoned, produced an aggregate of 617,052 barrels of whale oil during the five summers of 1905 to 1910. This was considerably more than the aggregate 362,550 barrels obtained from the southern grounds in the summers of 1904–05 to 1909–10. But whereas a catcher might produce an average of only 42 whales (an aggregate 1,270 barrels), those in the Southern Ocean caught an average of 168 whales, from which 4,075 barrels of oil were obtained. Jens Andreas Mørch calculated that 106,800 barrels of oil were produced from the whales caught by the 14 whale catchers employed off South Georgia in the sixth season of whaling – more than the global output of whale oil rendered only three years previously. That was why growth in the Southern Ocean became so explosive (Ellis 1992, 354–355).

The whaling companies had a profoundly different attitude from the Falkland Islands Government. The companies assumed that as the yield fell off, they would either find more productive waters or leave the industry altogether. The priority for the Governor and the Colonial Office was to maintain the productivity of the Dependencies' whaling grounds for as long as possible. The Colonial Government had to avoid at all costs the fate that had befallen the Newfoundland industry, where several species had, within only a few years, been 'well-nigh exterminated'. The first modern shore station in Newfoundland had been opened as recently as 1897. Income had risen from $1,581 in 1898 to $36,428 in 1900, and to $125,287 in 1902. There were 18 stations by 1905, but, as Charles Boardman Hawes wrote of the industry, 'the enthusiasts had gone too far'. Over 4,000 whales were killed in the first ten years. A fleet of 14 catchers caught an aggregate of only 481 whales in 1907 (Hawes 1924, 327).[1]

The priority for company shareholders was to secure the capital required for the next season's whaling to proceed. Allardyce also wanted to maximise whaling income, but in a way that safeguarded the longer-term productivity of the grounds. 'The prospects for the future development of this Colony' would otherwise be bleak. Far from issuing more licences, Allardyce warned, in July 1907, it might become necessary

> *to restrict the catch of the Norwegian Companies, otherwise the whales will suffer in the same way as they did off Newfoundland, and it would be a short-sighted policy now that the Falklands – hitherto a land of but*

> *one industry, i.e. sheep farming – have been fortunate enough to establish a small lucrative whaling business to permit it to be strangled in the course of the next two or three years owing to the issue of too many whaling licences.*(2)

Allardyce apologised in July 1908 to Sir Francis Hopwood (the Permanent Under-Secretary) for the trouble 'a small Colony such as this' was causing, but the development of 'a highly remunerative industry, must necessarily raise new questions'. The companies had done extraordinarily well at South Georgia, where the waters were simply teeming with whales. They could be counted, as 'old [Lauritz] Larsen' phrased it, 'not by the hundreds but by the *thousands*'. The Argentine Company had caught 800 whales, including 69 right whales – something 'unique in the annals of whaling'.(3)

Allardyce continued, 'I cannot help feeling that the Norwegians and the other Whaling Companies are "chortling" to themselves' on the returns made from their investments, but then, so too was Allardyce, who acknowledged, in his own words, being 'mildly prejudiced in favour of my protégés', both in

> *the leasing of South Georgia – whereby the Government receives revenue in excess of my salary (£1,200) – and the Whaling Ordinance, whereby several hundred pounds per annum have lately accrued to the Treasury, were creations of my own, as this Government did not receive a penny from either source when I came here in 1904.*(4)

Grindle readily acknowledged in the Colonial Office how the whaling companies had been given 'a paying monopoly and must submit to some restriction'. It was only when any limit was put 'into figures we get into difficulties'.(5)

Allardyce concluded in October 1908 that 'the most equitable method' was 'to restrict each licensee to two catchers, and for companies to pay an additional £100 for every catcher permitted over and above the number specified in their leases'. With so many companies 'keenly anxious at present to get a footing in South Georgia, the Whaler's El Dorado', Allardyce did not anticipate, in a despatch to the Secretary of State, any difficulty 'in getting one and all to accept the proposals I have submitted'. There was in truth no alternative if stocks were to be conserved. While 20 companies were permitted to operate in the first ten years, Basberg has calculated as many as 160 applications had to be rejected. Any move to expand 'the killing fleet' further would have meant there would be no 'remunerative industry ten years hence'.(6)

Even so, the annual catch continued to increase. The Norge Whaling Company recorded, upon its expedition's return from the South Shetlands in 1910–11, the largest catch ever of 586 whales (429 humpback, 110 fin and 47 blue) and production of over 17,000 barrels of oil and 50 tons of whalebone. Thirteen thousand barrels of first-class oil had already been sold for £23 10 shillings per ton, making for an overall return of £70,000, giving rise, as Allardyce reported, to 'a proportionate excitement and pleasure' among the small number at Stanley with shares in the company. Hubert Cowell (a Second-Class Clerk) thought them 'excellent results from a small licence duty'. Darnley reckoned them 'better than the pearl fisheries of Ceylon'. The rents and fees, totalling some £1,300 a year, were hardly excessive when set against an aggregate turnover of more than 'a quarter of a million' pounds. But as Grindle also pointed out, the highly profitable seasons had 'to pay for such catastrophes as the total loss of an uninsured (and uninsurable) expedition'. Allardyce also warned, in March 1911, that 'all the Captains and Managers I have seen are of the opinion' that, although whales continued to be 'extremely plentiful', the individual expeditions were having to venture farther and farther south, searching for 'new and good harbours' in the Belgica Strait and Graham's Land. The Admiralty charts were far from reliable. One catcher had run onto uncharted rocks at full speed and had sunk. Two floating factories, which had become grounded, were re-floated only with considerable difficulty. The captain of a catcher had been killed when 'a huge wave' struck the bridge. The vessel would surely have foundered had the wave struck bow on. The absence of good harbours in the South Orkneys made them especially difficult to work.(7)

Allardyce soon came to regret that there had not been closer regulation from the very beginning. It would have been better all round to have set a smaller catch. As he recounted,

> *You should have seen [Lauritz] Larsen plunge his hand in his breast somewhere and produce 250 golden sovereigns at the very thought of getting a station in S. Georgia . . . [T]hese fellows have been let off too cheap with a licence of £100 and no royalties. Believe me they are laughing at us. Why Nansen, the ex-Norwegian Minister was a large shareholder himself in the Sandefjord Whaling Company.*(8)

'The phenomenal number' of 5,521 whales, which had been caught in the summer of 1910–11, had produced 'a furore and excitement not unlike that of the mining boom'. Darnley agreed 'the whalers have had a phenomenal season, and the gross value of the product of the six companies working was over £400,000, and nearly double the value in the previous season'. But far from encouraging companies to show greater prudence, the size of the catches had reached such proportions they seemed to confirm, in shareholders' eyes, what some 'interested parties' had always asserted – the whales were as plentiful as ever.(9)

The industry had attained so high a level of profitability that Allardyce pressed for a harbour to be designated in South Georgia, alongside Stanley, as a Port of Entry where, upon the departure of the respective expeditions, an export duty of 3d a barrel (six barrels a ton) would be imposed to cover the costs. The whaling companies had after all 'done extraordinarily well in South Georgia' – it was 'simply teeming with whales'. The revenues would mean the Colonial Government could provide at last some of the basic amenities which the 2,000 or so Falklanders had so conspicuously lacked. Allardyce wrote, on 6 January 1913,

> *Too much emphasis cannot be laid on the part played by the whaling industry in lifting the Colony out of the rut in which it had remained for decades, in its subsequent development, and in the many improvements to be seen now on every side; and I feel convinced that the Colony's progress will be largely dependent, for some years to come, on the prosperity of this industry.*(10)

Malgosia Fitzmaurice wrote in *Whaling and International Law*, that a permit system was beginning to emerge, designed both to limit and regulate whaling in what proved to be its 'pre-conventional era' (Fitzmaurice 2015, 6). It might also be remarked that however embryonic in character, that same regulatory system had also provided a foretaste of the practical challenges – and not least the fact that the industry had already been allowed to attain its present proportions. Pesca would assuredly have demanded heavy compensation for any move to reduce the output from its pioneering lease. The Colonial Government would have no alternative in that instance of entering into what Ian Hart calls a 'gentleman's agreement', whereby *Pesca* simply undertook to make greater use of each whale brought ashore.

## 4.1 A resident magistrate

Every lease and licence holder was required to call at Stanley (or later Grytviken) both at the beginning and end of the season – in the latter case to report the summer's catch. The Governor invariably copied the detail as to species, sex, number and length of whales caught, together with his comment, to the Colonial Office, where it was copied to the British Museum (Natural History). The Museum Trustees responded by forwarding to the Secretary of State the observations invariably made by Sidney F. Harmer, their Keeper of Zoology (Burnett 2012, 42).[(11)]

Allardyce had become increasingly worried about the risks to law and order posed by the workforce employed at each shore station. Hubert Cowell (a Second-Class Clerk) accordingly ridiculed Allardyce's hesitancy at appointing a magistrate and constables because of the cost. The seven shore stations at South Georgia were expected to provide an aggregate income of £1,750 a year once all were operating – for which the Colonial Government provided nothing by way of material support. Gilbert Grindle was far more concerned at how little the magistrate and constables might achieve. They would have no practical means of keeping every shore station and floating factory under scrutiny, let alone maintain law and order over so large and remote an island. As Kirkpatrick Dorsey remarked, the finer points of any regulations are most likely to be overlooked in environments defined by hardship (Dorsey 2013, 30). Little would be achieved until the Colonial Government was willing to incur the extra expense of providing a boat whereby the magistrate could move between stations and keep watch over individual expeditions.[(12)]

James Innes Wilson was appointed the Magistrate for South Georgia in November 1909, with a constable joining him the next year. Although vested with wide-ranging powers, Wilson was instructed to use them with 'the greatest of caution'. As Grindle wrote, 'Our claim to some of these dependencies is very slender', and there was little material protection, if any, which could be afforded them. Both the Governor and those in the Colonial Office were impressed, nevertheless, by what Wilson managed to achieve as Grindle reported to the Patronage Department, having received only his very first report of February 1910, Wilson was a man to be noted (Hart 2001, 200–202 and 2006, 99; Tatham 2008, 562; Dudeney and Sheail, 2019, 55–56).[(13)]

Perhaps the most important aspect of Wilson's office was his ability to provide the firsthand information that the Governor and Colonial Office had so conspicuously lacked previously. He had instigated straightway a population census for 31 December 1909. Whereas no one had been recorded living on South Georgia in the census for the Falkland Islands of 1901, there were now 363 company employees. Grytviken had 84 ashore and 45 afloat; Leith Harbour 112 and 20, respectively, and Fortuna Bay (Ocean) 82 and 20. There were 103 working aboard the transport ships; 79 on Sandefjord's floating factory *Fridtjof Nansen* at Stromness Harbour; 88 on Tønsberg's *Bucentaur* at Husvik Harbour, and 81 on *Admiralen*, which Bryde and Dahl now operated from Godthul Harbour. As to their occupations, 30 were flensers and 20 coopers. Ten were captains and another 10 harpooners; seven were whaling managers. Wilson additionally counted four visitors, including the wife of the manager at Leith Harbour, and a woman and girl at Grytviken, as well as three doctors; a 'weather observer', and Wilson himself. Some 45 per cent of the Island's population were between 21 and 35 years of age, 24 per cent between 35 and 60 and 20 per cent between 18 to 20 years old. Four-fifths were Scandinavian (Dickinson 1986). Despite having no boat of his own, Wilson had somehow accomplished a 100-mile tour of all the shore stations and floating factories during December 1909. The *Geographical Journal* published an extract from his account of South Georgia, with sketch maps incorporating numerous amendments to the Admiralty chart (Anonymous 1912).(14)

The post of magistrate quickly became 'a symbol of British claim and authority encompassing the roles of local postmaster, shipping master, collector of customs and coroner, being above all the symbol of British occupation and authority' (Husby 2018, 87). Colonial Office officials were particularly attracted to Wilson's generally positive approach. South Georgia as he reported, was 'very mountainous, and for the most part is certainly bleak and desolate', but patches of green vegetation could be seen on the mountainsides up to about 1,500 feet. Although heavily glaciated, tussock grasslands covered fairly large areas of the well-sheltered valleys. Wilson's first whaling report, of 15 November 1910, recorded 3,516 whales being caught by the six companies in the 1909–10 summer: 3,391 humpbacks, 58 fin, 37 right, 26 blue and four sperm whales. The three shore stations so far erected had dealt with 1,990 whales. Pesca had accounted for 997; the South Georgia Company 527, and Ocean 466. The three 'floating stations' had processed the remaining 1,526 whales, of which 717 were caught by Sandefjord, 405 by Tønsberg and 404 by Bryde and Dahl. Although the fin and blue whales remained fairly plentiful during winter, the stormy weather had prevented much whaling.(15)

There was no longer doubt of 'the highly remunerative nature of the whaling industry' at South Georgia. Wilson wrote, in May 1911, that it had exceeded even 'the hopes of the most sanguine of whalers' – 'the enormous number of 5,521 whales' had been caught. The average return of just over £75 per whale would have been even higher if the companies had made better use of the whole carcase. Wilson's overriding priority had become one of achieving that fuller use. Ocean emerged as among the 'very good lessees', followed by Tønsberg, whose productivity that year was boosted by the large number of right whales caught. Pesca had made little improvement since Ernest Swinhoe had first reported its 'tremendous wastage'. It had obtained only 29,956 barrels of oil from the 956 whales it caught in the 1908–09 summer – only one barrel per whale better than in its first season. Colonial Office officials particularly regretted Pesca's having been granted a lease to Jason Harbour, its purpose being largely to exclude rivals from that particular embayment (Hart 2001, 123–127).(16)

Allardyce had toured the Salvesen factory at New Island in West Falkland in September 1909 and had seen for himself how it was possible to process the entire carcase without any part being abandoned to the sea or beaches. Its value might be increased by half as much again if every part was passed through the digesters. There was invariably greater wastage on the floating factories. Despite much investment and some marked improvements in these factories, Allardyce wrote in June 1911 that the ideal floating factory had still to be designed and built. Although fixed tanks had been installed in mineral oil tankers for some 20 years, it was not until 1910 that the first were fitted in floating factories: Salvesen's vessels the *Orn* and *Bombay*. Not only did they save space, but there was no longer the need to clean, weigh, fill and load each individual wooden barrel. Although there was now more space in which to install digesters and the machinery required for grinding the whale bones, Allardyce was sceptical of whether the floating factory could 'ever be as efficient as a land factory, and capable of utilizing in the strict sense of the words, the whole carcass of the whale'.(17)

Bonuses had meanwhile continued to be paid for the largest possible throughput of blubber yielding the highest quality oil. As the catchers became more powerful and skilfully deployed, the shore stations and floating factories became increasingly overwhelmed with carcases whenever sea conditions were good; consequently, an increasing proportion had begun to deteriorate before reaching the flensing plan. To save time, it had become the practice to discard the rest of the carcase, once the oil-rich blubber was removed. As Wilson commented, in his report of 18 May 1911, 'The more whales the more waste', making for 'a most gruesome spectacle' and showing 'very forcibly the great waste by merely flenching these huge animals'(18) (Figure 7).

It was to reduce such wastage that Allardyce wrote, in a despatch of 8 June 1911, that his government was 'endeavouring to arrange for the erection of a land factory at Deception Harbour (latitude 63° South), to utilize the whole carcass of the whale', the overriding purpose being to remove entirely the offal all too frequently abandoned outside 'the place where the whale is flensed'. But even if this could be accomplished for Deception Island, he remained 'distinctly doubtful' whether 'the severe climatic conditions' would permit further shore stations to be built 'at certain harbours in Graham's Land'.(19)

**Figure 7** The legacy of the early days of whaling when flensed carcases were left to rot. The picture shows piles of whale bones on a beach in Mikkelsen Harbour. The remains of a "water boat" can also be seen in the background. These boats were used to transport water to the whale factories lying offshore as it was needed for rendering the blubber down to oil. ©John Dudeney.

Whatever improvements were sought of the companies, the Magistrate remained physically powerless at South Georgia to keep the individual companies under close scrutiny. Allardyce pointed out,

> *He has no conveniences for visiting the different stations by sea, while the glaciers and the general climatic conditions prevent him making a tour of the Island by land. The natural, in fact the only highway is the water.*[20]

Whereas the whaling managers might generally offer him passage on one of their catchers, they were far less likely to do so at the height of the season when 'everything is at very high pressure'. They were in any case never likely to provide him with passage explicitly to report on their own particular shortcomings (Hart 2001, 124–125 and 461).

Allardyce had himself become increasingly impatient at his inability to witness at first hand what was happening in the dependencies, writing on several occasions how regrettable it was that, as the head of administration, he had been unable 'to visit South Georgia in one of His Majesty's Ships'. By witnessing the wastage at first hand, the Colonial Government would be in a much stronger position to threaten that, without improvement, the Crown might be forced to take repossession of the land on which the shore station was built. Should it be impossible to provide 'one of His Majesty's Ships', the only possible way by which Allardyce could 'personally visit the different whaling stations' and 'advise the Magistrate', was to accept the offer, made privately, of one of Pesca's steamers. Robert Wiseman readily acknowledged in the Colonial Office that no governor or senior figure of the colonial government had ever 'set foot on any of the Dependencies', but he also thought it 'a needless obsession in Mr Allardyce's mind'. Any acceptance of Pesca's offer for transportation would be tantamount to confessing that the Governor 'can only get to South Georgia through the good offices of an Argentine Company'. And the Salvesen company would charge 'a prohibitive rate of passage money'.[21]

However anxious the station managers might be to conceal their shortcomings in the processing of the whale carcase, they made the fullest use of the Magistrate in combatting strikes on their shore stations. Wilson's first report, for February 1910, described how he had been rapidly taken to Salvesen's strikebound station at Leith Harbour, where he personally warned the strikers they were breaking their two-year contracts. It was pointless to fine them, even if that had been physically possible, because the men were already so heavily in debt. Nor did he have the means to imprison them. Of the 50 strikers, 12 were 'the worst type of alien that can be ushered into any community'. Wilson was shocked by Salvesen's 'laxity' in appointing them. Their removal from Leith harbour aboard one of the Salvesen vessels brought the strike to an immediate end. Grindle thought Wilson had 'done extraordinarily' in the circumstances. Even 'if his legal decisions might have been questioned in a superior Court', they had achieved 'a very salutary effect'.[22]

Wilson was again called to Leith Harbour, on 22 March 1913, this time to resolve a station-wide strike centred on four labourers operating the 'dryers in the guano factory'. Wilson insisted that everyone should return to work, however legitimate the particular grievance might be. As he wrote later, the majority might well have done so but were afraid of the strike-leaders. Wilson arranged for the appointment of special constables from the crew of the Salvesen vessel *Horatio*, which was fortuitously in port, who in turn arrested 'the eighteen principal ring-leaders or agitators'. There was neither obstruction nor demonstration as Wilson and four 'policemen' escorted them aboard the steam whaler *Harpon.* With the vessel's departure for Buenos Aires, work was resumed immediately.[23]

Darnley thought Wilson had again done very well, and that 'a short Ordinance might be passed empowering the Magistrate at his discretion to require persons dismissed by the Whaling companies to leave the Dependency at the first opportunity'. Grindle regarded Wilson's action as having been 'wholly illegal' in terms of the judicial powers available to him, but 'justifiable in the very special circumstances of the Dependency'. Sir John Anderson (the Permanent Under-Secretary) acknowledged 'how peaceable and manageable the Norwegians are', but he too warned of how 'the next time Mr Wilson tries something of the sort', heads might be broken. The Governor should be warned of 'the illegality of Mr Wilson's action' and how, 'with no force to sustain him', there could have been a very different outcome. Lewis Harcourt (the Secretary of State) agreed that 'we have had a very lucky escape'. There were far worse industrial troubles in the immediate post-war years (Dudeney and Sheail 2019, 94–98).[24]

## 4.2 The first Antarctic magistrate

South Georgia was open all year and had a resident population. The other Dependencies might be ice-bound for half the year, making them 'one of the most interesting nomadic communities in the world' (Chatterton 1925, 243). The Colonial Government continued to depend on what the whaling companies chose to reveal of 'these remote and little known Archipelagos and Lands'. The return of the *Nor* and *Orn* expeditions from the 1908–09 summer was a chance to learn more of conditions in these remote places. Although confirming that Admiralty Bay was popular with the companies operating in the South Shetlands and therefore an obvious location for a shore station, the annual snow-melt would soon sweep everything before it – destroying the shore station, wharves and moorings.[25]

Whaling began in the South Shetlands typically in December and ended in February when the pack ice and violent gales made it 'most imprudent if not impossible to remain there longer'. Allardyce cited the experience of Peder Michelsen (the manager of the Southern Whaling and Sealing Company), in a despatch sent to the Colonial Office on 18 March 1909, in adjudging prospects in the even remoter South Orkneys. Whereas ice conditions had prevented him from reaching the archipelago the previous summer, Michelsen had passed that February through the Lewthwaite Strait between Coronation and Powell Island, where he encountered 'a phenomenal number of whales, so many indeed that he had to stop going full speed otherwise he would have struck against them'. The sea 'simply smoked' when they spouted . . . the atmosphere for miles was thick with vapour'. Most were humpbacks, with a few right whales.[26]

William Moyes accompanied Hektor's floating factory *Ronald* to the South Shetlands in the 1911–12 summer. Alongside his inspector's report, he also kept a diary. His report of 15 March 1912 recounted that between 130,000 and 140,000 barrels of whale oil had been obtained from the 4,000 to 5,000 whales caught. Although humpbacks were far less plentiful than in previous summers, many more blue and fin whales had been taken. The expeditions had intended to leave by 28 March, but the arrival of two transport ships with cargoes of coal and empty barrels meant that four could stay longer. Another report dated 8 April 1912 recounted that eight of the licenced companies had operated from Port Foster (in the flooded caldera of Deception Island) and two from Admiralty Bay. Most of the whales had been caught within 40 miles of Deception Island – many close to the harbour entrance. There was never any certainty when the whales might appear, being often scarce for a few days and then more plentiful. The season generally ended when all the freshwater aboard the floating factories was exhausted, the days were noticeably shorter and the gale-force winds made it impossible to handle the carcases in the sea.(27)

Allardyce drew on both Moyes's reports and his conversations with the individual whaling managers in explaining to the Colonial Office, in a despatch of 7 April 1912, that Deception Island was essentially 'the crater of a huge volcano', with continuous underground activity. Many areas were so hot that it was impossible to hold one's hand less than a few inches above the ground surface. From the recent increase in 'subterranean seismic disturbances' and the appearance of 'large volumes of steam', it was

> *quite within the bounds of possibility that the entrance may be found to be blocked before next whaling season owing to an earthquake and the downfall of the huge loose boulders which guard the channel. Geologists, scientists, whaling masters and others who have visited the locality do not hesitate to express the opinion that it is only a matter of time before a cataclysm occurs.*(28)

Grindle feared 'we are perhaps on the eve of an appalling disaster', but the Colonial Government was powerless to stop whaling because of the volcano, adding 'we can only hope the eruption will come in the winter'.

Moyes bitterly complained that more than 3,000 whale carcases had been left on the shores of Deception Island that summer, joining the many more abandoned from previous summers. The whales had been so plentiful that a profit could be made from simply stripping off their blubber. The larger part of the carcase would only be used when there were too few whales 'to keep the men on the floating factory fully employed for the full working day'. Irrespective of how many whales awaited flensing, the catchers were kept 'almost continuously whale hunting'. They were never allowed to be idle for a moment longer than could possibly be helped.(29)

It was not until October 1912 that 'the first Antarctic Magistrate' (as David Tatham calls him) was appointed to the seasonal post of magistrate, postmaster and coroner for the South Shetlands and Graham's Land. Edward B. Binnie (having briefly stood in for Wilson as Magistrate at South Georgia) was assigned to Deception Harbour, departing aboard Hektor's floating factory *Hektoria* in October, together with a prefabricated hut for his accommodation. Binnie completed his end-of-season report the day of his return to Stanley on 29 March 1913. Thirty-six licensed catchers had taken part – two more than the previous season. The weather up to New Year's Day could not have been better, but storms soon after meant it was impossible to keep whaling. A record number of 5,044 whales had nevertheless been caught, from which 198,410 barrels of oil were extracted, worth £599,886. As many as 70 to 80 barrels were obtained from each of the very large blue whales caught earlier in the summer. 'Some of these monsters' were almost 100 feet in length (Hart 2006, 159–160 and 2009, 113–140; Tatham 2008, 82–85).(30)

Binnie reported that the better equipped *Orn* and *Bombay* had obtained as many as 400 barrels per day compared with 300 barrels from the older floating factories. They had achieved this in part by using their pressure boilers to process the blubber, despite those boilers being primarily intended to deal with the tongue and jaw bones. It was in reporting such instances of abuse, as well as the noticeable shift of activity toward the even remoter waters of the Belgica Strait, that Binnie warned of the need for both closer and wider vigilance over the individual expeditions. Binnie especially deplored the absence of navigation lights along the narrow entrance into Port Foster, particularly on the darker nights of February and March. He said that only by positioning their vessels in relation to 'the high dark cliffs on the starboard side of the entrance' (where the water was deepest) could the ships' masters avoid the shoal, which ran 'for a considerable distance on the port side' and where many wrecks had occurred.(31)

William Moyes spent the summer of 1912–13 in the South Orkneys aboard Rethval's floating factory *Falkland*, during which time it was his practice to hand out copies of the Whaling Regulations to every ship's master he encountered. The *Falkland* had been the first to reach the ice pack on 20 November, joined by two more floating factories in mid-December and another on New Year's Day. Not until 8 January was the catch sufficiently large for all the ship's boilers to be used. The continued presence of pack ice caused Moyes to record that 'the South Orkneys must be considered as an exceedingly good whaling ground' but one very different from the other Dependencies. Although the whales were most plentiful in the early austral summer, ice prevented the individual expeditions from using either the few ports or more sheltered waters. More than a fifth of the catch might be obtained while the vessels were still 'waiting for the pack ice to break up'. So little could be flensed from each carcase that it was 'like taking only the cream and throwing the rest of the milk away'. However much it might pay to remove only the blubber when whales were plentiful, Moyes wrote that something must be done 'to compel the companies to be less wasteful'. He was sure that, if compelled to do so, a balance of floating factories and catchers could be devised whereby the industry was put on a sound and profitable basis for both the licence holders and the Government.(32)

Moyes warned 'there were few, if any, safe anchorages for floating factories when it was blowing a gale'. Hektor's floating factories *Ronald* and *Hektoria* had been the first equipped with wireless sets, provided by Telefunken and licensed with the Falkland Islands Government under the Wireless Telegraph Regulations of 1912 (Hart 2006, 150). It was the receipt of a wireless message from those aboard *Falkland* that caused Allardyce to emphasise, in a dispatch to the Colonial Office of early January 1913, 'what a hazardous enterprise' whaling in the South Orkneys could be and to warn that companies might well decide not to renew their licenses. Darnley thought Allardyce was being excessively cautious, noting on 27 February 1913 that 'the ball has started to roll'. Moyes's diary entries continued nevertheless to provide graphic accounts of both extraordinary catches and the hazards of working those waters. January had been 'an exceedingly good month, with most whales caught within 'a few miles of the factories', making for a total catch of 61 whales in a single week – 29 fin, 27 blue and five humpback. Some of the blue whales were 'extremely large and fat', two of them 93 and 95 feet in length. February proved a poor month with 'almost continuous bad weather', when the atmosphere was very thick with snow or fog or the strong winds created a heavy swell at sea. Few whales were seen and 'no sign of whale food'. More calamitously, the gale-force winds across Signy Island blew the floating factory *Tioga* onto the rocks of a small harbour on the night of 5 February 1913. A report by the managers of the three other floating factories confirmed there was no alternative but for Moyes to auction the wrecked hull and its cargo. The captain of *Normanna* bought both for £225. As Moyes suspected and was soon able to discover, the managers had agreed 'between themselves to share the salvage of the *Tioga* equally' so there would be 'no competition at the bidding'. Although many barrels of oil were recovered from the wreck as opportunity afforded, the priority for the other companies remained one of taking every advantage of the weather to continue whaling. Some of *Tioga*'s crew helped with the salvage but others returned with the now redundant catchers to Stanley. Of the four whales *Tioga* was about to flense, 'the *Normanna* and *Thule* each bought two of them'.[(33)]

Moyes's enquiries of the various ship's masters revealed a highly complex pattern of ownership. As Ian Hart writes, a countless number of whaling companies had been founded in Norway, often by the same owners or groups of shareholders, thereby making it increasingly difficult for the Colonial Government to allocate licences in any rational or even-handed way (Hart 2006, 133). Although *Tioga*'s licence was made out to 'Ingvald Bryde of Sandefjord for the Pacific Whaling Company', Moyes discovered that the floating factory and its catchers *Corral* and *Fyr* were actually owned by the Corral Whaling Company, whose principal investor was a 'Mr Jebsen, of Bergen'. Moyes also learned that Bryde had 'nothing whatever to do with the Corral Whaling Company . . . the probable explanation' being that he had sold the licence to Wilhelm Jebsen, without informing the Colonial Government. It was a further indication of the scale on which licences were being bought and sold in Norway. The opportunity was taken by the Whaling Regulations of July 1913 to prohibit any transfer of licences. Every licensed vessel had to be 'either owned or *bona fide* chartered by the licensee'. No licensed company could catch whales for another company, without the Governor's consent.[(34)]

Any close surveillance of the industry became ever more complex and demanding of resources, as the whalers ventured further into previously unexploited grounds. Despite Carl Larsen having written a damning report of his exploratory expedition to the South Sandwich Islands in the summer of 1908–09, others followed. Christen Christensen sent 'a specially built and innovative combined floating factory and catcher, with an auxiliary factory *Tulla* and store ship *Herfren* for the summer of 1911–12. The loss of the store ship to the ice, on 3 December 1917, meant the expedition had to be abandoned. Kosmos, as one of the six companies granted licences for the 1912–13, had straightway applied for a lease to erect a shore station at Southern Thule. Both Wiseman and Darnley acknowledged a shore station would make better use of the carcase and 'assert more vigorously our sovereignty' to the South Sandwich Islands. The difficulty was adjudging the length of the lease and how it might be adjusted to accommodate further experience of the grounds and the particular site where it would be built. Grindle ruled that any lease must be limited to ten years and terminable without compensation at a year's notice. Nothing came of it; no licences were requested for those whaling grounds in the 1912–13 summer (Hart 2006, 151–152).[(35)]

## 4.3 Towards closer regulation

The industry and Falkland Islands Government operated on very different time scales. The overriding concern of the companies was to have the capital to mount a further remunerative expedition the following summer. The Governor's objective was to ensure the longer-term productivity of particular whaling grounds. Every expedition strove for the greatest volume of oil possible, however many whales that might require; the Colonial Government's priority, in conserving whale stocks, was to maximise the monetary return from each whale caught.

Allardyce remarked, in a despatch of August 1911, that 'the question of the utilization of the carcase of the whale is quite a modern innovation – if I may so term it – and was practically unknown, or at any rate not taken seriously, in these seas until within the last three years'. The Whaling Ordinance of 1908 had made no mention of it. The first four leases had been awarded before the importance had been recognised of the companies having 'the necessary plant for utilizing the whole of the whales and thus avoid the waste consequent upon using the blubber only'. The oil would remain the primary product, but the fullest use must be made of the meat and bone residue. Oil was usually extracted from 'the tongue, lower jaw bone, and a piece of the tail' as well as from the blubber, but Allardyce remarked, in a despatch of April 1912, that even these 'titbits' might be thrown overboard with the remainder of the carcase when whales were especially plentiful. Rather than adjudging productivity simply by how many whales were killed, the greatest possible revenue must be raised from every whale slaughtered. The Colonial Government had to give much closer consideration to the number, disposition and efficacy of the individual shore stations and floating factories in terms of their cost-effectiveness in rendering down the entire whale carcase. It was through such calculation and husbandry practices that the permanency of the Dependencies' whaling grounds would be assured.(36)

Wilson wrote encouragingly of 'a great advance' in productivity through 'the utilization of waste whale carcases'. Whereas 5,516 whales were caught and 100,316 barrels of oil were produced in 1909–10, 6,531 whales and 184,363 barrels of oil had been obtained in 1910–11, and a further 6,535 whales and 205,187 barrels in 1911–12, during which time the output of

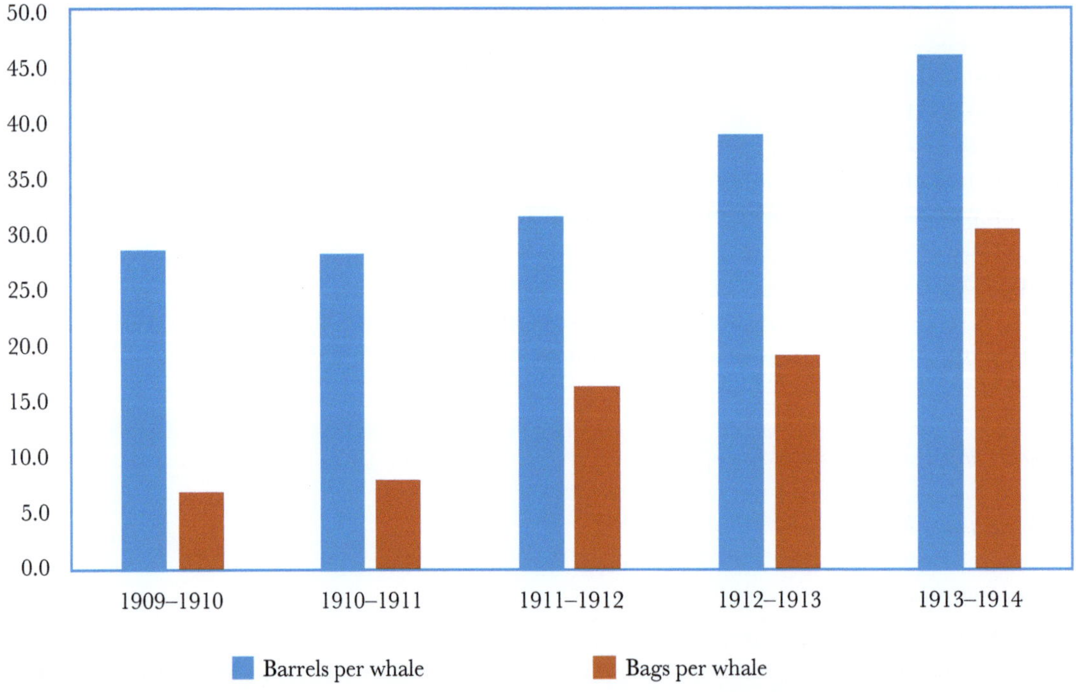


**Figure 8** This illustration shows the number of barrels of oil per whale and the number of bags of guano per whale at South Georgia leading up to World War I, illustrating the improvement of productivity as a consequence of investment needed to comply with the ordinances requiring the use of the whole carcase. The data come from Innis Wilson's annual reports from 1913 and 1914. There are small inconsistencies in the exact numbers given in the two reports that do not materially alter the interpretation, but note that the mix of whale species captured changed through the period from predominantly smaller humpback to larger fin and blue whales. This will partially account for the apparent increase in productivity. Data source CO78/126 and CO78/131.

guano had increased from 3,130 to 12,559 bags, and to 18,737 bags in 1911–12. The average yield of oil per whale had risen from 22.5 barrels in 1909–10 to 31.2 barrels in the 1911–12 summer. Although productivity continued to vary from one company to another, this too had generally improved, with Pesca, the South Georgia Company and Tønsberg each achieving an average of 30.2 barrels per whale over the three years, and the floating factory operated by Bryde and Dahl as many as 26.8 barrels[(37)] (Figure 8).

There had also been improvement in the general management of the floating factories. Never before had so many secondhand vessels been purchased and fitted out as in 1911–12, the number stabilising around 22. Invariably old merchant ships, many of them hastily and cheaply adapted, their efficiency remained scarcely half that of the smaller shore stations (Basberg 1985, 86 and 89). The most obvious refinement was the replacement of cookers mounted on the main deck with between-deck cookers, with top lids flush with the deck level. This way much less heat was lost during the cooking process. Some companies had begun to install the Hartmann rotary cooker, which the German manufacturers claimed to have the capacity of five ordinary pressure cookers. Along with considerable economies in space, labour, coal and freshwater, they were especially good at extracting oil from the whale meat, producing a reasonable quantity of Number 3 oil and good quality guano. First installed aboard A/S Odd's *Guvernøren* and used off the South Shetlands and Graham's Land in the 1914–1915 summer, the ship's log recorded 5,000 barrels of oil and 3,000 sacks of guano from the two units operated (Hart 2006, 114–118).

But beyond the technical improvements, there was need to organise the individual expeditions better. Albert Newing (aboard Odd's expedition to the South Orkneys) reported both the remarkable catch of over 3,000 whales and how, in coping with so large a number, even some of the whale blubber had been cast into the sea. Allardyce stressed in a despatch of May 1911 that the most urgent measures must be taken to ensure, in so far as local conditions permitted, that the whole carcase was always used. Odd's floating factory *Sobraon* had begun to show the way. It 'had merely followed the whaling expeditions' in the 1910–11 summer, flensing and boiling down those carcases that would otherwise have been abandoned. A further 7,000 barrels of oil were obtained, thereby adding some 30 shillings to the value of each carcase flensed. Drawing on such reports as those of Newing, Allardyce wrote, in August 1911,

> *It is my opinion and that of my advisers, supported by overwhelming evidence from South Georgia and the South Shetlands, where the wastage of whale flesh is enormous, that none of the floating factories can utilize the whole of the carcasses of several large whales daily. The only satisfactory preventative against this scandalous state of things is an adequate land factory with sufficient digesters, &c., &c., and boiler power.*[(38)]

The companies themselves had continued to lobby for a shore station at Deception Harbour. Hans Krogh-Hansen wrote from aboard A/S Hektor's floating factory *Ronald*, and later from Tønsberg in May 1911, describing how the carcases, when whales were plentiful, were left to 'drift around the harbours'. Despite 'certain risks and obstacles', he was confident a shore station would succeed. A clear-water frontage would be needed, together with free access to all the carcases abandoned by other companies, especially when there was a shortfall in the company's own supply. The Colonial Government responded by inviting tenders for a lease of 21 years for a shore station, working in conjunction with a floating factory, with the assurance that they could process an aggregate minimum of five blue whales per diem, or the equivalent number of fin or humpbacks (reckoned at 1½ fin, and 4 humpback per blue whale). The Secretary of State formally welcomed the proposal (Hart 2006, 103–104)[39] (Figure 9).

The Crown Agents for the Colonies had, by 27 July 1911, received tenders from Finn Bugge (Hektor), Peder Bogen (Sydhavet); Tarald Dannevig (Laboremus) and Salvesen. The other licensees to the South Shetlands whaling grounds had either declined to tender or had simply failed to reply. Hektor was adjudged to have the most capital resources and was awarded, in April 1912, both a 21-year lease for a shore station and a licence to operate a floating factory and two catchers, with the 'exclusive right to discarded carcases in Deception Harbour'. The Whaling Regulations were necessarily amended to enable the Governor to modify and withdraw the licences granted to the other companies, in the event that they ever seriously impeded the fulfilment of Hektor's contractual conditions (Tønnessen and Johnsen 1982, 195–196; Hart 2006, 69 and 104–107).[40]

Allardyce saw the lease as 'a decided step in the right direction', but he was anxious lest a daily throughput of five whales was sufficient to cope with the expected daily catch, as well as the carcases abandoned in the harbour. Darnley had also begun to have serious doubts, minuting that 'I think the Governor has managed badly here and failed to show as much technical knowledge as we could reasonably expect'. It was now however far too late to specify how 'many pressure boilers of such and such measurements &c' must be fitted.[41]

Hektor indicated its intention to install as many as 36 digesters and two large dryers. Although continuing to confess his difficulty as a layman in comprehending the technical

**Figure 9** The remains of the Hektor whale factory established in Whalers Bay at Deception Island in 1912. A key condition of the lease for this factory was that it would be used to process all the rotting whale carcases littering the beaches at Deception Island. It closed in 1931. © British Antarctic Survey, UKRI.

detail, Darnley became more confident that the shore station and floating factory, working together, would be able to cope. The first whales were hauled onto the flensing plan on Christmas Eve 1912. The further building work of 1913–14 provided accommodation for 89 workers; no winter working was anticipated. Through the combination of the shore station and Hektor's floating factories, output increased from 14,700 barrels in 1911–12 to 36,800 in 1912–13, and to 67,086 barrels in 1915–16. The output of guano rose from 900 sacks in the first summer to 7,000 in 1913–14 (Hart 2006, 104–108; Dibbern 2010, 214).[42]

Allardyce had come to acknowledge, in his despatches, the merit of administering South Georgia separately from the other Dependencies. Whereas the well-equipped shore stations could be relied on to make a greater use of the carcase, the same could not be said of the floating factories in the other Dependencies. Allardyce recommended each licensee be charged £500 for a floating factory and two catchers. Alternatively, a third catcher might be permitted where companies could prove they were sufficiently equipped 'to treat a large proportion of the carcasses of the whales caught'. Darnley agreed that the obvious course was to improve the ratio of catchers to the capacity of the floating factory to process the carcases. The question was what the ratio should be. It had become obvious the Governor could not be relied on to make these calculations. Any adjustment must avoid 'a waste of money and plant' being substituted for 'a waste of whales'. Grindle minuted 'we must not blame the Governor – both he and we have been gradually gaining experience in dealing with an entirely new industry about the technical details of which we had neither information nor precedents at the start'. Sir William Harcourt (the Secretary of State) was less forgiving, observing that 'the Governor living in the midst of whaling might have acquired more technical knowledge somewhat sooner'.[43]

As the industry grew in importance, Husby (2018, 86) writes of a discernible increase in involvement by the Colonial Office in setting regulatory policy. Whereas the Governor and his officials had learnt much from the expedition managers, an obvious course for the Colonial Office was similarly to speak to the leading figures of the respective companies. Theodore Salvesen had continued to profess that his floating factories were well able to process the whole carcase. Darnley acknowledged that Salvesen might well be right, but the subject was 'a difficult one', and it would be quite out of the question simply to follow his *ex parte* representations. A two-hour meeting proved nevertheless useful for Salvesen's keenness to impart 'a great deal of useful information' on the various combinations of pressure boilers (digesters) that might be installed. Hans Krogh-Hansen was also keen to show Darnley round his floating factory *Benguela*, berthed at Liverpool, in December 1912. As with Darnley's meeting with Salvesen, the most interesting aspect proved to be Krogh-Hansen's assessment of the processing plant currently in use. Although 12 pressure boilers might be optimally fitted, 'the blubber factory' could cost altogether £12,000 to £15,000. As Darnley wrote of his visit, 'we had better watch events, and gather figures of cost and output if we can', before insisting on the provision of a guano-making plant.[44]

Another despatch from the Governor, of September 1912, relayed Wilson's appraisal of the six floating factories, which had operated the previous summer off South Georgia. The Southern Whaling Company's *Restitution* had achieved sales of £16,996 in 'whale products' from a catch of only 136 whales – an average of £125 per whale compared with £85 from Bryde and Dahl's floating factories. The *Restitution* had even outclassed the productivity achieved by the shore stations operated by Ocean and Tønsberg. Despite catching far fewer whales than other companies, the 'most modern machinery' had attained a much greater productivity – a goal that Wilson thought every shore station and floating factory should be compelled to achieve.(45)

Sir William Harcourt initialled Gindle's acknowledgement of Darnley's close study of 'the technical side' and approved the Permanent Under-Secretary's instruction of November 1912 that 'the provision of adequate factory provision for treating the whole carcase' should be made a condition of each licence. A licence should be neither granted nor renewed where this was adjudged to be inadequate. The continued use of a third catcher should likewise be made contingent on the licensee having sufficient plant for treating the additional carcases. The present licence fees of £100 for a floating factory and two catchers and a further £200 for a second floating factory could be consolidated into one of £200 plus £100 for a third catcher, in those instances where it was permitted.(46)

## 4.4 The further ordinances

Rowland Darnley wrote, in February 1913, that 'the ball has started to roll; it will be a good while before we can finish the matter'. Companies continued to be warned that it was forbidden to deploy

> *any vessel or floating factory used for the purpose of flenching, rendering, boiling down, treating or manufacturing the blubber or any part of the carcase of any whale in Colonial Waters unless he shall have been expressly authorised to do so by a licence under this Ordinance or by the lease of a site on land for whaling purposes.*(47)

But it was one thing to issue regulatory ordinances but quite another to ensure they were followed. It was probable that 'these rules were skirted quite routinely' (Husby 2018, 60–61).

By further regulations of 22 January 1912, a non-refundable charge of £100 was imposed on every annual licence awarded from 1 October of each year to operate a floating factory and two steam whalers; an additional sum of £100 was to be paid when a third whaler was allowed. More particularly, no more than ten of these licences would be awarded in any one season for the whaling grounds of the South Shetlands and Graham's Land; a maximum of seven licences would be granted for the Falkland Islands, the South Orkneys and the South Sandwich Islands, making 31 licences in all. Each licensed vessel was required (as a condition of those licences) to proceed at the end of the season to a designated Port of Entry where the required detail of the catch was to be given to the collector of customs. Any breach of these various regulations, whether by the licensee or employees, would risk immediate forfeiture of the licence and denial of any future licence. Robert Wiseman, in the Colonial Office, particularly welcomed confining individual expeditions to particular whaling grounds. Darnley especially welcomed the Governor's retention of the power to decide whether a third whaler should be permitted.(48)

Allardyce forwarded to the Colonial Office in May 1913, a further draft ordinance, devised following discussions with Edward Binnie and William Moyes as well as the managers of the various expeditions on their return to Stanley. This recommended that every floating factory should be fitted with a minimum of 12 pressure boilers, each with a capacity of 385 cubic feet, for the purpose of the 1913–14 summer. In doing so, account had to be taken of the increasing difficulty for the companies in raising the necessary capital. 'The disappointing returns obtained from other whaling grounds' had meant 'the investing public and the banking establishments of Norway' were reluctant to venture further capital in 'the whale fisheries'. Companies should not be compelled to invest 'more than was absolutely required'. Allardyce remained particularly sceptical of whether tunnel driers should be required for drying the guano from the remnant carcases. However effective they might be, transport costs meant the guano was hardly profitable.[49]

The Amended Ordinance of 1913 prohibited the use of any vessel to tow 'the body of any whale in Colonial Waters', except where 'expressly authorised by licence or otherwise to do so'. It was Edward Binnie's intervention that had caused Pesca to protest, in February 1914, that

> *as the whale is getting shy and scarce the whalers are very often obliged to leave afloat the whales shot in order to work sufficiently rapid[ly] and get within range of others. In cloudy weather it very often happens the whales thus left adrift attached to small buoys are lost or – in the better cases – found after a day or more searching.*[50]

Darnley, in supporting Binnie, saw strong objection to the use of what he called 'tow boats', which enabled the catchers, having made a kill, departing immediately in pursuit of another.[51]

Allardyce had pressed, as early as July 1911, for King Edward Cove to be designated, alongside Stanley, as a port of entry and for a customs officer to be appointed to assist the magistrate in raising an export duty of say 3d a barrel; this was to cover the administrative costs and to build up a substantial reserve 'while whales are plentiful' for 'the lean years which are sure to follow'. Previous governors and their officials had been 'severely handicapped at every turn by lack of funds' that would allow them to provide the public works so badly needed by those living in the Falkland Islands. There had therefore been an increasing reliance on the goodwill of the Falkland Islands Company. Even the health officer 'going off to give practique' was compelled to make frequent use of the company's motor-launch, often putting the Colonial Secretary in the awkward position of having to aid and abet the company's interests. Allardyce could not

> *too strongly urge that provision be made for still further strengthening the accumulated assets of the Colony against a possible collapse of the whaling industry during the next decade, so as to prevent any repetition of the occurrences.*[52]

The annual revenue, based on the catches of the previous six summers, would be about £4,500.[53]

Allardyce had also begun to press for Deception Island to be made another 'port-of-entry and departure' and for a levy of 3d per barrel (18d a ton of oil) to be imposed to cover his Government's administrative costs. Darnley had meanwhile begun to warn that as the whaling industry further developed, 'we are going to have a good deal of trouble with lessees, and we may have to tax them to pay for the policing they appeared to require'. He hoped the convenience of not being required to call at Stanley would make the whaling managers more amenable to this payment.[54]

The Secretary of State formally approved the designation of the two ports of entry and the imposition of a levy of one-half per cent of the expected value of every barrel of whale oil exported, thereby raising an anticipated £2,500 a year, of which South Georgia was likely to contribute £1,500. At the close of season, every licensed vessel would be expected to proceed directly to a Port of Entry and report the 'full particulars of the catch, including the number of barrels of oil and the quantity of baleen and guano obtained'. Both Darnley and Grindle saw the consequent increase of £50 in Wilson's salary as 'well deserved' for the 'extra work and responsibility'. The same ordinance that designated the additional Ports of Entry also prohibited, as a licence condition, the killing of any whale calf or female accompanying a calf, on pain of a fine of £10 for each offence and immediate forfeiture of the licence.(55)

Revenue from an export duty might also overcome an increasing embarrassment to the Colonial Government. Whereas it had initially suited all parties that the magistrate should live and have his office at Grytviken, where Carl Larsen had helped Wilson to learn Norwegian, this location had become an increasing embarrassment as further shore stations were opened. It put the magistrate 'in a most delicate and difficult position'. The obvious location for the government office was a kilometre or so away – coincidentally alongside the weather station located at King Edward Point. Two English-style buildings were eventually built from the proceeds of the export duty, erected on the raised spur of morainic material overlooking King Edward Cove. Their design contrasted sharply with the otherwise Norwegian-style buildings of Grytviken (Headland 1984, 12).(56)

Theodore Salvesen lobbied hard for Leith Harbour to be made the port of entry for South Georgia. Not only was Stromness Bay much more used, but, he claimed, the requirement to call at Cumberland Bay would endanger 'life and property on that uncharted and storm-swept coast'. Both Allardyce and Wilson pointed out that Cumberland Bay was generally reckoned 'the safest, best known, and best charted opening along the whole coast line'. As Grindle minuted, any officer stationed at Stromness Bay 'would have to live on the shore station, which was just what Salvesen wanted'. Lewis Harcourt (the Secretary of State) thought he should nevertheless meet Theodore Salvesen, who demanded both the withdrawal of the whale tax and designation of Leith as the port of entry. Harcourt refused. Grindle pressed the propriety of using some of the tax proceeds for, say, the installation of a 'wireless telegraph; hydrographic survey, and the improvement in postal service', together with acetylene lights, and the buoying of the shipping channels. Each of these would be of direct benefit to the whaling industry. Sir John Anderson (the Permanent Under-Secretary) agreed, but cautioned that whaling was 'an industry which in the nature of things must have a short life – if a merry one'. Care must be taken to avoid, when 'whaling comes to an end', being 'saddled with a considerably increased establishment which it will be difficult to reduce'.(57)

Every whaling company had been invited to suggest how a portion of the revenue might 'most advantageously be spent for the benefit of the whaling industry in the Dependency in which your Company is operating'. Wilson wrote dismissively that 'the nature of the country precludes roads', and there was no need for public works. Others saw the principal need to be 'a wireless telegraph station', and for an acetylene navigation light on the northwest passage into Cumberland Bay and the repair of the slipway at Stanley. The priority for Salvesen was to repair the inadequacies of the Admiralty chart and thereby provide greater protection for the capital that companies had invested in their shipping. Darnley was supportive, despite Allardyce's outrage at the estimated cost of employing an Admiralty surveyor.

In acknowledging this, Grindle reasoned that 'we take a lot of money off the whaling companies and might do this for them'. The proposal was nevertheless abandoned when no whaling company was prepared to provide a vessel at what was also the height of the whaling season.[58]

It was a letter that Wilson wrote privately to Allardyce, in June 1913, that was most discouraging for any use of the fiscal revenues for local works. Having warned that the cost of any public works would be 'quite out of proportion to the value of the services to be rendered', Wilson recalled,

> *I have often remarked on the transient nature of whaling, the chief aim of the companies being to make as much out of it as possible while it lasts, and it has often occurred to me that Government should adopt a similar policy. In my opinion large sums of money should only be expended when a people are to receive a more or less permanent benefit or when good and permanent returns may be expected at some future time for such an outlay; but neither of these results can be expected at South Georgia. Here there is not any real local or Colonial interest. When whales are numerous it has often appeared to me like one great wild rush, with absolutely no room for any other real thought or interest but 'whales', and at the end of every season when 'the harvest has been gathered in' a joyous exodus to* gamle Norge *until the beginning of another season when the Viking again descends on his prey; and so the same wild rush goes on year after year.*[59]

## 4.5 The political dimension

The Colonial Government at Stanley issued the whaling leases and licences, but it was the Colonial Office in London that ultimately determined their number and allocation. Theodore Salvesen's goal was nothing less than unfettered access to the whaling grounds. Far from recognising the need for any even-handedness in their apportionment, not least for the Norwegians having established the industry, Salvesen contested every administrative move for its failure to eliminate his company's rivals.

Salvesen had contemptuously rebuffed the offer of an annual licence for the South Shetlands and Graham's Land, insisting on a 21-year exclusive licence, claiming it to be in the best interests of both the fishery and colony. In a letter of May 1908, he ridiculed the way in which 'Great Britain derives very little benefit' from an industry which continued to function largely under 'a Foreign flag'. Far from the position improving, even more foreign companies were being formed 'to carry out whaling operations in the Antarctic Seas', further threatening the depletion of the whaling grounds and British interests. The French government had by contrast granted a French company a monopoly of the grounds off the Kerguelen Islands. A similar monopoly should be granted to Salvesen as 'the only British firm in Great Britain carrying on the whaling industry on modern lines'. The Colonial Office went no further than to forward Salvesen's application to the Governor on the premise that he might be offered an annual licence, for which £100 would be charged.[60]

There followed, in April 1910, a considerably more modest application for an annual licence to operate a floating factory and probably three catchers in the South Shetlands for the 1910–11 season. It arose, as Salvesen was forced to acknowledge, from 'the almost total failure of the whale fishing at the New Island station' of the Falkland Islands. Whereas an average of 4,000 barrels was obtained per whaler in the South Shetlands in the 1909–10 season, 'our miserable result', Salvesen confessed, was little more than '280 barrels per whale of blubber oil or

419 barrels for blubber and carcase oil'. The shore factory had 'run at an enormous loss . . . for the want of whales'. The Falkland Islands were simply off the main migration routes. Salvesen finished his letter of application by more typically remarking that

> *we know that you are anxious to promote British enterprise. So far as we know, no vessel presently licensed flies the red ensign at the South Shetlands whereas 90 years ago about half of the vessels then engaged in sealing at the South Shetlands belonged to Great Britain. You will be aware that we are the first and only British whaling firm on modern lines both at the Falklands and at South Georgia, and we think all Britishers will be proud to hear of British enterprise being revived also at the South Shetlands.*(61)

There was more than an element of humbug in Salvesen's proud boast of being a British company. Although shortly to become the largest and best financed company in the industry, it would continue to rely on a largely Norwegian workforce, manning vessels largely refitted in Norway (Elliot 1998). Nor could the Colonial Office rely upon Salvesen's adherence to the terms of any licence granted. William Moyes's intervention had further soured relations. As Salvesen set out, in a long letter of July 1912, a spare catcher had been sent from the New Island station to assist the three licensed catchers working with the floating factory *Neko* at Deception Harbour. Moyes had ordered its immediate withdrawal as a flagrant breach of the licence conditions, threatening a £300 fine and confiscation of the oil obtained by its assistance (Hart 2006, 142–143). Salvesen's Scottish lawyers threatened legal action to recover the 'very serious loss' of earnings caused by Moyes's intervention. Grindle endorsed the action, minuting that

> *Messrs Salvesen are a grasping firm. We have gone out of way to give them exceptional advantages, allowing them, under the colourable pretence of a limited company, two leases instead of one in South Georgia, and they continue to ask for more. It goes very much against the grain to be bluffed by them into conceding further favours.*(62)

The legal assistant to the Colonial Office feared a British court might decide in favour of Salvesen. The company was nevertheless informed that 'Mr Harcourt was advised the action of the Customs Officer was justified'.

Far from turning the industry over to the kind of free enterprise sought by Salvesen, both the Governor and the Colonial Office were increasingly concerned that the industry did not outstrip the resource. Colonial Office officials took full advantage of Wilson's presence in London on home leave to discuss the optimal number of licences that should be granted. They agreed there should be no more, despite this effectively precluding any newcomers to the industry. Profitability could no longer be assumed. Pesca recorded its first trading loss in 1912–13. The whales were having to be hunted farther and farther from the shore. Not only did those longer journeys consume more coal, but the quality of the whale oil might have begun to deteriorate by the time the whale was on the flensing plan. Of more immediate concern was the decline in humpback numbers. Whereas they had only recently comprised nine-tenths of the catch, as few as 504 were caught in the 1912–13 season – 800 fewer than the previous summer. The overall output of oil had fallen from 46,795 barrels in 1911–12 to 23,622 barrels in 1912–13. Pesca's losses would have been even more embarrassing but for the additional 5,712 barrels of seal oil obtained from its separate lease for sealing (Hart 2001, 154–158).(63)

Darnley found a discussion with Hans Krogh-Hansen especially informative. While British yards had built all but two of the floating factories in use, the Norwegian yards had provided most of the catchers and installed the boilers, digesters and dryers during the re-fitment of the floating factories. Most of the provisions and stores were Norwegian, but almost all coal – some 90,000 to 100,000 tons in 1912–13 – was taken aboard at British ports. Krogh-Hansen remained highly sceptical of whether 'many English people would want to start a whaling company', even if a license was offered, warning that the issue of more licences could only result 'in an early killing of the companies already existing down there'. As he told Darnley in confidence, Irvins were 'pretty sick of whaling', and would 'only be too glad to dispose of their South Georgia business if they could get their money back'. Hektor and Kastor had both rejected the sales price offered for the Southern Whaling and Sealing Company as being too high.[(64)]

John Irvin had already met William Harcourt (the Secretary of State) at the House of Commons, and Sir George Fiddes and Darnley at the Colonial Office, when he wrote, on 10 January 1914, of the 'immense difference it would make in our prospects', if the company was granted a second licence, or allowed 'to employ a third whaler'. As the last company to be granted a lease to South Georgia, the Southern Whaling and Sealing Company had been given 'the least favourite site for our harbour' – one so remote from the best whaling grounds that its two catchers could not possibly achieve the productivity of other companies. Darnley thought the greater part of the company's difficulties stemmed from poor management. The catchers had been so late in their arrival from African waters that their catch was bound to compare badly with, say, A/S Ocean, whose sales had enabled them to pay a 40 per cent dividend. Irvin and his partners had worsened their financial position by insisting they should provide the entire capital required. It would be 'quite impossible to help them further without compromising our whole whaling policy'. Although feeling 'sorry for the Company', Harcourt agreed. However regrettable the Norwegian dominance of the industry might be, the two British companies had proven very difficult to work with. 'What with Messrs Salvesens' asperities and Messrs Irvins' moans, Harold Allen minuted that 'one is almost tempted to wish that the whaling industry was entirely Norwegian'.[(65)]

Licence applications had meanwhile been received from more than 20 aspiring entrants to the industry, including those of the Salvesen's Olna Whaling Company, a sheep farmer in the Falkland Islands, the manager of a marine engineering company in Sunderland and the partner of a Glasgow oil and paint company. Lord Crewe (the previous Secretary of State) lobbied Harcourt personally, writing, 'Dear Loulou, there is something rather picturesque in Cheshire farmers engaging in whaling in the southern seas, that they should be able to make such investments is gratifying to a landowner', such as himself. Harcourt responded by pointing out that it had become necessary, in 'the interests of whales and whalers alike, to stop the grant of further facilities in all the Dependencies'. Twenty-one whaling steamers had already been authorised for South Georgia alone – it was doubtful whether 'in the history of whaling there has ever been a locality in which as intense a pursuit has been carried on'. Having gone into 'all the technical and local advice available', Harcourt recounted that, a year or so previously, he had been compelled 'to take immediate steps to prevent the industry from crippling itself by its own assertions, and I think that if you had been here still you could hardly have acted otherwise'.[(66)]

It was the support given by the Bristol Members of Parliament that made an application from the British Antarctic Whaling and Oil Company so worrying politically.

As Sir William Davies (one of those MPs) complained, in a letter of December 1912, there were as many as 15 Norwegian whaling companies, but only three British companies (two of which belonged 'to a gentleman of Norwegian extraction and connections'). The whaling industry had become 'to all intents and purposes a Norwegian monopoly'. Discussions had therefore opened with 'the Bristol Dock and Harbour Board for the erection of a [whale-oil] refinery'. Avonmouth was both close 'to the great coal port of Cardiff' and 'practically in the dead centre of England' – unlike the present refineries of Glasgow and Liverpool – for marketing the oil.(67)

Darnley ridiculed such posturing of 'a wholly-British concern', whose first requirement would be 'a competent Norwegian manager' and largely Norwegian crew. Not only was 'this Norwegian trade' beneficial to the Falkland Islands but also for both British shipping and shipbuilding, and for the cities of Glasgow and Liverpool in respect of the whale oil trade. Sir John Anderson (the Permanent Under-Secretary) pointed out that any licence granted to the British Antarctic Whaling and Oil Company would mean 'Bristol would fair reap where it has not sown'. The present licence holders could not 'in common fairness' be ousted – nor was it in the public interest that they should be. Sir William Davies was accordingly reminded that 'the Norwegian companies are experts in the business' and had 'practically created the industry'. Considerable capital (much of which was British) had been invested. It was 'of course open to capitalists in any town to erect refining works', but the Secretary of State could hardly show preference for 'one port over another in carrying out the general policy to which he is committed'. That having been said, should a Norwegian licence be abandoned, every effort would be made 'to secure favourable consideration for British applicants for the vacant concession'.(68)

Another letter from a Bristol MP considerably widened the debate by quoting 'an expert' who had just returned from the South Shetlands and had 'an intimate and expert knowledge of its conditions during the past eight seasons'. He had found the whales to be 'larger and more plentiful this season than ever before'. There was so little need for 'any policy of preservation' that Bristol Members intended to raise the matter 'in the House of Commons at the first opportunity'. Darnley acknowledged that

> *most of the local experts think the whales are as plentiful as ever, and owing to the lack of information as to the ages to which whales live and the rate at which calves are produced, the question would be difficult to argue. The history of North Atlantic whaling is the strongest reason we have for fearing a practical extermination of whales in southern waters.*(69)

Harcourt had already noted, 'I have been attacked orally by several MPs over the subject and told them all that I will do nothing'.(70)

The British Antarctic Whaling and Oil Company wrote again, in January 1913, protesting at the way the Colonial Office, far from encouraging a British industry, 'penalizes British subjects, who desire to engage in it, by forbidding the harbours of a British Colony to vessels flying the British flag and engaged in a peaceful industry on the high seas'. Not only had whaling 'in the territorial waters of the Falkland Islands' become 'practically a Norwegian monopoly', but the Norwegians had begun to erect oil refineries of their own, causing further 'serious loss to British Investors'. The letter concluded by asking whether, should an expedition be mounted to the South Atlantic, the Secretary of State

> *would employ H M Naval forces to remove or confiscate the ships, or would the owners, their employees, or both be liable to penalties, if so, what penalties, and under what authority or Statute would such penalties, if any, be enforced?*[71]

The Colonial Office responded in the sense of 'clearing up some misapprehension'. The Secretary of State's decision to adopt 'a policy of limitation' had been taken in 'the general interests of whaling as a whole'. The Norwegian predominance was a consequence of its citizens having been the first and often the only applicants for licences. Since whaling could rarely be profitable 'without the use of harbours for reducing the carcases and of territorial waters for towing them in, if not for the actual captures', the industry had been brought under what might be termed 'municipal law' in a similar manner to that adopted in both Irish and Scottish waters, and indeed wherever whaling was pursued in the dominions.[72]

The British Antarctic Whaling and Oil Company wrote again on 3 February 1913 protesting that it was British industry, rather than whale stocks, that was the more threatened. The Norwegian dominance had become 'a very real danger to the British whaling industry and its dependent industries in the United Kingdom'. Having secured 'the sources of the business', the Norwegians were now clearly intent on capturing the entire industry. Their vessels would soon be supplying the Norwegian refineries now under construction and so bring 'British whale fishing to a speedy and complete extinction'. Norway would have become the sole supplier of oil. The letter finished by warning that 'we think the great body of British opinion' would perceive the Colonial Office's present policy as one of 'preference and protection for the foreigners'. It was upon 'these broader grounds that we appeal to the Secretary of State' to give a British company an opportunity both to revive whaling as a British industry, and to retain the refining industry.

> *We feel confident that if this matter were submitted to the judgment of the House of Commons, and the correspondence with the Colonial Office published, that the refusal of our application for permission to use the territorial waters in question would be condemned by a large majority, but this would entail delay.*[73]

A licence was required straightway if an expedition was to be ready for the 1913–14 summer (Dudeney and Sheail 2019, 55–57).

Colonial Office officials were fully aware of the obvious imbalance. British companies had taken only 21 per cent by value of the South Georgia catch in 1912–13, and only 7.5 per cent in the other dependencies. There was similarly no obvious way of preventing the Norwegians from erecting oil refineries of their own. But, as Owen Williams minuted, it would be a major breach of faith if the Secretary of State were to prevent those companies from renewing their licences – they had invested so heavily on the premise of their continued renewal. There would pretty certainly be retaliation in other sectors of 'our trade with Norway'. Harcourt agreed that 'we must stick to our prohibition of any new licence', but always on the understanding that, should any Norwegian licences fall vacant, they might be reserved for British whalers. But even if that eventuality should occur, Colonial Office officials pointed out that the Secretary of State would have to consider those British applicants who had sought a licence long before the application of the British Antarctic Whaling Company.[74]

CHAPTER 5

# The safeguarding of whale stocks

The origins of what became the Discovery Investigations and what they represented for the survival of whale stocks may be discerned in discussions immediately before the First World War. The magistrate's report for 1913–14 was a measure of what the whaling companies had achieved by way of developing the whaling potential of South Georgia. 2,710 whales had been caught and 126,457 barrels of whale oil produced. The average yield per whale had risen from 30.3 barrels in 1911–12 to 38.2 barrels in 1912–13 and 46.6 barrels in 1913–14. The increase, however, was a function not only of greater productivity but also a proportional increase in the number of the larger species taken, particularly blue whales. Humpback numbers had fallen from 5,299 in 1910–11 to 5,100 in the 1911–12 summer, 2,251 in 1912–13 and to 474 in the 1913–14 season. Whereas only 97 fin whales and 75 blue whales had been caught in 1910–11, their number had risen to 1,724 fin and 212 blue whales in 1912–13, and 1,417 and 651, respectively, in 1913–14.[(1)]

No one was sure why humpback numbers had fallen. Had the 'continued killing' of them thinned the population drastically and frightened the remainder off? Had they simply followed a different migration route and would in time return? Shipmasters continued to report 'schools of humpbacks three or four hundred miles off the land'.

> *A very reasonable contention is also held that the Humpback whale while foraging along his usual ocean track finds plenty of food on the way; this will delay, if not altogether prevent, him from touching at his usual resorts in these latitudes.*[(2)]

Darnley acknowledged, in September 1914, that there was a prima facie case for their greater protection, if 'we have the means of doing it'.

However successful colonial policy was in safeguarding stocks to the extent of avoiding the crashes experienced off Finnmark and Newfoundland, the Colonial Office was assailed, on the one hand, by those demanding the removal of all constraint on the industry and, on the other, by those wanting closer regulation. Both claimed to speak with authority, whether from an industrial perspective or, in the case of the British Museum (Natural History), from its standing as the acknowledged expert adviser to the colonial empire (Stearn 1981; Hodge 2011). Each lobbied both directly and through the publicity arising from public lectures and articles published in newspapers and journals. Theodore Salvesen used his lecture to the Royal Society of Arts, in March 1912, to emphasise the bountifulness of the dependencies' waters. A mere 19 catchers had caught 7,000 whales around South Georgia, producing 200,000 barrels of oil – about half as many more whales and 'nearly 40 per cent more oil than the total result of the combined efforts of the whole whaling fleet operating in the northern hemisphere'. The 22 whalers off the South Shetlands had 'bagged' a further 3,500. It was largely through the productivity of those southern waters that a global record of 22,500 whales was achieved, yielding 620,000 barrels (103,000 tons) of whale oil – twice the previous summer's output (Salvesen 1912).

The scientific journal *Nature* published a précis of another of Salvesen's lectures, which focused more explicitly on the Dependencies. Salvesens had, through their workforce of 3,500, made the dependencies 'the largest whaling business in the world' (Anonymous 1915). Salvesen's stance, however, was largely that of a goldminer in assuming there would be other lodes to work (or in Salvesen's case whaling grounds) as each was exhausted. The Colonial Office, and more particularly the Falkland Islands Government, had no such leeway. That was not to say that every whaling company was of the same mould. Salvesens were particularly notable for their greater use of the carcase. Theodore Salvesen boasted that it had become the practice at Leith Harbour to extract as much of the oil as possible and then to turn the 'absolutely fresh flesh' into 'a nutritious and wholesome cattle feed, with the remainder rendered down into what he called guano.

Others lobbied from profoundly different standpoints. It was through Einar Lönnberg (the head of the vertebrate department of Naturhistoriska Riksmuseet in Stockholm) that Eric Sőrling was able to spend the 1904–05 season at Grytviken and publish a monograph on the South Georgia fauna (Headland 1984, 64–65). Lönnberg was particularly worried by the slaughter of those sea elephant populations close to the shore stations. Among other moves, he appealed to the ornithologist, Edmund G. B. Meade-Waldo at the British Museum (Natural History), for the 'powerful legislation' required to prevent their 'speedy extermination'. Meade-Waldo handed a copy of the letter to Lord Crewe (the Secretary of State), in the course of a deputation led by the Society for the Preservation of the Wild Fauna of the Empire. Lord Crewe, in copying the letter to the Governor, thought a clause should be introduced to every lease and licence 'specially prohibiting the killing of seals'. Its enforcement became another reason for appointing a resident magistrate to South Georgia.(3)

Henry Cowell recorded, some two years later, the increasing concern for whale stocks. In warning of the 'total extinction of the whale in South Georgia waters', one correspondent had particularly attacked the Colonial Government for its failure to set a strict limit on the numbers that could be killed. Although the need for a numerical limit was strictly correct, Darnley wrote the rationing of both the number of licences and of catchers per licence had had a similar 'practical effect'. The really important question was what constituted 'a sufficient supply of whales' to maintain a sustainable species population. Any attempt to establish that figure would be useless without knowing more of how 'whales changed their haunts in the sea'. The Norwegians had the most experience from their 10-year prohibition of whaling along the Norwegian coast(4) (Figure 10).

Whaling historians have particularly focused on the alarm expressed by the British Museum (Natural History). The part played by Sidney Harmer has been particularly emphasised, not least for the relative ease with which the Museum's archives may be consulted. Harmer had been a Cambridge graduate, lecturer and superintendent of the University's Museum of Zoology before being appointed the Keeper of Zoology at the Natural History Museum in 1909. His initial research had been on the taxonomy and systematics of the polyzoa, but he

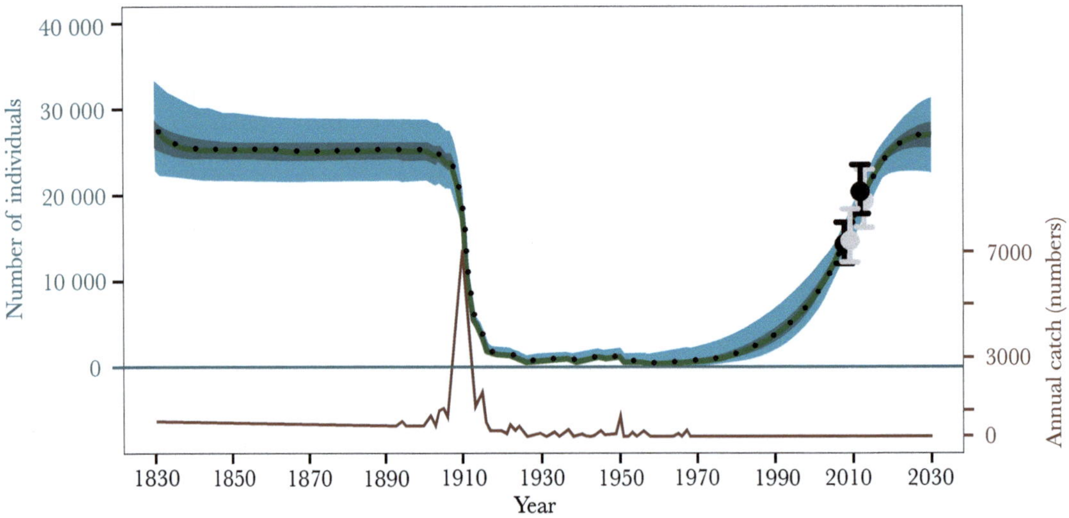


**Figure 10** A modern analysis of the modelled population trend of humpback whales in the Western South Atlantic (solid dark line) with the shaded region indicating the likely range. The model is fitted to population estimates made in 2008 and 2012 (black dots with uncertainty range as indicated). The red curve shows the number of whales killed. These data clearly suggest that the whalers had effectively wiped out the population of humpback whales that had traditionally inhabited the waters around South Georgia. Figure reproduced from A. N. Zerbini et al. (2019).

now found himself responsible for what was probably the largest museum collection of cetacea in the world, which he continued to expand. Harmer's personal interest in whales had begun in the 1890s when he co-authored a paper on a Sowerby's beaked whale washed up on the Norfolk shoreline. Years later, he pioneered 'a stranded-whales scheme', whereby coastguards reported any strandings to the Museum (Calman 1951).

Harmer was the obvious person for Jens Andreas Mørch (a whaling engineer from Christiana) to lobby. Mørch was the author of 14 patents between 1906 and 1908, one of which was designed to help load 'dead whales upon vessels . . . of relatively large size'. There followed a technical review of the 'Manufacture of Whale Products' in *Scientific American*. He attacked, in a further paper, the practice of simply stripping off the most accessible blubber 'blanket' and casting the 'huge masses of meat and bones' into the sea, asserting that the only reason the British government could tolerate such a practice must be its ignorance that it was happening. It should be made a condition of every whaling licence that the full carcase must be used (Mørch 1908, 1909 and 1911; Tønnessen and Johnsen 1982, 343–344).

Mørch recognised nothing substantive could be said about the industry's effects on stocks until they had been scientifically assessed. He accordingly published what became a pioneering (but very much overlooked) paper in the *Proceedings of the Zoological Society of London* titled 'On the Natural History of Whalebone Whales'. The southern industry had already, in its first six years, begun to surpass what the northern industry had taken 40 years to achieve. An ever-increasing number of whaling steamers 'brought oil in volumes' undreamt of only a few years previously. It was all the more reason for collating what was known of the baleen whale. So little commercial use was made of the carcase that zoologists had been given little opportunity to discover more, say, about the length and timing of gestation of the different species. There was much to learn from the observational data obtained off the Portuguese West African coast, where 'the great bulk of the Humpbacks from South Georgia' spent the austral winter. Mørch went on in his paper to speculate on the part played by the Gulf Stream of the North Atlantic. The intermingling of cold waters with the warmer, adjacent currents seemed to encourage 'the production of masses of plankton' upon which the whales depended. There was similarly much investigative work to be done on the currents of the southern hemisphere (Mørch 1911, 661–670).

Mørch sent two lengthy memoranda to Harmer, in June and July 1910, warning of the threat to whale stocks and emphasising the responsibility of the British government, as the colonial power, for protecting them.

> *Each whaling steamer under the licence of the British government should be furnished with a journal wherein should be noted down every whale killed, place where killed, species and sex; for females also if with calf. Also presence of plankton. (Savours 2001, 176–177)*

Harmer was particularly attracted to Mørch's idea that the catch data collected by the whaling companies should be submitted to 'some British scientific institution' capable of producing the large-scale maps required to provide a 'graphic view' of what was happening year by year. The cost could be covered by the licence fees paid by the whaling companies. Almost a further year elapsed before the Museum Trustees copied Mørch's correspondence to the Colonial Office, admitting how little definite information was held by the Museum (Tønnessen and Johnsen 1982, 344; Marsden 2005, 161–162; Burnett 2012, 31–32).[5]

Rowland Darnley's immediate response was to say there was little more the Secretary of State could do. Gilbert Grindle was also negative, noting that 'we limit the number of companies and the number of vessels they employ, and we must be content to let it go at that'. Allardyce pointed out that 'the many authorities with varied, extensive and practical experience of whaling' thought there was neither a significant risk to the whales, nor indeed evidence of their numbers being unduly reduced. Not only were there comparatively few whales caught globally, but the present-day whaling grounds were very small in relation to 'the vastness of the oceans in which these animals are found'. The Secretary of State's formal response, of 3 June 1911, indicated how little scientific information was available on which policies might be based. Perhaps the Museum might help rectify this?[6]

The Museum Trustees were far from reassured by the reports the Governor continued to receive from South Georgia and the South Shetlands. The Trustees found them impossible to read 'without very serious misgivings' at the lack of 'foresight and caution' being shown by those locally involved. The South African shore stations were not only increasing in number, but humpbacks formed a large proportion of their catch. It was by no means impossible, the Museum's Secretary wrote on 27 October 1911, that

> *the schools of Humpbacks which visit the South African coast during a certain part of the year may be identical with those which are found off South Georgia at other times. It is thus greatly to be feared that the persecution of the Humpback is being carried on at two very different localities, and that the species may be quite unable to resist so vigorous and determined an attack.*[7]

Darnley acknowledged that with so little known of the distribution and abundance of whales, 'our policy should be one of great caution'.

In January 1913, the Trustees forwarded an outline of the kind of investigation they believed was required. Darnley was sufficiently convinced to minute, 'There is a good deal to be said for sending a zoologist to study the industry on the spot', his costs being covered by 'the proceeds of whaling taxation'. The Colonial Office accordingly recommended, on 15 April, a sum of £600 to be spent for this purpose. The Trustees nominated Major Gerald E. H. Barrett-Hamilton of the Fifth Royal Artillery Rifles (then on manoeuvres in Ireland), their letter of 3 June 1913 describing him as a mammologist and 'constant worker at the Museum', who had 'a special interest in marine animals'. He had also visited the seal rookeries of the Pribylof Islands in 1896–97 as one of the two British government 'expert' advisers appointed to the Alaska Fur Seal negotiations (Burnett 2012, 54–64).[(8)]

Barrett-Hamilton accepted the appointment and secured a further £100 from the Colonial Office for Percy Stammwitz (the Museum taxidermist) to accompany him (Stearn 1981). Allardyce had meanwhile continued to argue that there were sufficient data already, provided through both his own officials and the whaling companies. A far greater priority was to discover the migration routes taken by the different species of whale. He tacitly agreed to the Colonial Government meeting the costs of employing Barrett-Hamilton and Stammwitz for a single summer. They arrived, aboard Bryde and Dahl's floating factory *Thor* at Leith Harbour on 15 November, where they began straightway to measure, describe and sample an eventual 294 whales.[(9)]

Barrett-Hamilton reported, on 2 December 1913, there had been only 25 whales to sample – little more than a whale a day. Those at the whaling station remained confident 'the big herds of Humpback' would still arrive. The shore station manager was 'one of the most intelligent people on the island'. Barrett-Hamilton hoped he would be able to visit the South Shetlands, but this would depend on the companies providing him with a passage. Another letter, three days later, contained 'a good many details of the financial and business side of the whaling industry, which Darnley might find useful'. Another to Harcourt of 2 January 1914 (who had asked to be kept personally informed) relayed how so very few whales were being caught that without a substantial increase in the daily catch, the companies were likely to incur 'a serious loss of money'.

> *I find many people here nervous about the prospects of whaling in these waters. Large sums of money have been sunk in shore stations and factories, and the working expenses of concerns situated at not as such a long distance from home are very high. Therefore a big catch – vastly bigger than in waters nearer home – is absolutely indispensable for a profitable season.*[(10)]

Barrett-Hamilton had, by the time of his letter of 16 January 1914, begun to attribute the continued absence of humpbacks to their excessive killing – their decline was a warning of how the blue and fin whales might soon also be 'decimated' (Headland 1984, 73).

Barrett-Hamilton died the next day of heart failure, bringing the investigations to an abrupt end. Martin A. C. Hinton at the Natural History Museum synthesised the mass of data that Stammwitz had brought back, including the taxonomist's own measurements, descriptions and samples (Headland 1984, 73). The Museum Trustees were in no doubt that the figures showed a clear decline in humpback numbers through 'excessive slaughter'. As Grindle acknowledged,

'A calf [born] in alternate years' did not 'point to rapid reproduction of the numbers caught'. Darnley's difficulty was not so much making a case for continuing with Barrett-Hamilton's investigations but how account might also be taken of the consequences for the shore stations along the West and South African coasts (Matthews 1931, 141–142). Barrett-Hamilton's data (and more particularly Hinton's synthesis of them) proved so relevant to the post-war investigations of the 1920s that rather than remain 'buried in the obscurity of a Confidential Office Print', they were published and made quite accessible (Hinton 1925).(11)

## 5.1 Foreign Office enquiry

The United Kingdom Government had signed, on behalf of the Canadians, a Sealing Convention in July 1911, along with the Japanese, Russian and United States governments, banning pelagic sealing in the North Pacific and the use of any port for that purpose. No seal skins could be traded without the official marking and certification of the respective breeding grounds, which were to be closely patrolled (Young and Osherenko 1993; Dorsey 1998). The expert advisers to the convention had commended the treaty as a model for other marine species. It was this recommendation that prompted the Foreign Office to write to both the Board of Trade and the Colonial Office on 14 August 1911 (Dorsey 2013, 31–32).(12)

The Board of Trade's Commercial Relations and Treaties Department had received about the same time what was perhaps the most succinct appraisal made of the whaling industry. The Board's lack of concern for its content was signified by a further ten years passing before the attention of the Colonial Office was drawn to its content! The author, Arthur F. Bearpark, had first worked as a chemist at the Saldanha whaling station and thereafter in various managerial roles until, by the 1930s, he ran his own technical and commercial consultancy specialising in whale and fish oils, with an office in the City of London (Hart 2020, 369 and 371). He acknowledged, in his letter of 1911, how any move to prevent 'the killing of any kind of whale' would be largely of benefit to posterity, and that he must therefore leave it to others to adjudge how justifiable it might be 'to seriously interfere with a present-day industry'. He was inclined to think every species should be preserved, but he was also 'distinctly of the opinion that the extinction of any kind of whale through the operation of the whale fisheries' was impossible. Whales were so widely distributed that the whaling industry would have to be 'very considerably extended to cause any serious or general diminution in the total number of whales present in the sea'. Even where they had become rare, they might simply have learnt 'to frequent other and safer areas'. The evidence seemed to point in that direction. Even if the industry expanded to the point of causing 'a serious diminution in the number of whales', there could be no risk of extinction. A whale fishery was 'a purely commercial undertaking', entirely dependent upon 'a plentiful supply of whales'. Should the industry expand to the point of drastically reducing numbers, the business would cease to pay and exist. Free from harassment, 'their numbers would again assume great proportions as at present'. Bearpark expected a cycle to develop, whereby there was sufficient recovery of stocks upon the abandoned grounds, to preserve both the whale and whaling as 'a paying industry'. Should some form of regulation continue to be contemplated, whaling might be prohibited upon the subtropical breeding grounds. Whereas a general ban, for example, on the use of floating factories would serve no good purpose, it was reprehensible that 'the enormous bulk of the whale', once the oil had been extracted, was left for the sharks to devour. It would not be difficult for governments to insist 'the blubber, flesh and skeleton of the *largest* number of whales' should be worked up. Regulation would not be difficult.[(13)]

The Marine Department of the Board of Trade did not answer the Foreign Office enquiry until 10 May 1912, indicating there was likely to be little, if any, objection to any protective measures contemplated. The delay, on the part of the Colonial Office, was more a case of not knowing what to say. Allardyce thought the greater priority was to impose a ban on pelagic sealing in the Southern Ocean similar to that of the North Pacific. The rookeries were still affected by poaching. As with the North Pacific, the most objectionable aspect was the wastefulness. Over half the seals shot at sea from the sealers' boats were never recovered, with the pups left to starve ashore. It was however physically impossible to mount any kind of regular patrol of those remote waters, even if the 'value of the property' had warranted it (Dickinson 2007, 95–104). Rowland Darnley dismissed out of hand any thought of simply extending the North Pacific sealing model to the southern whaling industry. The effect would be to impose a ten-year ban on all whaling and for the floating factories to be permanently outlawed. The cost would be ruinous in lost revenues and the heavy claims for compensation from the shore station leaseholders. As Sir George Fiddes (an Assistant Secretary) minuted, 'I don't see how we can go further' than the procedure of furnishing the Natural History Museum with the latest statements and statistics.[(14)]

Robert Wiseman (a Second Class Clerk) was deputed to begin preparing the Colonial Office response. His memorandum, dated 7 November 1911, acknowledged that there was no limit set to the number of whales a lease or licence holder could take, nor any distinction made between the species that could be caught. Any attempt to do so would have required 'the payment of enormous compensation' to an industry already with an annual turnover of almost £500,000. Nor would the Colonial Government have had the physical means to enforce any catch limit or protection of a particular species. The only course presently open to the Governor was either to refuse further leases and licences or to postpone any kind of action until there was irrefutable evidence of a decline in stocks. Any move, however pressing, would have to contend with 'all sorts of vested interests'.[(15)]

Lewis Harcourt's intervention provides an outstanding example of how even the briefest minute, written by the Secretary of State, could determine policy. Harcourt wrote, on 28 January 1912, 'In this matter I am on the side of the whales and seals. I think we ought to take care that no species of fauna are exterminated for commercial purposes.'[(16)]

As he wrote soon afterward, 'We ought certainly to try to get an International Agreement about the fur seal' and, in respect of whales 'we should certainly decline to issue any further licences for the present', other than ordinary renewals of existing licences. In so informing the Foreign Office, 'the magnitude of the vested interests of the whaling companies operating in the Falkland Islands and their dependencies' should be emphasised. The Colonial Office must 'proceed very carefully'.

It was accordingly not until 20 April 1912 that the Colonial Office responded, summarising the action that had already been taken 'to limit the pursuit of whales in the Falkland Islands and their dependencies, without the delay inseparable from international action'. As well as warning of the heavy claims for compensation and the essential part played by floating factories in those instances where it was physically impossible to erect a shore station, the letter emphasised, more positively, the need to maximise the use of every carcase. Should the Foreign Secretary agree, Harcourt recommended the appointment of a small interdepartmental committee to formulate more definite proposals, as advised by the industry. There should be consultation with the 'self-governing Dominions' before any international conference was called. Although not 'absolutely essential', the Norwegian and South American governments might be contacted.[17]

The Foreign Office indicated, on 2 May 1912, that Sir Edward Grey was 'of course in favour of doing all that can be done to preserve rare and valuable species from extinction'. Grey had himself amended the draft reply to suggest the first step should be to consult the dominion governments; the second to discover (through the respective embassies and legations) how other governments were likely to respond. Grey's third point was to enquire how far whaling could be pursued entirely on the high seas and therefore free from government regulation. The Colonial Office wrote to each dominion and colonial government asking whether and how their industry was regulated. It was obvious that foreign governments would also have to be involved because 'whales migrated over long distances, and that accordingly the industry carried on from one country at a time is liable to injury from excessive pursuit of whales carried on from another country at another time'. Of the various embassies and legations written to, the Japanese response proved particularly relevant because its entire national fleet was already restricted to 30 whaling steamers. The Ambassador, Sir Claude MacDonald, was asked for further detail of the scale of capital investment, the number and location of ships involved and value of the catch.[18]

Darnley circulated a paper in June 1913, drafted on the basis of the replies, in which he proposed that the United Kingdom Government should initiate an international agreement to limit the size of the global whaling fleet. There were as many as 60 catchers in the dominions, of which 24 were South African. The Falkland Islands government had licensed 80, of which 61 were Norwegian, 12 British, four Argentine and three Chilean. If the three vessels operating from St Helena were included, there were as many as 154 licensed whaling steamers in the British Empire, of which one-half was foreign owned. Norway had by far the largest number of licensed vessels (169), followed by Japan (30); United States (about 15), and perhaps a handful under the Mexican and Chilean, and the French, Danish and German flags.[19]

There would be a need for the utmost diplomacy in any negotiations. The British Empire played so dominant a role there might be 'some awkwardness' in proposing that those other governments should limit their number of licences. Some restriction would be needed, however, if there was to be 'an adequate degree of protection of the whales'. Darnley accordingly proposed that an agreement should be sought whereby governments that had awarded 20 or more licenses already should not exceed their present number, and that those with fewer might operate a sliding scale whereby they increased their present number to, say, no more than half the difference between that figure and 20. Whereas there were 130 steamers operating in the South Atlantic and about 45 in the North Pacific, there were thought to be as few as 60 steamers and 24 sailing vessels in other waters. The whaling grounds of the South Atlantic and North Pacific would therefore have to be dealt with separately. Darnley doubted whether the Secretary of State would wish to commit himself to any particular scheme at that stage.[20]

## 5.2 The Whaling and Protection of Whales Committee

The initiative for investigating whale stocks came not from industry but from the Imperial Government in the form of departmental initiatives and interdepartmental discussions in the guise of committees, sub-committees and government reports. The Foreign Office readily agreed to the appointment of a small interdepartmental committee to draft 'definite' proposals for an international agreement both to limit the size of the whaling fleet and to make greater use of the whale carcase. Darnley used a letter from the German government to warn, in January 1913, how easily the initiative might be lost. The letter spoke of the possibility of 'the complete extermination of the whales' off Germany's Southwest African Protectorate. Grindle assumed that Darnley would represent the Colonial Office in the interdepartmental discussions – he had 'made the subject his own and knows far more of it than any of us'. His knowledge of the Norwegian language gave him ready access to the Norwegian literature.[(21)]

Darnley saw the purpose of any international conference as determining what more was required to protect the whales, predicated on the assumption that it was impossible to process the carcase without recourse to a harbour or sheltered waters, and therefore the jurisdiction of the relevant government. There was principled support from both the dominion and colonial governments for the conference. The Minister of Marine Fisheries at St Johns attributed, in the course of correspondence, the sharp decline of the Newfoundland industry to a combination of overfishing, low market prices and consequent over-capitalisation. The Governor-General wrote more generally that whales were 'intelligent creatures' and had learnt to avoid those waters 'where they are subject to attack'. The Governor-General of Australia relayed the Federal Government's concern that 'modern appliances for the capture of whales' had become so efficient they might easily wipe out stocks. Although whaling seemed to have failed on the East Coast, three Norwegian shore stations had been licensed on the Australian West Coast. The Union Government of South Africa had licenced as many as 24 land stations – nine were already operating. More than 1,120 whales had been caught the previous season off Cape Province. Of the 800 brought ashore in Natal Province, 770 were humpbacks. The senior British naval officer for the West Coast of Africa reported that recent whaling expeditions had been mounted for humpbacks from 'the South Polar Seas on their way to the Bay of Biafra, where they breed each year'. The 'wholesale butchery' would probably have exhausted stocks within six years, so 'feverish' were the companies to catch the greatest number of whales possible for their oil, the carcases were 'ruthlessly thrown away'.[(22)]

Darnley took advantage of an invitation to tour the *Benguela*, the floating factory operated by Kastor of Tønsberg. It had recently returned with 8,000 barrels of Number 1 oil; 1,500 barrels of Number 2, and 4,000 barrels of Number 3 oil, making for a profit of 12 to 15 per cent, if account was taken of depreciation costs and need for a reserve fund. The consul at the British Legation in Christiania forwarded 'a most interesting report' compiled from Norwegian sources. Three times as much capital had been invested the previous year in the 30 or so companies comprising the Norwegian fleet, making for a fleet of 196 catchers, 47 'boiling ships' and 26 transport vessels. [(23)]

From further enquiries made of the respective embassies and legations, Japan was found to have caught 1,620 whales in both its territorial waters and the North Pacific in 1911. Two companies operated from the German colonies and as many as seven in the Portuguese waters of southwest Africa. The information for the United States was both vague and conflicting. The Chilean company Sociedad Ballenera of Punta Arenas operated off the Chilean coast and under licence off the South Shetlands and Graham's Land. The British Legation at Buenos Aires described Pesca as exceedingly well managed. Whereas the whales had initially been caught close by, they now had to be hunted some 40 miles offshore. Unlicensed whalers were known to operate in Argentine waters. Despite an executive order, of September 1913, requiring them to be licenced, no effective way had been found of policing a coast of such length.(24)

Darnley adjudged that sufficient data had been collected by March 1913 to advise the Foreign Office that an interdepartmental committee might be convened to prepare the way for an international conference. Although a Colonial Office initiative, the Board of Agriculture and Fisheries had also recognised both the urgency and considerable difficulty of framing regulations that would be sufficiently effective to secure the acceptance and enforcement of governments. 'The good effects of prohibition or limitation of whaling in one locality' might so easily be 'neutralised by overfishing in another'.

> *When it is remembered that whales produce only one young at a time and probably not every year, the necessity of greatly restricting the operations of whalers appears ever more urgent. No doubt many of the companies recently formed to exploit the southern whale fishery will fail, but the successful companies will probably, if they go on at the present rate, reduce the stock almost to vanishing point.*(25)

In emphasising the global scale of the problem, Darnley cited the instance of an American harpoon having been found recently in a whale caught in Norwegian waters.

Henry G. Maurice (the Fisheries Secretary of the Board of Agriculture and Fisheries) was appointed chairman of what was styled the 'Whaling and Protection of Whales Committee', charged with considering what measures might 'be laid before an International Conference with a view to the protection of whales and the regulation of the whaling industry'. The Board of Trade, Colonial Office, Foreign Office, Scottish Office and the Department of Agriculture and Technical Instruction of Ireland were all represented. Darnley attended all six meetings of the committee, chairing two of them. As the Board of Trade member remarked at the first meeting, on 12 August 1913, the industry's success and the large number of whales taken were themselves an indication of the need for protective measures. Maurice sought more detailed evidence from the Trustees of the British Museum (Nature History), the attendance of Johan Hjort (the Norwegian Director of Fisheries) and a nominee of the United States Fisheries Commission as witnesses before the committee. Detailed evidence was to be obtained from the Governor of the Falkland Islands and the relevant departments of the Irish and Scottish governments about their experience of enforcing their respective regulations. A further four months elapsed before the committee met again on 10 December 1913.(26)

Sidney Harmer, the first witness, pressed for more studies of the kind undertaken by Barrett-Hamilton, together with what he called 'proper investigations' of the whales' environment and the effects of hunting. Little was known of their breeding habits; there seemed to be 'no definite period' for mating. There was, as Henry Maurice concluded, no 'guiding principle' by which a 'closed season' might be imposed. Harmer was likewise unclear as to whether whales migrated by some definite route or indeed why they should choose to do so. When challenged by Maurice to prescribe a 'satisfactory method of protection', he emphasised that this must await further biological examinations of the kind Barrett-Hamilton had begun, ideally from a number of shore stations at the same time. When pressed by Maurice whether there was definite evidence of a serious depletion of whale stocks, Harmer cited both the increase in the catch size in the Southern Ocean and observation of whaling in the northern hemisphere, which had shown that, once reaching 'a certain point, the catch fell off rapidly'. Harmer took solace from the knowledge that weather and ice conditions would always prevent whaling in some parts of the Southern Ocean.

Governor Allardyce, on home leave in London, testified to what the 'dozens and dozens of different whaling captains' had told him when calling at Stanley. Whereas Harmer was alarmed by the destruction of whale stocks as the whaling fleets grew larger, Allardyce was much more concerned lest the whales were 'frightened off from entering Dependencies' waters. 'A very clever mammal', the humpback was able to sense danger and avoid where the whaling fleets operated. While the Southern Ocean was 'a very large place', the whaling grounds were comparatively small. There was an obvious risk of the same herds being attacked from different points along the coastlines in the course of their yearly migrations. In an allusion to Barrett-Hamilton's work, Allardyce ridiculed the notion of sending a single 'man to South Georgia, or indeed to the South Shetlands, to study dead whales'. The essential first step should be to charter a vessel to follow the whales and learn more about their habits. The cost might be £5,000 a year, but the monies would be well spent – providing they came from the Imperial Exchequer rather than the Colonial Government. From the knowledge gained of the migration routes and of 'the habits and peculiarities of the whale food', there would be a more informed basis on which to regulate the industry. However obvious it was that the greatest use should be made of each carcase, every whaling captain had protested to Allardyce that because the summers were so short and the weather was often so bad, their overriding priority must always be to flense and cook the blubber – their 'shareholders at home' expected no less. Even when conditions permitted the greater part of the carcase to be used, transport costs meant there was often little or no profit to be made from the guano. Allardyce and his Legislative Council continued to believe the most effective form of regulation remained one of refusing more leases or licences, except where any fell vacant.

Maurice found both Harmer and Allardyce so evasive in their evidence that he feared there would be no whales to protect by the time a 'satisfactory method of preserving them was found'. Captain L. Bruun, the Norwegian manager of the Irish whaling station at Belmullet (but with experience of both the Antarctic and Portuguese colonial waters) warned the committee of the heavy costs of designating any kind of whaling reserve, most obviously in the levels of compensation the whaling companies would require. Sanctuaries were not needed as the greater part of the Antarctic coast was effectively a whale reserve. There was already abundant opportunity for depleted stocks to recuperate. The whaling companies would require not only very long notice of the closure of any particular ground but, as Charles O. Johnson (the co-founder of Joseph Crosfield and Sons Ltd of Warrington) warned from a trading perspective, the closure of particular grounds would cause the market price for whale oil to fluctuate. There had to be a regular supply of whale oil at the cheapest possible price for it to compete successfully with other animal and vegetable oils and fats. Rowland Darnley assured him from the chair that this would 'carry much weight with the Committee'.

Each witness brought, as intended, his own personal agendas to the committee, and none more so than Charles Johnson, the co-owner of Irvin and Johnson, which had so recently invested in a new floating factory *Restitution*, equipped with enough meat boilers to produce

13,000 barrels of oil and almost 9,000 two-hundredweight bags of dried guano. The number of whales previously required to produce the same financial return had been halved. Johnson saw the main impediment to protecting whale stocks to be the commonly held fear that the whaling grounds would soon become exhausted and that everything must be done to retrieve the investment and achieve profitability before that occurred. Johnson thought a greater effort should be made to persuade companies that through a better balancing of oil and guano production, their greater productivity would ease the pressure on whale stocks and thereby secure the longer-term prosperity of the industry. It would be far more effective than limiting the number of vessels permitted on particular grounds.[27]

Denis G. Lillie was perhaps the most relevant witness for the committee's purpose. He had been a member of Scott's Antarctic Expedition of 1910–13, for which he was awarded the Polar Medal. A Cambridge graduate and conscientious objector, he was employed as a military bacteriologist during the First World War. But for his catastrophic mental breakdown from which he never fully recovered, he might well have become a key member of the later Discovery Investigations. Lillie described how he had carefully sampled the waters through which Scott's *Terra Nova* had passed, recording the increasing quantities of whale feed as the vessel moved into the much cooler waters of the Antarctic (Lillie 1910). He had since continued his studies of the whale while working at the New Zealand shore station of Whangamumu, close to the Bay of Islands on North Island, and then aboard the floating factories of the New Zealand Whaling Company. There, he had found the whales 'to hang around the coast', particularly where the rivers brought down the nitrogenous material that encouraged plankton growth. Whereas a number of whaling companies operated off the West Australian coast, he was told they would not 'touch' the East Coast for its fewer rivers and therefore little plankton. But even the whaling captains, most familiar with Australian waters, acknowledged the Antarctic grounds to be the most remunerative, particularly those off South Georgia and the South Shetlands.

Lillie had witnessed firsthand what he described as the humpbacks' obvious 'breeding migrations' along the New Zealand coast. The only explanation for their northward migration away from the summer-feeding grounds, and their much poorer condition when they returned, was their need for warmer waters for giving birth. Many heading southward were accompanied by young. Examination of those caught suggested the gestation period could be 13 months, with the whales which had gone north to be impregnated, returning with a 2.5 inch embryo which, by the time of their next northward migration, might have grown as much as 14 feet and be ready for birth in June and August. As to their actual routes and destination, the herds might range as far south as 67° south, appearing off the Bay of Islands toward the end of April, with most in May, continuing to move northward as far as Norfolk Island, where the whalers contended they bred. The first could be seen returning southward past the Bay of Islands in mid-September, forming a continual stream just as they had when moving northward. The Bay of Islands shore station had closed by late December. Lillie reasoned that the humpbacks' preference for shallower waters went far in explaining why they had provided the bulk of the initial catches. There was a correspondence between the reports of the herds passing the Durban whaling station of South Africa and those travelling southward in New Zealand. There had been no reports of them moving in any direction other than north and south. Blue and fin whales were generally found in their greatest numbers between November and March, by which time it was too dark to see whether any remained. Their preference for the deeper waters had made it especially difficult to adjudge whether they both gathered and migrated specifically to breed. Although the whaling captains thought so, Lillie reckoned 'it would be a good bit of work' to confirm this by anatomical work at South Georgia.

Lillie contended that there was nothing to show whether the whales were becoming scarcer or were simply 'driven from place to place' by the 'tremendous slaughter'. While there might still be 'any amount of whales in the Southern Hemisphere' there was no way of knowing whether they belonged to particular grounds or had simply been driven there. Stocks might, however, be 'just beginning to get appreciably fewer'. Whaling captains, although averse to regulation and sceptical of what policing might achieve, were beginning to acknowledge in their conversations the only way to secure their longer-term livelihood might be to close particular grounds for a number of years, and then to re-open them for perhaps a further five or six years, before imposing 'another close season'. As developed by Henry Maurice from the chair and by Lillie in the course of his evidence, a routine might be devised for some kind of rotational management, alongside measures to prevent 'a tremendous slaughter' on those grounds that remained open. The companies would have to be given plenty of notice. The outstanding question was how the companies might survive commercially during those periods of closure.

The outbreak of war caused the committee to be dissolved, but for Darnley, sufficient evidence had been obtained to confirm 'the general soundness' of present British policy. The companies would always require a sufficiently large catch to cover the cost of the expedition, but the annual catches must not be great enough to threaten the survival of the whale population. A closer investigation was needed to determine whether the Southern Ocean constituted, as was widely believed, 'a vast reservoir of fin and blue whale', of which only the dependencies' waters were presently 'tapped'. If this were found to be true, those other waters might one-day offer 'permanent and adequate protection' through the designation of a reserve protected by international agreement.(28)

## 5.3 The wider fisheries context

Deep-sea fishing provided an essential context. Johan Hjort, the Director of Fisheries at Bergen, was the sixth and most important witness to appear before the Whaling and Protection of Whales Committee. The detailed memorandum that he handed Maurice at the commencement of his evidence showed his unrivalled knowledge of whale numbers, their incidence and behavioural patterns (Hardy 1950).(29)

So important was the cod harvest industry to Norwegian life that Georg Ossian (the son of the distinguished zoologist Michael Sars) had been commissioned as early as 1864 to investigate the striking differences in the landings of cod made from one year to the next around the Lofoten Islands (Smith 1994, 1). Johan Hjort, whom Vera Schwach describes as 'a talented researcher, but not an easy character', had trained at Christiania and Munich before spending a year at the Zoologica Station in Naples and becoming director of the University Biological Station at Drøbak in southern Norway in 1897. It was from there that he organised cruises as far north as Spitsbergen and Jan Mayen aboard the *Michael Sars* – an English trawler purchased and refitted by the Norwegian government as a deep-sea research ship. Already a member of the Norwegian Board of Fisheries, he became director in 1906 (Schwach 2002, 40).

Hjort's goal was to modernise the industry through economic reform, based on scientific inquiry, thereby considerably extending the mapping programme that Georg Sars had initiated, plotting, for example, 'the abundance of eggs laid by the cod along the Norwegian coast, using their concentration to identify the more productive spawning grounds'. He also used a range of statistical techniques to determine the age of fish by their scales and otoliths which indicated striking differences in spawning behaviour from year to year. Hjort's paper 'The fluctuations of the great fisheries of northern Europe' took 'fishery research to a higher plane'. It was an exemplar of how cetacean studies might develop (Hjort 1914; Maurice 1948; Andersson 1949; Hardy 1950, 167–168; Schwach 2002, 40; Schwach and Kjesbu 2019).

It was a time of increasing interest in marine research generally. The Sixth International Geographical Congress, organised by the Royal Geographical Society in London in August 1895, was remarkable both for the stimulus given to Antarctic exploration (Dudeney and Sheail 2019, 18) and oceanographic studies. A Swedish report on studies made in the Baltic was followed by a resolution acknowledging both the scientific and economic importance of what had been discovered and the need for similar studies to be made of the North Sea and North Atlantic. So positive was the response that the Swedish government convened a follow-up conference extending to the Arctic Seas. When the British complained that the discussions were too heavily weighted toward 'the scientific side', the Swedes countered with the inclusion of sessions on the 'many practical questions with respect to the conduct of fisheries at sea' (*The Times*, 13 May 1899; Lee 1992, 3–6).

A second meeting, attended by representatives from Denmark, Finland, Germany, the Netherlands, Norway, Russia, Sweden and the United Kingdom, was held at Christiania in May 1901, during the course of which the International Council for the Exploration of the Sea (ICES) was founded. A twofold division also emerged quickly in its proceedings, with Hjort and his group focusing on the natural variability of the annual catches, and the British and German delegations concentrating on overfishing (Lee 1992, 35–40; Schwach 2002, 40–43). The British contribution was strengthened by the Board of Agriculture and Fisheries Act of 1903, whereby fishing responsibilities had been transferred from the Board of Trade to the Board of Agriculture and, more particularly, to Henry G. Maurice in his appointment as Fisheries Secretary in 1912 (Lee 1992, 51–53, 67, 71 and 74–75; Jenkins 1927, vii). It was following Maurice's attendance at the annual meetings of ICES that Hjort and Sir John Murray developed a proposal, whereby the Norwegian government loaned the *Michael Sars* and its scientific crew for a four-month summer cruise, which Murray financed. This left Plymouth in April 1914 and followed the European coast as far south as the Canaries, from where it headed, by way of the Azores, to Newfoundland. From there, the vessel returned via Ireland and the Faroe Channel to Bergen. Those aboard had throughout that time pioneered the use of tow nets and pelagic trawls drawn simultaneously at various depths. Never before had so detailed an insight been obtained of pelagic life in the different water layers. The findings of the 11,500-mile voyage became the source for the jointly authored volume *The Depths of the Ocean. A General Account of the Modern Science of Oceanography Based Largely on the Scientific Researches of the Norwegian Steamer* Michael Sars *in the North Atlantic* (Murray and Hjort 1912; Andersson 1949).[30]

It was within that considerably broader context of fisheries research that Henry Maurice's committee endeavoured to identify and project a policy for safeguarding both the productivity of the whaling grounds and continued prosperity of the industry. Was enough known of the whales' global distribution and behaviour to devise a regulatory regime? How effective were the measures being taken by the Falkland Islands Government? How relevant might the concept of a whale sanctuary become, as the industry further developed? How much of the evidence taken by the committee was hearsay, speculative or simply special pleading on the part of interest groups? How reliable was the frequently voiced view that the Southern Ocean was so large and densely populated with whales that the industry could afford to lose a few isolated populations? Could a much greater resilience be expected of the Southern Ocean than from the more accessible and confined waters of the northern hemisphere?

Hjort began his evidence to the committee by emphasising that practically everything known about the whale had come from the whalers themselves. There was a need for scientists to interpret the evidence gathered. A study of some 40 whalers' logbooks, which had been made at the request of the Norwegian government, had highlighted the 'immense pelagic animal life' in the Dependencies waters between November and April/May, which could go far in explaining what might otherwise have been conjectured as purely 'a breeding migration'. No way had been found so far of identifying the age of individual whales.[31]

Hjort, in presenting his evidence to Maurice's committee, spoke both as Director of Norwegian Fisheries and as a spokesman for the industry itself. It was perhaps significant that Theodore Salvesen was content to let Hjort speak on behalf of the whole industry; there was no need for Salvesen himself to appear before the committee. Although uniquely placed through his executive responsibilities and stature to speak as a scientist, Hjort was also spokesman for the commercial aspect. He was noticeably very reticent in his answers to Maurice's questions of whether more regulation was needed. In acknowledging the considerable concern expressed 'among scientific people about the extermination of the whales', he personally doubted whether this could happen for as long as the industry was so restricted in where it could operate. More positively, Hjort saw a considerable potential for designating reserves 'like Yellowstone Park for the bison'. The priority was to obtain a lot more information. Some form of marking might be devised to help trace whales in their movement. Much more might be learnt from what was written in the whalers' logbooks. Many more carcases should be dissected, and studies made of their foetuses and 'all that sort of thing'. Hjort offered to act as a broker by arranging with the various whaling companies how this information could be gathered for the forthcoming 1914–15 season.[32]

The committee had been expected to issue an interim report in the latter part of 1914, but the outbreak of War led to its eventual abandonment. Darnley summarised the Colonial Office's position for the Secretary of State in June 1914. The Committee had shown that

> *the conditions which have led to the practical extermination of the Greenland and North Atlantic right whale do not exist in the case of modern whaling in general, and that the fact that now-a-days it pays to hunt whales only when they are in considerable numbers is an ample safeguard against extermination.*[33]

'The whole Antarctic area' might be acting as 'a vast reservoir of fin and blue whales, which is only being tapped at one point' – the Falkland Islands Dependencies. While it was commercially impracticable at the present time to operate over the entire Southern Ocean, there was no assurance this would continue. There would be obvious difficulty in enforcing protection at the many points along the humpback's West African migration route, but any failure to do so would considerably reduce the value of any Antarctic reserve.

Any long-term and adequate form of protection was likely to require an internationally agreed on 'whale reserve'.(34)

Darnley believed the evidence taken by Maurice's committee had 'strongly confirmed the general soundness of our policy as regards whaling in the Dependencies'. The question had become more one of whether enough was known of the whales' global distribution and behaviour for the present measures to be effective. The French Government was adopting a very similar policy for the Congo. The Committee had shown both a breadth of vision in investigating the whales' ecology and the importance of assessing the industry's impact on the individual cetacean species. It had begun to discuss the fundamental question of whether an annual catch of, say, 6,000 whales off South Georgia was sustainable. The close personal interest which the Secretary of State William Harcourt had taken in the enquiry was signified by his minuting, 'I believe it would be useful if we could insist on a return of the number of females killed with *foetus in utero* and its size . . . [W]e are still without clear knowledge of the length of gestation'.(35)

In the absence of any kind of infrastructure within which a research programme might be developed, every straw had to be clasped. Maurice's committee particularly welcomed Hjort's suggestion that Sir Ernest Shackleton be asked to allow observations to be made from the vessels taking part in his forthcoming Trans-Antarctic Expedition. It was for Maurice an opportunity not to be missed; there would presumably be 'a number of competent biologists' aboard the two ships. Owen Williams wrote from the Colonial Office in June 1914, impressing upon Shackleton how valuable it would be to collect evidence of both the presence and migration of whales 'round the verge of the Antarctic Continent' and whether there was any degree of coincidence with the plankton found. Shackleton readily agreed to make such observations.(36)

Although not appreciated at the time, some stocks might already have become so diminished that it would take a century or more for them to recover. In a lecture to the Royal Colonial Institute in 1895, Sir William Flower had reported that the Dundee and Norwegian expeditions of the 1890s had failed to find right whales, despite James Clark Ross having described them as 'very abundant in the month of December' in 1840. Flower wondered whether their absence was attributable to their recent and 'ruthless extermination' on the winter breeding grounds on the Australian and New Zealand coasts (Flower 1898).

As it was, Maurice's committee never reported. All constraints on the industry were removed to meet the wartime need for the glycerine so urgently required for munitions production. The Board of Agriculture and Fisheries published almost immediately following the War a bold programme, *British Fisheries: Reconstruction*, calling for an investigation into all types of fisheries; the building of research vessels, re-establishment of a marine research station at Lowestoft and continued support for the ICES (Board of Agriculture and Fisheries 1919; Lee 1992, 83–86). The Colonial Office also published a post-war Reconstruction Report, authored by an interdepartmental committee, in which whaling figured so large that some historians have mistakenly imagined its sole purpose was to complete what the pre-war investigation of Maurice's committee had set out to acheive. The focus of the further enquiry was post-war reconstruction, as it generally affected the Falkland Islands Dependencies. The post-war committee's primary concern was not therefore whale stocks per se but how to set the Falkland Islands' economy on a more sustainable footing.

CHAPTER 6

# *Weapon of war*

As the world's largest imperial, financial and trading power, the British declaration of war, on 4 August 1914, was bound to have global consequences. The first naval battles were fought not in the North Sea, as many had expected, but off the coasts of Chile and the Falkland Islands, where a German fleet threatened Antipodean and South American trade routes. British bankers and manufacturers dominated the global markets, enabling the United Kingdom Government to commandeer much of the productive capacity for the Allied cause (Strachan 2010; Larsen 2021, 3).

Governments were so astonished to find themselves at war and so confident of a swift conclusion that there had been minimal planning (Uekőtter 2018, 286–287). Whaling policy continued to be ruled largely by the uncertainties of the weather and whale numbers on any particular whaling ground. Harold Allen wrote, as a Second-Class Clerk in the Colonial Office, an account of the war's effects some five months before what proved to be Armistice Day 11 November 1918. The war had affected whaling in five particular ways: first, through the restrictions imposed on the sales of whale oil; secondly, the relaxation of whaling regulations; thirdly, the use of additional catchers, and fourth, the promises made as to the retention of the additional catchers after the war. The fifth reason was the use of floating factories for general shipping purposes when not otherwise required by the industry.(1)

The priority for the Colonial Office at the outbreak of war was to prevent whale oil from falling into enemy hands. Nearly all direct trade with Germany and the Austro-Hungarian Empire was stopped – the German mercantile fleet disappeared almost immediately from the seas. It proved much harder to sever the 'arteries' through which the neutral countries continued to trade with the Central Powers. The Ministry of Blockade had, by the second year of the war, shifted from a policy of searching for 'contraband' to one of rationing each neutral country to what was adjudged to have been its pre-war level of imports of the otherwise prescribed goods. Rather than relying on the oversight of the British legation and consul, the respective trade associations were made responsible, where possible, for keeping track of the volumes and destination of products (Fayle 1923, 301–302; Keilhau 1930, 306 and 402–403; Riste 1966, 89–93; Salmon 1997, 112).

The governance of whaling had become considerably more complex. The United Kingdom Government, in taking effective charge of the industry, was helped considerably by the Norwegian reliance on purchases of British coal and the British market for sales of its fish. This mutual dependency was a powerful inducement for companies to adhere to the Whale and Seal Oil Export and Sale Ordinance (Number 2) of 1917 and therefore possession of the special licence required for every sale of whale oil. Pesca was exceptionally allowed to continue its supplies of oil to Argentina, provided that three-quarters was re-exported to the United Kingdom and the remainder was disposed of locally (Fayle 1924, 47; Keilhau 1930, 310–317 and 402–403; Dudeney and Sheail 2019, 57–60).[2]

Among the desperately needed products sought for waging war on what was rapidly becoming an industrial scale was the glycerine needed for 'the Propellant Explosives' in the armaments industry. Although an attempt was made to substitute other products, such as sugar beet, an official history of the Ministry of Munitions, published shortly after the war, recalled that whale oil was the cheapest and most plentiful source. Not only was the existing stockpile of 10,000 tons requisitioned but the United Kingdom Government effectively took charge of the industry (Ministry of Munitions 1921, 78–80; Macleod 2018, 103).

So great was the military demand for whale oil that Darnley acknowledged, 'We might properly disregard the protection of whales'. As many whaling licences should be awarded as the industry could effectively use (Figure 11). The Governor, by the emergency powers granted him, was instructed to remove every constraint on the use of the whale catchers. There was no longer a need to make the fullest use of the whale carcase. Unlicensed vessels should be allowed both to buoy and tow carcases. Every vessel carrying whale oil was ordered to proceed directly to one of three ports nominated by the Ministry of Munitions: Lever Brothers at Port Sunlight, Ardol at Selby and Joseph Crosfield at Warrington, to provide the glycerine so urgently required (Ministry of Munitions 1921, 81; Dudeney and Sheail 2019 57–61).[3]

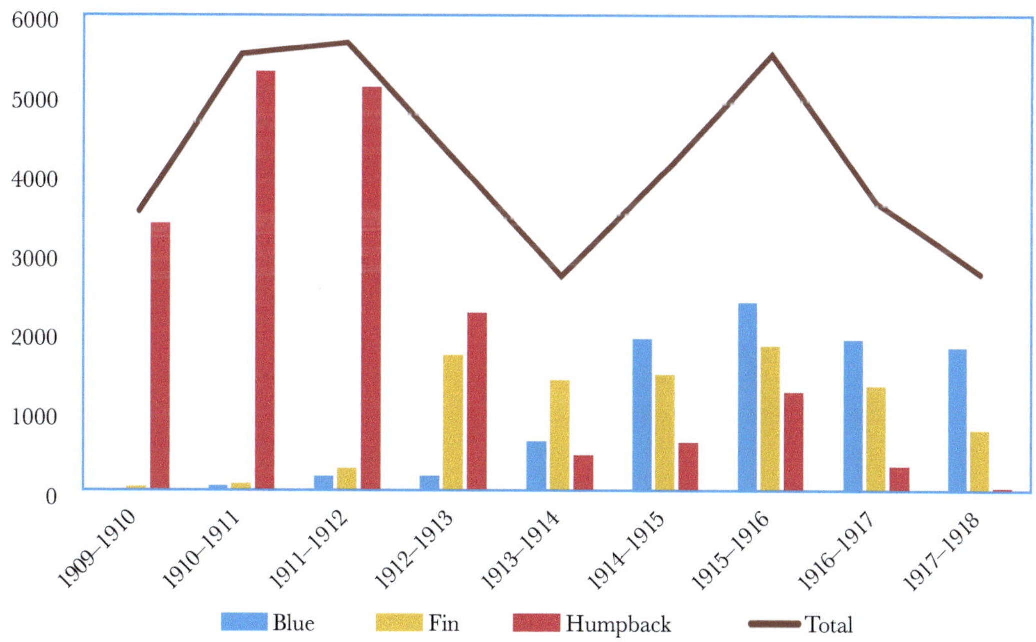


**Figure 11** The total whale catch at South Georgia (all species) for 1909 to 1918 and the breakdown by individual species for the main species targeted. Note the dramatic decline in humpback and the coincident rise in blue and fin whales. Note also the overall decline in catches just pre-war and the wartime peak as regulatory restrictions were eased. Data source: CO78/116, CO78/131, CO78/139, CO78/143, CO78/149.

As the military situation on the Western Front worsened, the British Legation at Christiania exploited ever more ruthlessly what had previously been a close alignment of interests – one which, after the war, was never fully restored. Matters were hardly helped by what the British Minister at Stockholm described much later as the failure generally to think through, and systematically apply, those hastily conceived wartime controls (Howard 1935; Dudeney and Sheail 2019, 59). Whaling became considerably more complex. Every company, except for Pesca, was required to enter a bond with the Falkland Islands Government, whereby its entire output was brought to the United Kingdom under penalty of the company losing its licence. Thor Dahl's refusal to sell his cargo of oil at the price offered him had already caused its compulsory purchase upon the insistence of the Ministry of Munitions. Dahl responded by paying the levy of £543 demanded of him the following year on the export value of his 18,601 barrels of oil at King Edward Point. But rather than deliver the oil to the prescribed British port, he diverted *Thor I* to a United States port, where the price paid for the oil more than covered his forfeiture of the bond. The lease holders of the shore stations might be threatened with their confiscation – except for Bryde and Dahl which possessed only a few sheds at Godthul.(4)

The British mercantile fleet was the largest in the world, but there was immediate pressure from the Admiralty and the Board of Trade for the release of those whaling vessels when not needed for their immediate purpose. Similar pressures were imposed on the Norwegian mercantile fleet. Their floating factories were also ideal for conveying mineral oil and other war materials across the Atlantic. However much this might suit their Norwegian owners, their reliance on 'best Admiralty coal' left them with no alternative. Such dual use of the vessels put even greater pressure on the Colonial Office. Along with all the uncertainties integral to whaling in the Antarctic, there now had to be an assurance of the relevant vessels being available and of their being adequately provisioned (Keilhau 1930, 347).(5)

It was not until a meeting at the Ministry of Munitions in January 1916 that the Colonial Office fully grasped the seriousness of the military shortage of glycerine. Darnley confessed that it was 'very difficult to say what would be the danger to the future of the industry' in terms of whale stocks, but the most recent letter from the Ministry of Munitions seemed 'abundantly to justify us in taking the risk' of removing all the constraints previously imposed upon the companies. While every preference should be given to the British companies in meeting the urgent need for increased output of whale oil, Darnley minuted it would 'probably be inexpedient to exclude those foreign firms which had the plant and ships and were in a position to act quickly'. Darnley warned Alfred Bigland MP (the Assistant Director of the Propellant Supplies Branch), that the main constraint on output was not likely to be whales but of shipping. While there might be little scope for increasing the output from the South Shetlands and Graham's Land, much more could be expected from the shore stations of South Georgia, provided the Norwegian companies were willing to increase their production: they continued to own two-thirds of the productive capacity.(6)

## 6.1 The Dannevig Pledge

Tønnessen and Johnsen, in appraising the impact of the First World War on whaling, rightly remark that a large book could be written on the protracted negotiations with the Norwegian companies, but they also exaggerate the extent to which 'Norwegian whaling' was entirely subject to the whims of the British (Tønnessen and Johnsen 1982, 294–296). The Foreign Office warned in May 1916 that there was 'no small political feeling' at the pressures being imposed on the individual companies. Nor were the negotiations entirely one-sided, in the sense that it would have been absolutely contrary to Allied interests to have alienated the Norwegian companies. They were the key to attaining the greatest possible provision of whale oil for British weaponry, and when not so engaged, for conveying the mineral oil required from the United States for the increasingly mechanised warfare along the Western Front. The overriding objective had to be one of making it worthwhile for the Norwegian companies to maximise their output of whale oil and of re-deploying those self-same floating factories and tanker vessels, when not so required, upon the trans-Atlantic run.(7)

Darnley attended a meeting on 13 April 1916 at which the Ministry of Munitions and the Board of Trade offered to purchase all the whale oil being currently conveyed to the nominated British ports, together with an assurance that these companies' whaling licences would be extended for three years. Den Norske Hvalfangerforening (the Norwegian Whalers' Association) would, for its part, accept the purchase price offered and agree that those vessels might be chartered for conveying mineral oil from America, when not otherwise engaged in whaling. The delegation (comprising Finn Bugge, Christen Christensen and Johan Rasmussen) agreed to put the offer to the respective shareholders. Andrew Steel-Maitland (the Parliamentary Under-Secretary) noted that it was '*very* important not to jeopardize the production of whale oil', but Andrew Bonar Law (the Secretary of State) saw the delegation's hesitancy in accepting the offer as a 'bluff to get a better price'.(8)

The companies' rejection of the offer caused Gilbert Grindle to draft, within the Colonial Office, what proved to be a mid-war conspectus of the industry. Four Norwegian companies (Bryde and Dahl, Ocean, Sandefjord and Tønsberg) operated around South Georgia and a further six companies were active off the South Shetlands and Graham's Land. Hektor/Hval operated eight licenced catchers; Ornen/Nor six, and the other four companies (Laboremus, Norge, Odd and Sydhav) three catchers each, making 26 in all. Although those companies that had invested so heavily in their shore stations were unlikely, even if dissatisfied, to abandon them, there was a real danger that other companies, with comparatively little material investment in the Dependencies, would do so. Rethval and Thule had already found it more profitable to charter their floating factories entirely for freight purposes. Although still making a profit, other companies might also withdraw, resentful of having to sell their entire output for what they claimed to be only a quarter of what they could obtain in other markets.(9)

A further delegation, comprising Tarald Dannevig, Hans Krogh-Hansen and Johan Hjort, returned to London under instruction to demand £40 per ton for the best grade of whale oil. The British side, in the course of renewed negotiations in May 1916, countered with £32 per ton, provided those same vessels were chartered at other times to the United Kingdom Government. Hjort asked for a guarantee that the eight annual licences currently held by the Norwegian companies would be extended beyond the war. Grindle (now an Assistant Secretary in the Colonial Office) saw no objection, provided the companies worked 'the fishery next season' rather than withdraw as their 'shareholders are inclined to do' for the more lucrative freight trade. Darnley thought an extension of up to five years sufficient. It was on that basis that Grindle gave what became known as the 'Dannevig pledge'. In a letter addressed to the Hotel Victoria, where Tarald Dannevig (manager of Laboremus and President of the Norwegian Whalers' Association) was staying, Grindle indicated that the Secretary of State for the Colonies was prepared to renew the licences for the duration of the war and beyond, provided those vessels were available for charter at other times of the year for such purposes as the United Kingdom Government required. The agreement encompassed the companies Hektor (with Hval), Norge, Odd, Ornen (with Nor), and Sydhavet, whose vessels had been chartered the previous season for a similar purpose.(10)

The Ministry of Munitions confirmed, in February 1917, that the demand for glycerine would continue for the duration of the war. David Lloyd George wrote, as Minister of War, he was confident that he could rely on the Colonial Office to foster output to the limit 'consistent with the maintenance of supply of whales'. Grindle responded, in a letter of May 1917, that every endeavour was being made to raise output, most obviously through the enlargement of the shore stations. Salvesens had to that end scoured the length of England and Scotland for a suitable plant to do so. The Secretary of State was reluctant to make the same request to the Norwegian companies lest it cause further competition for the scarce resources required and so increase foreign dominance of the industry. The shore station at Deception Island had already been closed following the charter of Hektor's floating factories *Beguela*, *Hvalen* and *Hektoria* by the Board of Trade indefinitely on the trans-Atlantic crossing. Nils Bugge rejected a Colonial Office proposal to re-open the shore station in an all-out effort to maximise oil output, pointing out that it had been designed primarily for 'the dissection of carcases'. It was hardly worth the coal, barrels and stores required for producing the inferior grades of oil that would otherwise be obtained. As Darnley noted in April 1917, 'We were trying on off-chance and expected no better'.(11)

As it turned out, only the 1915–16 summer exceeded the pre-war peak in the production of whale oil. An even greater military need had developed for mineral oil as well as a civilian need for the vegetable oils required to make margarine to counter what was becoming a marked deterioration in the civilian diet (Ministry of Munitions 1921, 78–80;). As Arthur Steel-Maitland (the Parliamentary Under-Secretary in the Colonial Office) recorded in May 1916, it was for the Ministry of Munitions to decide between 'lubricating oils from U.S.A.' and whale oil. Six Norwegian companies had meanwhile negotiated time-charters for the austral winter, modelled on those already granted the British floating factories. It was while operating those separate charters that the *Benguela*, *Capella*, *Falkland*, *Hvalen*, *Hektoria* and *Normanna* were all sunk by enemy submarines.(12)

## 6.2 Salvesen – 'the Sturdy Beggars'

*In English historical law, a sturdy beggar was a person who was fit and able to work, but begged and wandered for a living instead.*

A striking dichotomy soon developed in the Colonial Office's attitude toward the companies. The two British companies were given every facility to obtain whatever vessels and equipment they required to increase production. There was a considerable reluctance to grant the Norwegian companies the same concessions. As Darnley acknowledged in April 1916, 'We have granted no fresh concessions to foreigners'. Any greater profitability on the part of the British companies would bring higher taxable revenues for the Exchequer, whereas any fillip given to foreign companies would benefit their respective governments.(13)

Fourteen Norwegian companies had continued to operate under British leases and licences: the nine Sandefjord companies (Bryde and Dahls, Hvalen, Laboremus, Nor, Normanna, Ocean, Odd, Sandefjord and Sydhavet), the three Tønsberg companies (Hektor; Rethval/Thule, and Tønsberg) and the two Larvik companies of Norge and Ocean. Although acknowledging Hektor to be 'one of the largest and most successful' companies, Darnley refused its application for another catcher. Ocean fared no better despite being the only leaseholder limited to two catchers. Carl Larsen had left Pesca to become a farmer in Norway but almost immediately sought a licence for 'a large floating factory and three small steamers', without success.(14)

The wartime closure of the fishing grounds around the British Isles prompted John Irvin to claim that his companies would soon be bankrupt unless the Colonial Office transferred the whaling concessions previously awarded to the 'Norwegians or to any other Foreigners' to the Southern Whaling and Sealing Company. Darnley was unmoved, minuting that Irvin's company had 'got their lease originally by an *ad misericordiam* appeal' even though their losses had been largely due to 'mismanagement, overcapitalisation, the poor 1913–1914 season, and depreciation of investments in another (South African) company'. As to further British companies seeking a licence, only David Geddes and Son had some previous experience of whaling and, even in that instance, the company had failed to follow up its tentative application for a licence in the South Shetlands in the 1916–17 summer. One submission drew particular comment from Darnley. Edward Anderton wrote from a private address in Ilford, in April 1915, citing articles published in *Hvalfangst Tidende* (*Norwegian Whaling Gazette*) as evidence that the whaling grounds had never been so bountiful. Some 20 or more whaling captains had assured him 'the extinction of the whale under present conditions' was 'a practical impossibility'. Whaling was so expensive there must always be a sufficiently large catch per steam whaler to cover costs. The demise of the American 'New Bedford whaling enterprise' had shown how, despite its having become uneconomic some 50 years ago, sperm whales continued to be seen 'in large quantities all over the former whaling grounds'. Darnley noted that Anderton's application was the only one to have attempted 'a reasoned attack upon our policy of restriction'. Anderton was nevertheless told that from the information presently available, 'Mr Harcourt does not consider that he would be justified in departing from the existing decision' that no further licences should be granted, except for renewals to those already granted.(15)

Both companies Irvin and Salvesen had been invited, following the alarm raised by the Ministry of Munitions in February 1916, to apply for as many additional licences for the South Shetlands as they could use. For Theodore Salvesen, it was an outstanding opportunity to break free of all previous constraints, causing officials to point to the irony of Salvesen's being the principal beneficiary. 'So fond of waving the British flag', as one Colonial Office official observed, it was 'not unfair to draw attention to their Norwegian origin'. Another noted how Theodore Salvesen's resentfulness of the Southern Whaling Company entirely overlooked how his company's interests in 'Falklands whaling' were three times larger. But that aside, as well as the 'truculence and disingenuousness' of the Salvesen's letters, Darnley minuted, and Harcourt and senior officials agreed that 'it would seem to be bad national policy not to encourage the employment of British whale catchers during the war'.(16)

Salvesen persisted, further attacking the Colonial Office in the course of correspondence, in June 1915, for its continued failure to take advantage of 'the yearly nature' by which licences were granted. Those held by Norwegian companies could so easily be transferred to meet the Salvesens' long-standing anxiety to expand further. Darnley found the particular correspondence so 'full of errors' as to suggest that parts were intended deliberately to mislead. The effect was to make officials ever more wary – none more so than of an application, in July 1915, to deploy a further two catchers in South Georgia waters where no other company was operating. Although sympathetic to the professed intention, Harold Allen feared 'a trap'. Not only might the concession 'lead to increased killing' of whales, but it would be unfair to those 'other companies bound by their leases to the use of certain harbours'. However, the war situation had become so critical by March 1916 that both the Irvins and Salvesens were licensed to operate as many catchers as they could effectively employ.(17)

Far from being grateful, Theodore Salvesen protested that his company could hardly be expected to make the necessary investment without an assurance that their wartime concessions would be made permanent. Alongside the higher operational costs of wartime, the company had lost 10,000 barrels of whale oil when the floating factory *Horatio* caught fire at Leith Harbour in March 1916. Grindle protested that it was still far too early to decide on any post-war concessions. Much would depend on whale stocks and how the industry had survived the war, causing Sir George Fiddes (the Permanent Under-Secretary) to remark that Salvesens 'are sturdy beggars'. He felt that they should be informed that the Secretary of State was unable to approve any permanent increase in the number of catchers at that stage of the war.(18)

Theodore Salvesen revealed, in another letter to the Colonial Office dated April 1917, that Pesca had offered to sell its entire business to the Salvesens. He assumed the Secretary of State would agree to the transfer of the shore station at Grytviken to 'any British Company that we might arrange to take over the business'. Darnley saw no fundamental objection, as Salvesens were 'the most capable' of all whaling firms, but the lease originally drawn up for Grytviken must be aligned with the covenants introduced to the later leases. Salvesen was informed that the lease could be transferred to 'another reliable British Company'. Nothing came of the proposed sale, but their aggressiveness toward other companies continued unabated. As Theodore Salvesen wrote personally to Darnley, in May 1917, 'I am strongly of the opinion' that you should 'annul Bryde and Dahl's lease for Godthul'. Should Darnley agree, Salvesen proposed that the self-same lease and licence should be transferred to his company. Darnley responded by indicating it was the Colonial Office's intention to continue with the present licensing policy.(19)

Before the war, Salvesen had been allowed to let out any catchers not immediately needed at Leith Harbour, a practice considerably extended as a consequence of the wartime practice of granting as many licences as both Irvins and Salvesens could effectively use, whether directly or through 'a profit-sharing arrangement' with the other companies. Sandefjord, Tønsberg and Ocean had made considerable use of the facility. The hiring arrangement reached such proportions that Claude F. Condell (the Colonial Secretary at Stanley) cautioned the Governor, in June 1917, against allowing Salvesen to attain the same kind of dominance that 'Mr Lafone and his successor the Falkland Islands Company' had previously secured over the colony's cattle and then sheep industries. Salvesen would soon be in 'a very strong position' to attain such prominence in whaling. Darnley saw the danger but noted that the company had considerable capital resources, most obviously in its 'fleet of steamers not entirely engaged in whaling'. It was this separate mercantile fleet which had enabled Salvesen to take over 'the business of the Sandefjord Company', following that company's loss of all its transport vessels to enemy action.(20)

Matters came to a head when Condell discovered an angry letter, which Theodore Salvesen had written to Ocean, reprimanding it for applying directly to the Governor for a further catcher (for which £100 would have been charged by way of extension to its licence), rather than hiring the additional catcher from Salvesen, for which £2,000 would have been charged. Pesca had also raised with Condell the leeway given to Salvesen in subletting its licences. As Governor Young protested in November 1918, the disposition of additional catchers should be a matter for himself and not an individual company. While both he and Condell thought every opportunity should be taken 'to transfer licences, as opportunity offered, to British Companies', Young wrote, in August 1917, that none should be awarded to Salvesen, who 'had not shown the spirit expected of 'a British firm''. Darnley acknowledged, in February 1919, that Salvesen should never have been given so much leeway in 'trafficking' government licences.[21]

## 6.3 The wartime catches

There is no knowing what the pre-war Whaling Committee would have decided, but the reports from the whaling grounds had hardly suggested any kind of crisis. The Governor, in forwarding Edward Binnie's report from South Georgia for 1914–15, wrote of 'continued prosperity and increased yield'. The average output of oil per whale had risen from 38.2 barrels in 1912–13 to 46.6 barrels in 1913–14, and now to 52.94 barrels. Binnie warned, however, that the increase was due to the greater proportion of blue whales being caught than any greater economy in the use of the carcases.[22]

The most pressing question was why humpback numbers had declined (Figures 10 and 11). While 1,285 more blue whales and 59 more fin whales had been caught in the 1914–15 season, the number of humpbacks had fallen from 2,251 in 1912–13 to a mere 474 in 1913–14, and 631 in 1914–15. Some whaling captains thought 'the continued killing' had driven the humpbacks to other grounds, but others attributed their movement to the availability of more whale food in another part of the ocean. A further factor might be that whereas the humpbacks had been the most sought-after species for the comparative ease of catching and handling them, the catchers were now so 'much larger and stronger' that companies increasingly rewarded their gunners with much higher bonuses for catching blue and fin whales. Binnie took consolation from how the shift in preference toward the larger whales might enable the humpback 'to return to its old haunts'. Should that happen, and humpbacks were able to enjoy 'a peaceful time', the British Museum's fears of their being exterminated would have proven exaggerated'.[23]

The summer months of 1914–15 were notable not so much for the intrusion of war but for the hazards so integral to southern whaling. Ice conditions had prevented the whaling fleets from taking full advantage of the weeks when the blue whales were often most numerous. William Barlas found the returning tankers far from full. The floating factory *Governøren* was lost to fire (Figure 12), the catcher *Harpon* wrecked and another floating factory, *Benguela*, was badly damaged from running onto uncharted rocks. Arthur Bennett's report from the South Orkneys also focused not on the war but the need to ensure that the floating factories were fitted with sufficient pressure boilers for the carcases to be worked to their utmost capacity. Far from inflicting any loss to the companies, Bennett was confident they would make half as much profit again.[24]

**Figure 12** The partially submerged wreck of the *Governøren*. In 1915 she was the most modern floating factory ship in operation. She caught fire in Foyn Harbour (Enterprise Island) in January 1915 during a boisterous party and was deliberately sunk by firing two harpoons into her hull, apparently to stop the fire to allow some salvage. All the crew were saved, but 16,614 barrels of whale oil were lost. © John Dudeney.

Comparing the year-on-year statistics from South Georgia, Darnley found no steady development but a pattern of one 'bad year followed by two good ones'. Only 166,578 barrels had been produced in the 12 months from October 1913 to September 1914, but output then rose to 270,507 barrels in 1914–15 and to 346,209 barrels in the same months of 1915–16. James Innes Wilson had been the Magistrate for the South Shetlands for the two summers 1915–16 and 1916–17, and he too was struck by their variability. Some whalers claimed December and the first part of January to be the best time for whales, but the 1915–16 summer had been quite the reverse, both in 'the coming and going of the whale and in the weather and ice'. The abundance of whales, and more particularly blue whales', was 'astounding' in 1916–17 – whalers had never 'seen the like'. Catchers had commonly returned with six large blue whales in tow, as opposed to the more usual three or four. The floating factories had worked 'to their highest pressure', producing 300 to 400 barrels of oil without ceasing. Both the companies Nor and Norge had produced 'the finest cargoes of whale oil' ever to have left the South Shetlands: 75,406 barrels of Number 1 oil, 10,435 barrels of Number 2 and 6,919 barrels of Number 3.[25]

From its first expedition to the South Shetlands, the *Admiralen* had brought back only 4,200 barrels, and 7,500 barrels from the second expedition of 1906–07. Since then, there had been, in Wilson's words, 'quite a southern rush of Norwegian whalemen'. Whaling was now on 'an unprecedented scale'. The South Shetlands had become 'a whaler's perfect and real

El Dorado'. There were now nine companies operating, raising production from 93,900 barrels to 148,700 barrels in the 1911–12 season, to 204,700 barrels in 1912–13, and a peak of 229,600 barrels in the 1913–14 season. Production had thereafter slipped to 211,700 barrels in 1914–1915, and 212,336 barrels in 1915–16. For Wilson, the 1916–17 season had proved 'unique in almost every aspect'. An output of 92,760 barrels from only 4 companies had equalled what nine companies had managed to achieve in 1910–11, and double what the four companies had obtained in 1909–1910.[(26)]

Wilson had sought the closest involvement possible with the expeditions in the 1916–17 season. After only eight days at Deception Island, he had boarded *Bombay* for Mikkelsen Harbour (within what is now called the Gerlache Strait). Many expeditions, rather than operating close to the South Shetlands, were now venturing south along Graham's Land, where 'the hunting area is greater and the possibilities consequently better'. One had even gone as far as 65° south, taking shelter in the small harbour of Port Lockroy. But although the fitting of additional pressure boilers to process the carcase more fully benefited both companies and whale stocks, there remained the practical challenge of realising the potential of such investment, particularly where the summer was so short and the appearance of the whales so uncertain. Wilson's report included details of the South Sandwich volcanic archipelago, comprising some 'twelve large islands, separated by deep channels and a number of islets and rocks'. Almost every part was 'bold and rugged, rising abruptly from the sea in perpendicular snowy and ice-bound cliffs', except 'where here and there the naked face of a barren black rock protrudes itself from the glittering mass of snow and ice'. Everything had 'a bleak and dismal aspect, terribly forbidding and depressive'.[(27)]

Wilson thought the term 'floating factories' was a very apt description of the vessels capable of processing as many as 16,000 to 20,000 barrels of oil a season. Each had a crew of 60 to 100 – for the most part Norwegian – six to 10 open boilers for 'trying out' the blubber and eight to 16 press boilers for the bones and meat. The blubber took only 10 hours of steaming to boil down, but cooking the meat and bones took much longer, with the oil running off into tanks that could hold the equivalent of 500 to 2,000 barrels. The Number 1 oil from the open boilers contained no more than 2 per cent free fatty acids, and 0.5 per cent water or dirt. The other three qualities came from the press boilers. Number 2 oil contained up to 8 per cent fatty acids and 0.5 per cent water or dirt, Number 3 oil no more than 15 per cent free fatty acids and Number 4 oil 30 per cent. Number 1 oil could fetch £32 per ton, Number 2 oil £30, Number 3 oil £27 and Number 4 oil £25, that is, 40 per cent more than 'normal times'. Dividends of 30 to 50 per cent were being paid and even 100 per cent in one case.[(28)]

The catch of 2,751 whales at South Georgia in 1917–18 was significantly down from 4,048 in 1914–15 and 5,517 in 1915–16 to only 3,629 in 1916–17. Whereas 214,324 and 258,782 barrels, respectively, had been obtained in the first two years of the war, it had fallen back to 197,213 and 182,310 barrels, respectively, in the last two. The combination of a scarcer resource and higher prices had indicated how much more could 'be got out of the whale carcase than the companies in ordinary circumstances will admit'.[(29)]

Among the reports forwarded to the Colonial Office, Binnie described from South Georgia that there had been hardly any blue whales after January, their scarcity being attributed to a shift in the currents causing the 'the blue whale food' to be replaced by a 'red' plankton, upon which the sei whales appeared to feed judging from their large number in March. Arthur Bennett had taken Wilson's place for the 1917–18 summer in the South Shetlands, where the catch proved similarly disappointing. The five floating factories had produced only 47,605 barrels of oil, compared with 92,760 barrels from the four floating factories in the 1916–17 season. Not only had the expeditions arrived late, but the scarcity of whales was similarly attributed to the currents and therefore the lack of whale feed. There had been much ice to the south.[30]

CHAPTER 7

# *Research and the development of the Falkland Islands*

Much has been written about the part played by scientists and their respective institutions in promoting imperialism. Whether holding university or museum posts, or simply volunteering their personal expertise, they became a vital part of 'the workings of science and empire' (Pietsch 2011, 154). It is through the use made by government officials of those individuals and their learned societies that a clearer picture emerges of the development of the individual colonial territories. With whaling so largely in Norwegian hands, account had also to be taken of the further networking of expertise and experience represented by Johan Hjort and his colleagues.

Tønnessen and Johnsen (1982, 343–344) greatly over-simplify what they describe as the 'chain reaction' that began with Mørch's lobbying of Harmer; the appointment of the pre-war interdepartmental committee and the eventual establishment of what became known as the Discovery Investigations. The documentation held by The National Archives tells a much more complicated story of initiatives taken by officials in using the opportunities arising from post-war reconstruction to appraise the entire Falkland Islands' economy. Rather than simply picking up where Henry Maurice's Whaling Committee left off, the overriding purpose of the post-war enquiry was to identify ways in which the entire Falkland Islands economy might become more prosperous.

The United Kingdom remained the foremost imperial and naval power, determined to retrieve its overseas investment and outclass its competitors (Young 1997, 88–89). Rather than some kind of top-down imposition, imperial power arose from the local initiatives required for every colony to prosper. The whaling industry exemplified the approach – or rather the dilemma in obtaining it. It had fallen to Owen Williams (a Second-Class Clerk) to acknowledge Harmer's analysis at the British Museum (Natural History) of Edward Binnie's reports from South Georgia for the 1914–15 summer, in which Harmer claimed to have found the statistical evidence of a further reduction in humpback numbers and the beginnings of a similar decline in other species. The Museum Trustees and Colonial Office had no means of following up his analyses. The individuals assigned by the Governor to accompany the expeditions could do no more than report what they had observed from the decks of a particular vessel. They were in any case highly sceptical of some of the assignments identified by Harmer as being required. However interesting it might be to establish whether the migratory schools comprised whales of the same sex or of a similar physical size, experience had shown there might also be mixed groupings, as well as individuals on 'the outskirts of those large schools and sometimes after such schools had left the particular area'.(1)

Wilson was similarly sceptical of the practicality of Harmer's recommendation that the ban on killing calves should be extended to encompass '*all* small specimens'. Harmer contended that however difficult it might be to gauge the size of a whale when immersed in water, the gunners could surely be expected to stop when one small specimen was caught after another. Wilson agreed this might be expected of whalers when whales were plentiful, but hours and days might pass 'before anything turns up large or small'. Any move to extend the present prohibition to all small-sized whales would inflict serious injury upon an industry already beset by so many uncertainties. The 'one great factor' so commonly overlooked by 'outsiders' was the 'high degree of protection' afforded by the great cost of equipping a shore station or floating factory. As Wilson warned, from the evidence of the 1915–16 summer, 'the time must surely come at no very great distant date when whaling in the south must cease to be profitable'. An expedition must catch at least 150 whales for the season to be profitable. Whalers were already turning to the fresh grounds of the Bransfield Strait, Smith Island and the Belgica Strait, and, in so doing, enabling the stocks of the well-worked grounds to recover. As Governor Young summarised the position in his covering despatch of December 1917, the whale was already sufficiently protected – any further measures would unduly hamper the industry.(2)

Harmer acknowledged the heavy inroads made on stocks by expeditions that continued into the winter breeding season, closer to the Equator but insisted that the Falkland Islands Government was uniquely placed to take charge of the industry. Gilbert Grindle wondered whether the Museum Trustees had ever considered the practicality of imposing the kinds of restrictions they variously sought? It would be the officials accompanying the individual factory expeditions who would have to face the wrath of the crews. Grindle remained sceptical of whether sufficient account was taken of both the whales' apparent ubiquity and their marked habit of frequenting the pack ice, where whaling was generally impossible. Grindle went no further than formally to acknowledge the Trustees' submissions, promising they would be laid before the pre-war interdepartmental committee on whaling 'as soon as war exigencies admitted, and its sittings resumed'.(3)

Claude Condell (the Colonial Secretary at Stanley) volunteered a more local and altogether different perspective. The Norwegian captains were 'as nice a set of men as one could wish to meet', being 'plucky, courteous, and kind hearted', but some were optimists and others pessimists. However exaggerated the fears of the industry's demise, there could be no doubting that a class of mammals, producing a single offspring a year, was being severely affected. It was a very weak argument to profess 'the matter will right itself' once the whales became too scarce to be hunted profitably. As Condell phrased it, 'We wish to preserve the industry as well as the whale'. There had to be closer regulation. It might be impossible to set any kind of limit at the beginning of a single season without knowing the weather or number of whales to expect, but it should surely be possible to impose one from the evidence collected over perhaps five years. Once determined, the regulations must be rigorously enforced.(4)

There was obvious need for scientific research, but Condell insisted, it must be directed 'by a man educated in the "Humanities" rather than 'a pure specialist'. He should be someone 'capable of understanding the reality and importance of general ideas – a man able to deal with the whalers and to appreciate them'. Some of the scientific writing on whaling read more like 'the denunciation of a prosecuting attorney – with the whalers as prisoners at the bar accused of assassinating the whale species!' There were others who, through their 'exaggerated commercial spirit', had 'a somewhat sneering preference for a dead whale rendered into fertiliser' than for 'the whale as a living animal'. So long as the restrictions imposed were reasonable and applied equitably, Condell was confident there would be little objection from whalers as a body. Much would be achieved by employing those 'Government officers, local Managers, and Captains with any taste in that direction' as 'observers for some man of scientific learning'. He in turn should have the perspective that extended beyond 'the ascertaining of lengths with a tape measure'. He must also appreciate 'the necessity of fostering an important industry as well as the desirability of obtaining scientific information'. There were those who contended there were still 'vast spaces in the Antarctic where whales are undisturbed'. Condell thought it just as plausible that the present whaling grounds were 'a favourite gathering place owing to their abundant supplies of plankton during the summer season' and that, without them, whales would be comparatively rare. Until such time as one or the other theory was shown to be correct, it was better to protect every species, wherever found, rather than risk inflicting 'irreparable harm' should any particular theory prove false.

There were invariably objections to any 'remedial measures' for the difficulty of making them effective and universal. They ignored, in Condell's experience, the fact that a very large part of the whaling grounds was within the Falkland Islands Dependencies and that it was impossible to exploit those grounds without recourse to territorial waters. It should accordingly be within the grasp of British diplomacy to devise some workable convention. An international committee might be formed to investigate and collate what was known of the whale, drawing on the large store of information obtained from those engaged in the industry. This would enable future research expeditions to have 'more definite aims and less of a general roving commission' and thereby avoid the risk of their coming to resemble the labours of Sisyphus, who, in Greek mythology, was condemned to roll a huge stone up a hill, from where it always rolled down again. It was Condell's experience that every newly appointed enquiry rolled a further stone up the hill, during which it gathered 'much moss of knowledge', but with the odd exception or two, the stone was left to fall not so much to the foot of the hill but into a pigeon-hole, from where nothing emerged, except 'perhaps the trace of the easiest track to the summit'.[5]

## 7.1 Realising the Colony's potential

Jonathan Haslam has observed that those using official papers have to tease out 'the unwritten assumptions' made by officials hastily drafting whatever was required 'for purely operational purposes' (Haslam 2012, xii). His observation is particularly apposite to the reconstruction programme drawn up as accepted policies came under challenge, both publicly and within government. More than usually, officials were required to write down what might otherwise have been left implicit. The emergence of a further enquiry into the southern whaling industry had begun with the appointment of a Cabinet Reconstruction Committee as early as March 1916. Sir George Fiddes (now the Permanent Under-Secretary at the Colonial Office) had asked for a 'skeleton memorandum' indicating 'the heads of subjects in which we are interested' as well as others 'as they occur to us, or we feel it safe to handle them'. Rowland Darnley used the opportunity to appraise the longer-term prospects of the whaling industry within the larger perspectives of both the Falkland Islands and imperial development generally.[6]

*A Preliminary Memorandum on Post-War Problems* was circulated within the West Indian Department (of which the Falklands were part) in August 1916. Darnley saw two priorities.

One was British Guiana – that 'magnificent province' whose great resources of gold, timber, bauxite and water power were almost untouched. The other was 'the British Antarctica' where the Falkland Islands and Dependencies had already provided whale oil to the value of £20 million. A steamer might be provided to survey their geographical, geological, fishery and whaling potential, as well as one or two aeroplanes (following the Russian example) both to study ice conditions and to scout for whales. The cost could be offset in part by a whaling tax. Not only would this be far beyond the capability of the Falkland Islanders, but the Imperial Government must fully engage with other governments in the wider development of the mineral resources of the Antarctic continent and 'a million square miles of sea rich in animal life'.(7)

Among the never-ending succession of deputations before the Secretary of State, one of November 1916 had pressed for individual studies to be made of the respective colonies. *The Times* had called for 'a generous scheme of imperial assistance for the development of our resources'. While Charles Strachey (a Principal Clerk) thought this was to ignore the 'annual fight with the Treasury' for any kind of grant-in-aid, the Chief Clerk to the Colonial Office felt some kind of post-war imperial development board would nevertheless be needed. More fundamentally, however, both *The Times* leader and articles had been seriously flawed by assuming that every colonial territory was open to exploitation for the benefit of the Imperial Government in London. Far from that being so, it was a fundamental tenet of Imperial Government that every colonial 'territory was administered entirely for the benefit' of its inhabitants and those entrepreneurs who, through their various activities, would advance the wellbeing of those living within the respective territories.(8)

In acknowledging this, John F. N. Green (a First Clerk in the Colonial Office) warned nevertheless that the Treasury's antagonism towards any investment in science must continue to be a major constraint on development. Every one of the dominion's governments had instigated 'great Geological Surveys over their undeveloped regions', whereas the obduracy of the Treasury had prevented any geological survey being made of the tropical African colonies. British Guiana continued to be the most undeveloped colony for its size, largely because the Colonial Government had insisted it should decide how the monies were spent. There was neither need for grand committees nor large funds, but for 'small special bodies' free of Treasury control. Gilbert Grindle cautioned there must always be a degree of Treasury involvement in any public expenditure. Arthur Steel-Maitland (the Under-Secretary of State) saw no risk of 'our principles as regards the natives' being jettisoned where support was given to such figures as Sir William Lever, where the purpose was to identify the geological, meteorological and, where applicable, marine potential of the territories which, when taken together, might provide a basis for the further entrepreneurial development of the particular colony. Walter Long (the Secretary of State) rounded off the interdepartmental exchanges by confidently remarking how the immensity of the post-war problems would call for some kind of fresh approach.(9)

It was in the context of these considerably wider discussions that Darnley drafted a note on the economic resources of the Falkland Islands Dependencies. He did so in the context of the recently revised Letters Patent of March 1917 (see Figure 1, page xx), whereby the Dependencies were defined more closely as the lands and territorial waters lying between 20° and 50° west longitude and south of 50° south latitude, and between 59° and 80° west and south of latitude 58°, covering an aggregate 3,100,000 square miles, or 1.5 per cent of the globe. Although largely ice-bound, more than a million square miles were accessible for whaling, fishing and sealing. Their harbours were open for the most part in summer and, in the case of South Georgia, all the year round. As well as whaling, there were abundant opportunities for fisheries development 'if the ease with which the penguins and seals get their living' was any guide to their richness.[(10)]

Christopher Addison had, upon being appointed Minister of Reconstruction in July 1917, asked what preparatory work had been done. It was the Chief Clerk's reply that 'as yet very little definite action' had been taken, that caused Darnley to protest there had already been interdepartmental discussions on the development of the Falkland Islands Dependencies. Among the typewritten notes to survive in the Colonial Office files, there was warning of the enormity of the marine area and the inherent difficulties of studying whales in sub-Antarctic conditions. There could however be no doubting 'the great value of the marine resources'. The few remaining fur seal colonies would 'require careful nursing and protection from poachers'.

Darnley had already begun to investigate the feasibility and 'probable cost of equipping and maintaining' a vessel for scientific work. Among those he consulted, Darnley described William Speirs Bruce as 'the highest scientific authority on the Dependencies of the Falkland Islands', having led the Scottish National Antarctic Expedition of 1901–04 (Bruce 1894 and 1915). Darnley had also met and corresponded with Robert N. Rudmose Brown, a member of that expedition and now a university lecturer at Sheffield, serving with the Admiralty. It was Rudmose Brown who first suggested, in February 1917, sending a research expedition consisting of a wooden whaler with an auxiliary steam engine. As this type of ship was no longer built, the *Balaena, Terra Nova* and *Aurora*, or perhaps the French vessel *Pourquoi Pas?* might be borrowed for the purpose. The alternative would be to use an iron ship well-sheathed with greenheart (one of the densest and strongest woods on Earth). A vessel twice the tonnage of a whale catcher would be needed to accommodate the scientific staff, laboratories and necessary equipment. Both the analysis and publication of the research findings might be undertaken by a Scottish university reputed for its marine biology.[(11)]

Sidney Harmer cordially 'welcomed what he took to be the pre-war Whaling Committee', now charged with devising research expeditions for such purposes as mapping the routes by which the different whale species migrated, with winters spent investigating the impacts of the shore stations on the tropical African coast. Harmer prepared a seven-page appraisal detailing what would be achieved if funds were available at the British Museum (Natural History) for curating the research materials obtained. This had been done for the earlier *Southern Cross, Discovery* and *Terra Nova* Expeditions. The Trustees were well used to 'dealing with collections of this nature'. An annual sum of at least £1,000 would be needed for this purpose and publication of the research findings. Drawing upon the Museum's advice, and an evening spent with Bruce, Darnley wrote a detailed proposal on 21 June 1917, pressing the timeliness of such an initiative, remarking how, in the years prior to the war, the Falkland Islands were 'one of the major undeveloped estates of the Empire'. The fisheries potential might far exceed what the whaling industry had so far achieved. Darnley went on to sketch out a scheme for the employment of a research vessel both for study of the whale and the hydrographic work needed to make those waters safer for navigation. Meteorological and geological investigations could also be included. Both Grindle (who had recently been promoted to Assistant Secretary) and Fiddes were supportive, and Walter Long (the Secretary of State) wrote that he was 'in entire sympathy with Mr Darnley's admirably expressed proposals'.[(12)]

Darnley, in related correspondence with the Admiralty of August 1917, emphasised that the most pressing question from an imperial standpoint was the future of the whaling industry.

Both the Greenland and Spitsbergen industries had 'collapsed somewhat suddenly for want of whales, due to the development of more efficient methods of hunting and to the absence of any form of control'. Not only were present methods so destructive, but far too little was presently known of the feeding, migratory and reproductive habits of the whale to provide any reliable basis for a control policy. The Secretary of State had concluded that 'a fully equipped scientific investigation was needed to provide 'the basis of a really reliable control of whaling'. The Admiralty's advice was accordingly sought for commissioning such a research vessel specially fitted 'to navigate the pack ice and provided with scientific staff'. A further sentence was deleted, inviting the Admiralty to identify which of the tasks it might assist financially. The draft concluded by reminding the Admiralty that, alongside a half-dozen, fully equipped scientific expeditions recently to pass through those waters, only Bruce's Scottish National Antarctic Expedition of 1902–04 had been British. It would be 'very undesirable in the interests of the Empire that this state of affairs should be allowed to continue'. The Admiralty, responded by emphasising the need for hydrographic surveys of the Island groups, the collection of meteorological and magnetic data and a better knowledge of the tides and currents. Grindle thought the Colonial Government might well cover the costs of so doing.

The war had still many months to run, but Darnley had timed his initiative well. An Interdepartmental Committee was appointed comprising himself, John O. Borley (the Superintendent Naturalist Inspector of Fisheries in the Board of Agriculture and Fisheries), Captain Charles V. Smith (representing the Admiralty Hydrographer), and Percy C. Lyons (the Assistant Secretary of the newly formed Department of Scientific and Industrial Research (DSIR), whom the committee appointed chairman.[(13)]

## 7.2 The Committee's findings

Royal Commissions and Departmental Committees were a flexible and adaptive way of obtaining knowledge and opinion. As many as 1,400 were convened in the first 70 years of the twentieth century, with 45 appointed in 1919 (Cartwright 1975, 3–4 and 40). Darnley had drafted very comprehensive terms of reference whereby the purpose of the Interdepartmental Committee on Research and Development of the Dependencies of the Falkland Islands would be

> *to consider what can now be done to facilitate prompt action at the conclusion of the War, in regard to the preservation of the whaling industry and to the development of other industries in the Dependencies of the Falkland Islands; and to consider not only the economic questions above referred to, and the scheme for the employment of a research vessel, but also what purely scientific investigations are most required in connexion with these regions. (Parliamentary Papers (PP), 1920, 1)*

The Committee, which had met 21 times, made 36 recommendations. Although only 31 pages in length, the report signed by members on 27 August 1919 included a further 130 pages of appendices (Headland 1984, 80–81).

Tønnessen and Johnsen (1982, 292) rightly caution against exaggerating the impact of the war on whaling generally. The greater part of the evidence taken by the Committee would have been just as relevant to Maurice's pre-war Committee. The post-war appraisal was nevertheless timely. Colonial Office officials were incensed by an article published by the American naturalist,

Robert Cushman Murphy, implying that the United Kingdom had ignored the scientific advice proffered on whaling. An editorial in *Nature*, of 15 August 1918, compounded the error by calling for much greater use of scientific expertise (Burnett 2012, 116). But in publishing a rebuttal, Harmer (1918) acknowledged how very little was known about the whale scientifically. Those with relevant information should contact the British Museum (Natural History) on 'a matter which is not only of great zoological interest', but of 'supreme national importance'.

However supportive publicly, the British Museum Trustees were by no means uncritical of its handling of the whaling issue. Charles Fagan (the Museum Secretary) warned, in a letter of 22 January 1917, that whales were so migratory they could well be diminishing over their entire range.

> *Unless a real effort is made, and made promptly, to control Antarctic whaling, the responsibility for the depletion of the Whaling grounds – if that should be the ultimate result – will lie largely with those who are in a position to control the Falkland Islands at the present time.*(14)

For Harmer and the Museum Trustees, the purpose of a Dependencies Committee was not so much to reveal the risks to whale stocks, to the extent that these were already known, but to identify the safeguards needed.

Carl Larsen and Thoralf Sørlle (the station manager at Stromness Harbour) gave evidence on behalf of the recently formed Den Norske Hvalfangerforening (the Norwegian Whalers' Association). While the oil-processing industry was largely in British hands, oil production was largely Norwegian. There were 13 Norwegian companies (two of which were only nominally independent), compared with only three British companies (two of which were only nominally separate). Pesca was an Argentine company. A better guide to the national balance was the number of whale catchers: there were 37 Norwegian, 17 British and four Argentine. Whereas their number had increased, there had been a heavy loss of floating factories to German submarines; some years would elapse before the pre-war scale of operations could be restored. Harmer acknowledged the technical progress made by the whaling industry. Gunners now had the larger boats and the stronger tackle required for pursuing the blue and fin whales. It was more than ever important to discover whether there was a large 'reservoir of whales' whereby stocks might be replenished. The whalers' contention that the depleted whaling grounds would be restocked within ten years had still to be proven (PP 1920, 6, 51, 85–86 and 91).

Any adequate investigation, although focused on the Dependencies, must encompass the entire Southern Sea and indeed wherever whales were to be found. The Committee reprinted a long paper by Johan Hjort summarising what was known of the cetacean species as an essential context for adjudging what further might be achieved through the kind of investigation made by Barrett-Hamilton. Among others to give evidence was the recently ennobled Lord Rothschild who, as Walter Rothschild, had established his own zoological collection and become closely associated with the Natural History Museum. He was sure the greater number of whales bred at a specific time of the year, for which purpose they migrated northward. Assuming the southern migration was driven by the abundance of whale feed in those southern latitudes, the Committee pressed for further studies of the whales' stomach content on the flensing plan. Studies should also be made of the abundance and character of the plankton at different times, depths and localities, together with the part played by light, temperature and the chemistry in its distribution (PP 1920, 9–10, 83, 96 and 105).

The overriding purpose of the Dependencies Committee was therefore to safeguard and stimulate what Graham Burnett (2012) calls 'a home-grown industry' and therefore the promotion of the survey and research required to secure the greatest possible income for the Falkland Islands, free from the risks of over-exploitation. Every endeavour should be made to maintain and extend the statistical returns required of the whaling companies, as a means of determining what might be ascribed to natural causes as opposed to 'whaling operations'. How far might a shift in the currents, consequent upon the ice being much further south, contribute to the comparative scarcity of whale feed in a particular year? Theodore Salvesen, in his evidence, recounted that the catch was usually smaller in the milder seasons, when the 'fast ice' was well to the south, than when the ice-front was close by in colder summers (PP 1920, 5–6 and 12).

The Committee recommended that every whaling company should be required to provide, for every whale caught, detail of the date and location; its species, length and sex; and the reproductive status of the females. It should be possible by statistical analysis to determine whether there was any decrease in adult size as well as to determine both the distribution and migratory pattern of each species. The foetal-length data would further enable estimates to be made of the gestation period, growth rate and times of impregnation and parturition. Records of every female accompanied by a calf would provide further insight into the timing of the reproductive cycle and makeup of the migratory herds.

The interdepartmental report described the types of inquiry required to maximise the studies made on the flensing plan, to identify distinct populations within the particular species of whale, and the practicality of making dietary analyses of stomach content and measurements of the seasonal changes in the blubber thickness in relation to the reproductive cycle. Perhaps even more important, close observation was needed of live whales in determining their feeding habits, diving depth and duration, migration in all its forms, schooling behaviour and any predation by killer whales. Concurrent observations were needed of widely separated parts of the polar seas as well as within warmer waters. And as a final recommendation, the Committee commended the practice of putting observers on the whale catchers to gain a fuller understanding of hunting methods, and similarly at the shore stations and on the floating factories in looking for efficiencies in carcase use and the catch generally.

The Committee's eventual report was remarkable, even when judged in a more modern context, for the emphasis placed on data collection at a large spatial scale. Statistics should be gathered from wherever the whaling companies operated. Expressed in modern terms, the research requirements (although not structured as they might be today) saw the critical need for more information about the life cycles of those species being exploited and their relationship to food resources and predators. No reference was made to what would today be recognised as so critically important for managing any stock under exploitation, that is, the rate of recruitment of young whales into the fishery. Nor was there reference to natural and fishing mortality rates as the factors most directly affected by fisheries management. It is curious that these intuitively obvious questions were not addressed; the Committee may have thought them too obvious to mention or, more probably, too intractable to be tackled.

The Committee's report was signed on 27 August 1919, and published as a Command Paper to Parliament in April 1920. Both Sir George Fiddes and Leo Amery (then the Parliamentary Under-Secretary) recorded that Darnley and Harold Allen (as the Committee's Secretary) deserved the very greatest credit. The distinguished marine biologist, Sir Alister Hardy, wrote many years later that the Discovery Research Expeditions had emerged directly from 'the deliberations of a very unusual Government Committee' and 'the imaginative vision' of Rowland Darnley. It was through reading the Committee's report and more particularly the article that Darnley published in *The Nineteenth Century and After* that Hardy himself applied for and was appointed to the investigations (Hardy 1967, 37).

It is very exceptional for an official to publicise so personally the advice his department has given the Secretary of State, but Darnley was exceptionally placed to do so in respect of whaling. Anxious that the Committee's findings should not be 'buried' in a government 'Blue Book', he published an article under his own name in the monthly review *Nineteenth Century and After*, in which he defined the Committee's purpose as nothing less than 'A New Antarctic Expedition' – one far surpassing what even the *Challenger* Expedition had attempted. If successful, knowledge and understanding of those high latitude oceans would be transformed. Whereas the recommendations made by so many previous reports had 'died an early death from financial starvation' – from the refusal of the Imperial Treasury to support them – nothing resulting from the new report would be added 'to the burdens of the home Treasury'. The cost would be met largely by the whaling industry through an increase in the whale oil duty raised by the Falkland Islands Government. Although this might imply that the Piper must call 'a good many of the tunes', there was an explicit intention to range far beyond the immediate needs of the industry (Darnley 1923, 718).

With knowledge of the collapse of the Spitzbergen and Greenland whaling industries, Darnley wrote that any 'wise government' would seek to investigate their own whaling industries to ensure that such a fate did not befall them. But doing so is difficult. The Colonial Office and 'the local Government' of the Falkland Islands were at that time confronted by a 'mass of opinions', not one of which was supported by more than 'a slender basis of ascertained fact'. The obvious course was to reduce the number of catchers employed, but Darnley wrote that there was presently no 'firm basis of definite scientific knowledge' for deciding what that limit should be. As he explained to readers, a well-known biologist had been employed just before the war 'to study whaling on the spot'. His work had been of great value and promise. A Committee had since appraised both the potential of whaling and other industries, and it had recommended 'the employment of research ships', with their cost being met by a sufficiently large increase in the export duty on whale oil 'to enable the matter to be taken in hand seriously' (Darnley 1923, 721–722).

The intended investigation should be conducted on the broadest lines. One or more research ships would provide the platforms from which the requirements of the different whale species could be assessed in terms of 'their food; the temperature, salinity and currents of the water, and the ice and climatic conditions which might be encountered'. Prolonged voyages would be required, even as far as the tropics. A practical way of tagging (or marking) the whales must be found, using darts fired in the manner of a medieval crossbow. Close investigation would be made of the plankton on which the whales fed. A 'sound conservation policy for any living thing' required the most fundamental knowledge of 'the annual increment and the conditions of reproduction'. As Darnley (1923, 724) expressed it, the first would 'fail if the latter are not properly safeguarded'.

Neither the Committee in its deliberations nor the resulting report made any specific use of those who had been most closely involved in the supervision of the whaling grounds: the Governor's own officials. Not one of them appeared as a witness. Although Erik Hamilton and Edward Binnie welcomed the 'scientific and economic researches' the Committee had called for, it was left to their individual initiative to seek the correction of what they saw as major errors in the Committee's findings. None was more glaring than the assumption that British personnel would be inferior to those Norwegians presently employed. Binnie recounted that of 'the fairly large numbers of raw farm hands', which he saw engaged each year from Norway, some became quite expert. An average 'Britisher', with a 'power of observation and a willingness to serve his apprenticeship, might become similarly skilled', if given the chance. The Committee, in its anxiety to avoid prejudging the findings of its proposed investigations, brought about the paradoxical situation where those (and none more so than Harmer) who had previously pressed the urgency of intervention, now counselled caution and delay until at least the preliminary findings of the proposed investigation were available. Only two recommendations were made for immediate effect. One was to avoid entering into any further long-term leases, and the other to avoid employing more floating factories than absolutely necessary (PP 1920, 12–3).(15)

Sidney Harmer (now the Director of the British Museum (Natural History)) briefly reviewed the Committee's report in the *Geographical Journal* of July 1920 (Harmer 1920). The blue whale might already have 'passed its zenith', while the ever more 'intensive capture of Fin Whales cannot fail to affect the numbers of this species as well'. Henry Allen saw no alternative in the short term but to continue with the pre-war safeguards imposed on the industry. To go further – perhaps prohibiting the use of floating factories – would mean 'the almost total abandonment of the South Shetlands field', something that could only be justified by 'some conclusive evidence that operations on the scale now authorised were seriously threatening the stock of whales and consequently the continuance of the industry'. As the Trustees were informed, the Secretary of State (Winston Churchill) saw no alternative to continuing with present policy, subject to 'such modifications as might from time to time be found practicable and necessary'.(16)

CHAPTER 8

# *The heyday of Dependencies whaling*

Every whaling company had lost vessels to submarine warfare and the revenues they would have brought. Hektor sought an assurance that its lease would be renewed for the shore station at Deception Island before raising the capital to replace *Benguela* and *Hektoria*. An obvious course for companies was to look for cost savings. The merger of the Norwegian companies Ocean and Sandefjord meant the shore station at New Fortune Bay (Ocean Harbour) could be closed and its machinery moved to Stromness. Thoralf Sørle was confident the additional four open boilers and eight pressure boilers installed at his station could cope with the greater throughput (Tatham 2008, 508–509). The Colonial Office had little choice but to agree the arrangement.(1)

Bryde and Dahl responded to the Governor's refusal to permit its floating factory at South Georgia to have a fourth catcher by protesting that, while other leaseholders could use any sheltered water, it was confined to Godthul. The company should also have access to any unoccupied bay. Harold Allen acknowledged that Bryde and Dahl's catchers might have to tow the whale carcases longer distances, but it was also the only leaseholder not to have erected a shore station as required by its lease. Any concession would be unfair to those companies that had incurred the considerable expense. Bryde and Dahl was in any case 'hardly deserving of special consideration', having so abused the wartime regulations. Rowland Darnley quite agreed.(2)

Theodore Salvesen lobbied as hard as ever, setting out in considerable detail, in November 1919, how the whaling licences should have been allocated. Harold Allen responded that 'the Secretary of State is not accountable to Mr Salvesen', who should be told that 'the normal number of 21 catchers' would continue to apply. Far from closing the matter, Salvesen continued to demand that 'foreigners should be curtailed in their operations and British firms offered additional facilities'. Allen and Darnley found this letter particularly 'disquieting' for its 'wilful perversion' of what the Secretary of State had put in place. It was time Salvesen was 'given a broad hint that it was the Secretary of State who was responsible for the whaling policy'. Responding to another letter of complaint – this time from Salvesen's local Member of Parliament – the Parliamentary Under-Secretary, Leo S. Amery, wrote that Salvesen knew full well that any vacant lease or licence would be granted to a British company, but since every Norwegian company except one had 'met the wishes of H.M. Government during the War in the most praiseworthy spirit', it would be unfair to remove their licences. Far from wishing to award the additional concessions that Salvesen desired, Amery found it 'rather regrettable that circumstances should have rendered it necessary to allow the additional facilities' even for British companies, during the war. George Irvin had meanwhile continued to plead the financial plight of the Southern Whaling and Sealing Company and, more particularly, its entitlement to more catchers – even if that meant 'the dropping out of some of the Norwegian Companies who have already made so much out of the business'. As Allen minuted, what with Messrs Salvesens' asperities and Messrs Irvins' moans, 'one is almost tempted to wish the whaling industry were entirely Norwegian'.(3)

There was lobbying too from an entirely new quarter: the Chairman of Lever Brothers Lord William Lever, drawing attention to an associated company, the New Transvaal Chemical Company, which operated a fleet of whalers off the southeast African coast. His letter proposed two licences, currently held by Norwegian companies, should be transferred as a way of obtaining 'more effective control' of 'Whale Oil consumption in the United Kingdom'. Owen Williams acknowledged that this might be done if 'some Norwegian firm misbehaved itself', but he could see no other reason 'why the Norwegians should be kicked out at short notice'. Darnley agreed. Any move to confiscate their licences would cause deep resentment, particularly for the suffering caused in that part of Norway. There could be reprisals by the Norwegians. Lord Lever was informed that his application would be placed with the numerous other British applications for a licence.(4)

Through the purchase of their rivals, the Crosfield and Gossage companies, Lever Brothers had become the largest soap manufacturer, diversifying into other products and further capitalising on its West African palm oil and oil milling interests (Fieldhouse 1978, 31–32). Lever had also bought the Southern Whaling and Sealing Company reportedly for £750,000 in November 1919. Darnley found no reason to object other than 'some distrust' lest the company's increasing dominance of the market enabled it to charge higher prices for its products. While acknowledging the risk, Grindle noted, 'We need not be sorry to lose Messrs Irvins' (former owners of the Southern Whaling and Sealing Company).(5)

Critics of the whaling industry continued to warn that the southern grounds might soon suffer the fate of the northern grounds – prescribing in ever more detail and comprehensiveness how this might be avoided. James Travis Jenkins (Superintendent of the Lancashire and Western Sea Fisheries) provided an account of 'the natural world – the ecosystem within which whales flourished' and of 'the economic drivers of its human destruction'. There had, in every previous instance of whaling, been an initial abundance of whales and great prosperity. The turning point had come quickly as whaling technique improved. While acknowledging that economic pressures would cause the whaling industry to cease to exist long before any species became extinct, zoologists had a responsibility for both alerting and educating public opinion. It was only through the pressure of public opinion that governments would take effective steps to protect 'one of the most interesting groups of animals' from the rapacity of 'a small section of capitalists' (Jenkins 1921, 46–47 and 58).

Oil prices had quickly recovered from their post-war collapse, but there was never any confidence they would remain high. Bjørn Basberg describes those years as one of 'great Antarctic expansion' in a logistical sense, but never the kind of goldmine some historians have imagined (Basberg 1985, 89–98). Among the numerous reasons, Ian Hart writes that Pesca might well have been one of 'the very great whaling companies of the world' but for its lack of business acumen. Grytviken could well have closed, but for the astute business sense of Fridthjof Jacobsen (Carl Larsen's successor as manager), who later became the manager of Pesca itself in 1921. Jacobsen had secured the investment needed to achieve a throughput of as many as 20 whales a day, not least through improvements to the quayside and hydro-electric power supply (Hart 2001, 209–214, 232–235 and 241–248; Tatham 2008).

Companies generally were compelled to invest both in larger vessels and the oil-extraction equipment required to secure both their competitiveness and compliance with the Colonial Government's regulations. The number of patents to achieve these ends peaked in 1929, by which time they had transformed the industry. Christensen's *Admiralen* had set the dominant design. Second hand passenger ships were generally preferred to cargo vessels for the greater accommodation for crew and machinery. The year 1929 was the busiest for the installation of the Hartmann boiling apparatus, which in turn might soon be replaced by the Kvaerner cooker. There was much more innovation in the design and performance of the catchers, which might now be required to tow as many as six carcases more than a hundred miles to a shore station. Breech loading guns, using smokeless powder, replaced muzzle loaders. Ships were fitted with direction-finders and wireless telegraphy. Salvesen's catchers were typically up to 120 feet in length, with a beam of up to 25 feet and a moulded depth of 12 feet or more, their flushed deck cut away at both bow and stern. With engines of 500 to 750 horsepower, they could attain 12 knots. The tanks that had held the oil or coal for the outward journey from Europe could now be used, once thoroughly cleaned, for conveying the whale oil on the return journey. Apart from producing as much Numbers 0/1 oil as possible, judiciously mixing the remaining grades was found to increase the value of the overall catch. Centrifugal separators were being fitted to some stations by the later 1920s to remove even the smallest proportions of water and dirt. A caustic alkali could be used to neutralise the fatty acids (Matthews 1931, 128–129; Boyson 1924, 228–230; Basberg 1985, 101–102 and 1998, 24–25).

The uncertainties remained. The 1925–26 season had strikingly demonstrated how much continued to depend on the presence, size and wellbeing of the particular whale species in the waters where the expedition masters had adjudged they might concentrate. Binnie reported from South Georgia the largest catch ever made of 7,819 whales, causing the Trustees of the British Museum (Natural History) to warn ever more stridently of the inevitable and catastrophic diminution of the whale population and 'perhaps the lesser evil of the extinction of the industry'. There could be no denying, even among those working on the grounds, the obvious recklessness, extravagance and lack of regard shown for the future. Many whales, although harpooned and killed, were never retrieved from the water. Carcases would often break loose from the tow ropes or might be 'lost during the stormy passages required between the whaling field' and the floating factory or shore station.(6)

## 8.1 Local direction

However closely scrutinised by the Colonial Office, the Governor and his officials remained directly responsible for awarding the concessions, scrutinizing the catches and generally maintaining oversight of the fishery. Much was learnt and imparted during the brief hours the expeditions spent at their prescribed ports of entry and departure. As the Governor, Douglas Young, reported, in a despatch to the Secretary of State in February 1919, Victor Esbensen (now the manager of Ocean's shore station at New Fortune Bay) had admitted in conversation that the humpbacks had been so overhunted they were now rarely seen within 80 miles of a shore station. Both he and several other managers thought a closed season might be imposed from say 15 May to 1 October each year. Binnie saw the added advantage of encouraging the managers of the South Georgia shore stations to follow the practice of the South Shetlands by sending their entire workforce home for the southern winter. Not only might this give 'the whales a chance' and make for better labour relations, but workers on two-year contracts would be assured of a passage home and return to their jobs six months later. Morale would be higher, with less 'wages and food troubles'. Norske Hvalfangerforeningen (Norwegian Whalers Association) was supportive but pressed for a shorter closed season of 1 June to 15 September. Pesca and the Salvesen company thought the closure pointless – few whales were caught in winter. Binnie wanted to go further by prohibiting all pursuit of the humpback. It now comprised so small a part of the catch as to have negligible effect on oil production. The Colonial Office went no further than to re-impose the pre-war regulations.(7)

It was almost fortuitous that a government naturalist and assistant came to be appointed. The Governor had long wanted to appoint a geologist to prospect for workable minerals. The opportunity came when James Innes Wilson resigned to become a sheep farmer. Harmer succeeded in persuading the Secretary of State that a biologist should be appointed and, more particularly, John Erik Hamilton, who had previously been employed at the whaling station at Belmullet Bay in County Mayo. He immediately impressed those at Stanley with his diligence in providing much greater protection of the seal colonies around the Falkland Islands. It was readily agreed that he should be made Government Naturalist, with a salary comparable to that of the Chief Inspector of Livestock (Tatham 2008, 261–262).(8)

Since his arrival in the Falkland Islands in 1912, Arthur Bennett had risen through promotion to become the senior customs officer which, among his multifarious duties, had enabled him to publish a volume, *Whaling in the Antarctic*, giving a firsthand account of his eight summers as a whaling inspector (1913–14 to 1926–27). However rudimentary those early shore stations and floating factories may have been, Bennett emphasised in his book both the heavy investment and risks, physical and financial, they involved. As well as labour, coal and other operational costs, heavy expenditure was incurred in both refitting and maintaining the individual expeditions. Having now travelled some 8,000 miles through those southern waters in the course of his duties, Bennett was able to provide a firsthand account of this 'modern industry', writing of how every hour of delay, incurred for whatever reason. 'tells adversely on the quantity of oil'. Speed and adaptability were essential if the floating factories of the South Shetlands were to compete successfully with the shore stations. Working along the ice-edge was not dissimilar to operating close to land in calm weather, but 'the ice could be 'a deadly and treacherous enemy when it is angered' (Bennett 1931, 125–133; Palmer 1995; Tatham 2008, 72–73).

Bennett wrote confidently – as did Gunnar Isachsen (a not dissimilar Norwegian authority on whaling) – that the whales could never be exterminated (Isachsen 1929). There was, as Bennett expressed it, no reason whatever to fear the extinction of whales as species, much less as a family, because long before these animals became so thin in numbers as to be seriously threatened, commercial whaling would be numbered among those occupations that 'have been' (Bennett 1931, 35).

Gunners would be incredulous of 'any idea of a possible shortage of whales'. So few of those seen were actually killed. The much greater anxiety among those aboard the catchers was meeting the expectations of 'those at home' in providing for the 'dividends on their capital'. Bennett's greater worry 'was not the extermination of any species of whale, but how far 'the present death-rate of whales exceeds the economic limit' whereby sufficient stocks were left to maintain profitability. If only public opinion could be roused on the question, it might be possible to stay 'the slayer's hand' sufficiently to confer benefit on both 'the pursued and the pursuer'. Bennett was deeply concerned at how the numbers of humpback had fallen, not to their point of extinction but to where there was no profitability in pursuing them. Although numbers were bound to vary from year to year, there was now irrefutable evidence that they had either been severely reduced in population or frightened off the particular grounds. Although the shore stations had initially been very wasteful, the floating factories had proved even worse – 'millions of pounds worth of valuable material were allowed to rot' (Bennett 1931, 62–63 and 212–214).

As a self-taught taxidermist and naturalist, Bennett had made the fullest use of his postings to the South Shetlands. Harmer had secured the publication of his short paper in the *Philosophical Transactions of the Royal Society* on the incidence of diatoms upon the skins of whales. Bennett had found from firsthand observation the whales with the thickest covering of the yellow slime of diatoms had also the thickest layer of blubber, reflecting the greater length of time they had spent in Antarctic waters (Bennett 1920). Allen had met Arthur Bennett several times, when on home leave. He had been 'struck by Bennett's scientific knowledge' but also his 'distinctly rough' manner, meaning that Bennett would probably never have an opportunity for advancement beyond the Falkland Islands. Allen strongly supported Hamilton's commendation that Bennett should be made Assistant Naturalist, with 'a non-pensionable allowance of £100 per annum'. His 'Customs duties should always take precedence over other work'.(9)

Hamilton appointed himself the South Shetlands Magistrate for the 1920–21 season, during which time he identified four basic requirements. The first and most fundamental was for each magistrate to have his own vessel. Not only were the whaling expeditions venturing ever further, but the universal fitting of wireless on each ship meant there could no longer be 'surprise visits'. Each expedition leader knew where the Magistrate was at any particular time. The Colonial Government should additionally be given the powers to limit catch size and decide which species could be caught. There should be a prohibition on taking humpbacks until more was known of their behaviour. As only 159 were caught off the South Shetlands in the 1920–21 season – contributing as little as 1.08 per cent to the total oil output – overall profitability would hardly be affected without humpbacks. If indeed the survivors of this 'reportedly-intelligent species' were simply frightened off (as the whaling companies contended), those remaining might act as the decoy birds used in wildfowl shooting to lure others back to the hunting grounds.(10)

Hamilton similarly pressed for the more economical use of the carcase by requiring every floating factory to produce, as a licence condition, at least one barrel of press oil for every two-and-a-half barrels of blubber oil. Whereas it might seem that the most direct and obvious way of safeguarding whale stocks was to impose a limit on the number of whales an expedition might take, too little was currently known about the individual species for this to be presently possible. Hamilton personally thought the present industry was 'attacking merely the fringe of a vast herd'. He finished his report to the Governor by acknowledging the energy and courage of the townspeople of Sandefjord and Tønsberg. Their prosperity continued to depend on the respective expeditions. As well as finding the initial capital, each of the expeditions had to be fitted out and maintained annually. An ordinary expedition might cost as much as £175,000, and provision also had to be made for such risks as 'ice, weather, unknown rocks and other dangers of a like nature'. And after all these expenditures, the entire success of the expeditions would rest on the price paid for the oil at the point of sale. The Colonial Office forwarded Hamilton's report to the Trustees of the British Museum without comment. The Governor went no further than to prohibit the hunting of humpback, right and sperm whales, and to require at least one barrel of press oil for every two-and-a-half barrels produced of blubber oil.[(11)]

Hamilton was appalled at how little the Colonial Government knew about the market constraints within which whale oil was produced and eventually sold, and of the life histories and behaviour of the particular whale species. It must become better informed. Copies of the company balance-sheets should be obtained. But above all, the whaling inspectors must be given the flexibility and means to decide for themselves, as opposed to having to rely on decisions made by the whaling managers, where and when their presence was most required. Only then would they have the opportunity to see at first hand both the opportunities and hazards of attaining the industry's potential. A particular instance of the inherent constraints and hazards of operating in such climes was the loss of the catcher *Graham* during the 1924–25 summer in the South Shetlands. The vessel was one of a number of catchers that experienced two days of blizzard conditions shortly after leaving Admiralty Bay. Most managed to turn back, heavily coated in ice and listing dangerously. The *Graham* and its crew of 11 were thought to have foundered off Joinville Island – the most serious casualty in some 20 years of whaling.[(12)]

Each of the five shore stations at South Georgia was required by its lease to maintain a weather station. The Argentine Meteorological Service had provided instrumentation and the analysis of data obtained by Pesca at Grytviken. An obvious course was for the Colonial Government to appoint a meteorologist to coordinate and disseminate the data obtained from all the shore stations and floating factories wherever they operated. The Director of the Meteorological Office in London found the prospect 'very tempting' but warned, in correspondence of early 1927, that it would be extremely difficult, if not impossible, to run 'a Government Observatory' – most obviously for the difficulty, if not impossibility, of recruiting the 'succession of trained observers'. The decisive factor for the Colonial Government was that the costs would no longer fall to the individual companies and would have to be met by the Colonial Government.[(13)]

The single Government building at King Edward Point on South Georgia was now required to accommodate the Magistrate, his deputy and two customs officers, a police constable and caretaker as well as to function as the courtroom where 'all Government business' was done. Edward Binnie's marriage and the intended appointment of a wireless operator brought matters to a head. Together with the installation of electric lighting and a three-quarter-sized billiard table, the additional building was estimated to cost £2,300. The optimal location for the wireless station was found to be east of the existing building. The obvious course would have been to place the contract for its supply with the Norwegian company that had provided the original building and for it to have been despatched aboard a floating factory from Sandefjord or Tønsberg. Unemployment was so high in the British building trade that the contract was awarded instead to an Essex company. There followed weeks of anxiety lest the strikes crippling the building industry in England would prevent the order from being ready in time (Hart 2009, 201–218).[(14)]

A summertime steamship service between Stanley and the principal Dependencies had long been discussed, but it was Hans Borge (the manager of Tønsberg) who made the first substantive move by seeking a subsidy of £300 from the Colonial Government for every crossing made between Stanley and South Georgia. Borge had found a vessel suitable for the purpose in France and suggested six return-crossings a season. Allen, having met Borge in April 1924, pressed for the offer to be accepted, on the premise that the Governor would meet the cost of the subsidy. Despite Grindle's fear of objections in Parliament to what would be a mail contract placed in effect with a foreign company, it was 'perfectly clear' that only companies operating in those waters were in a position to provide the service. The Falkland Islands Company had no suitable vessel; the Southern Whaling and Sealing Company expected a considerable loss, and Salvesen prevaricated. Tønsberg was eventually offered the contract whereby *Fleurus* would make five voyages each season to South Georgia and one to the South Shetlands initially for a period of two years. Much of the 1924–25 summer had passed before the Colonial Government consented to meeting the cost.(15)

Tønsberg's offer was in practice conditional on it running alongside its whaling operations and, more particularly, the erection of a shore station on Signy Island in the South Orkneys. Edward Binnie joined the expedition both to agree a location and to the terms of the lease. The weather proved so bad as to prevent the erection of any building in the short time the expedition was there. A dispensation was accordingly granted whereby the floating factory *Orwell* and two catchers could continue to operate there, provided further pressure boilers were fitted. William Barlas accompanied the expedition in 1923–24, when the combination of winds and exceptionally bad ice conditions meant few blue whales were caught. Little more than the blubber could be removed. Although the operations were scarcely profitable, Grindle secured the renewal of both the five-year whaling lease and Tønsberg's agreement to provide the mail service for a further two years.(16)

## 8.2 Fraudulence

Other than John Quayle Dickson (the Colonial Secretary) in 1914, no senior Falkland Islands' official had visited the South Shetlands until Herbert Henniker Heaton (the Colonial Secretary) and the Colonial Surgeon made a 12-day visit to Deception Island in 1925. The Treasurer and Collector of Customs followed with a visit to Grytviken, where they found the Magistrate so overwhelmed with 'routine work' that he had little opportunity for 'supervising the operations of the different whaling companies and, more particularly, visiting the outlying whaling stations'. Although 'admirably conducted', the magistrate's office at Deception Island was so 'very comfortless' that the incumbent magistrate slept whenever possible on the floating factory *Ronald II*. Henniker Heaton had taken advantage of his visit to witness at first hand, from aboard a catcher, 'the actual capture of whales'. He had, while visiting Melchior Harbour and 'a comparatively well sheltered length of coast on the north side of Anvers Island', been able to tour the floating factory *Southern Queen*. His report to the Governor finished by praising the efficiency of those he had encountered and by relaying their confidence there had been no diminution in whale numbers. The whales were instead becoming 'much shyer than they used to be', avoiding those parts where large numbers had previously been caught.(17)

Governor Middleton's eventual successor, Arnold Hodson, became the first Governor to set foot upon the Dependencies, first during a brief visit to South Georgia in August 1927 and then on a more comprehensive tour in the 1927–28 season. As he recounted in a 50-page report to Sir Samuel Wilson (the Permanent Under-Secretary), he had boarded the *Fleurus* and called at Port Howard and Fox Bay in West Falkland before heading for Deception Island, where he remarked on the hazardous entrance to the flooded caldera and the tremendous depth and 'so much chain' being required to anchor the boats in the otherwise excellent harbour. It was in conversation with the various whaling managers that he had learnt that the floating factories, on encountering ice, would proceed with one catcher ahead and others each side, maintaining close touch with one another by wireless telegraphy. The summer had been exceptional both for the 'enormous number of whales' and for the ice pack remaining so close to the South Orkneys. As Hodson remarked in his report, whaling had been effectively pelagic that season in the South Orkneys and South Sandwich Islands. The risks to the floating factories had been considerable. The *Professor Gruvel* and *Southern Queen* had been holed and sunk, and the *Maudie* and *Pythia* badly damaged.(18)

Hodson had concluded his tour aboard the *Fleurus* with a visit to South Georgia and more particularly Grytviken, where the shore station managers had organised 'a salute of seventeen dynamite blasts carefully laid in the top of a small hill close by', making 'much more noise than a cannon'. All the whale catchers were 'gaily decorated with flags', causing Hodson to write, 'I much appreciated this graceful courtesy'. A sports meeting was organised, a rifle range opened and a ceremony held at Sir Ernest Shackleton's grave. The remainder of the day had been spent 'interviewing the different managers', whom he found 'all particularly nice men and intelligent' (Hodson 1929; Tatham 2008).(19)

However valuable the visits of the Governor and senior officials, the responsibility for ensuring the whaling companies' compliance with the whaling regulations remained with the officers aboard the floating factories. Erik Hamilton considered their presence essential, most obviously for the apprehension raised in the expedition manager's mind. Not only did their presence encourage a more productive use of the carcase but it also helped detect fraud. Two notorious instances of fraud occurred aboard the Salvesen vessels: the *Sevilla* (the secondhand steamer that had been refitted as a floating factory in 1922) and the *Saragossa* in 1925 (Somner 1984, 52). The whaling inspector complained of being physically threatened by the master of *Sevilla*. Governor Young used his home leave to warn Theodore Salvesen personally of the risk of the ship's licence being withdrawn, prompting Harold Allen to report that he had personally 'never trusted Mr Salvesen the least'. As Salvesen had to be given the opportunity of refuting the various allegations, the forthcoming season had necessarily to go ahead before any formal enquiry was held at Stanley. A new master had in the meantime been found.(20)

In the course of his analyses of reports made annually by every company, Sidney Harmer had become ever more suspicious of the statistical returns made by Salvesen. As he wrote formally to the Colonial Office in July 1924, there was considerable reason to believe the statistical information had been deliberately falsified. Salvesen countered by challenging the very basis on which the Colonial Office and Governor attempted to regulate the industry. Governor Middleton had continued to argue that the best way to measure 'the economic use made of the catch of whales' was by calculating 'the average production of barrels of oil per whale over a series of years'. Salvesen had responded with the 'astonishing proposition that the larger the proportion of the lower grade oils produced, the more economical was the use of the carcases'. As Middleton remarked, in his despatch to the Colonial Office, no manager would have the slightest difficulty in achieving that goal by simply leaving 'the carcases to decay and the processing plant dirty'. Matters hardly improved. The assigned whaling officer became so suspicious of the returns handed him by the master of *Sevilla* at the end of the 1924–25 season that a check was made of the ship's logbooks in which the individual catch was recorded as the basis for calculating the bonuses for each member of the crews. The correct number in that particular instance was found to be 410, rather than the 383 whales originally stated.(21)

But Henniker Heaton's report of his earlier visit to the South Shetlands had also revealed a conflict of priorities within the Colonial Government itself. Despite the *Sevilla* having obtained the lowest average yield of oil per whale in February 1925, Henniker Heaton was far more impressed by the captain's endeavour to dispense with Norwegian labour. In Henniker Heaton's judgement, the employment of an all-British crew more than compensated for 'his inefficiency as a Manager, the surpassing state of dirt on his ship and his low oil production'. Despite a report by the whaling inspector of more decomposing waste than found on any other floating factory, Henniker Heaton instructed that the captain should be encouraged for his 'genuine efforts' to employ British subjects – a goal Governor Middleton especially commended in his covering despatch to the Colonial Office.(22)

Erik Hamilton had meanwhile continued to press the priority of every inspector having a craft of his own rather than having to rely on what the individual company chose to provide. The Colonial Government might begin by purchasing one of the largest and most up-to-date catchers of the kind a British yard had recently built for a whaling company at a cost of £14,000. Hamilton calculated a further £5,000 would be needed to cover crew costs, an annual overhaul and the laying up of the vessel at Montevideo. A further £750 must be set aside for coal. Such detailed investigation was to no avail. Darnley dismissed all such proposals. The greater priority was to fit out a research expedition.(23)

## 8.3 The Ross Sea

The opening of the Ross Sea to commercial whaling marked a further important step in the absorption of the Southern Ocean into the global economy. Alongside the commercial challenges, the Ross Sea was so remote from the Falkland Islands that it was designated a Dependency of the Dominion Government of New Zealand. Carl Larsen, who had played so prominent a part in pioneering the whaling industry of the Falkland Islands, had turned his attention to the Ross Sea – that other large embayment of the Antarctic continent. A succession of polar explorers had used either the Ross or Weddell Seas to obtain access to the continent beyond. Both water bodies have a powerful ocean gyre, carrying pack ice and icebergs in a northeasterly direction into the Southern Ocean. Their character is otherwise very different. The west coast of the Weddell Sea – the Antarctic Peninsula – acts like a huge ice trap. The comparable coastline of Victoria Land does not extend far enough north to have a similar effect. Larsen had followed the edge of the ice pack until a passage was found through which his expedition could enter the open waters to the south as far as the Ross Ice Shelf – the largest ice shelf in the world (Ommanney 1938, 183; Jackson 2005, 178–179) (Figures 13a and 13b).

Carl Larsen had remained manager at Grytviken until 1914, when he purchased two holdings near Christiana as model farms. He had also invested heavily in A/S *Antarctic*. When it failed to raise the capital for an expedition to the Ross Sea, Larsen helped form Hvalfangeraktieselskapet Rosshavet explicitly to do so (Tønnessen and Johnsen 1982, 346–353; Tatham 2008). The Colonial Office had responded with an Order-in-Council of 30 July 1923, vesting the administrative responsibility for the Ross Sea in the New Zealand government, that is,

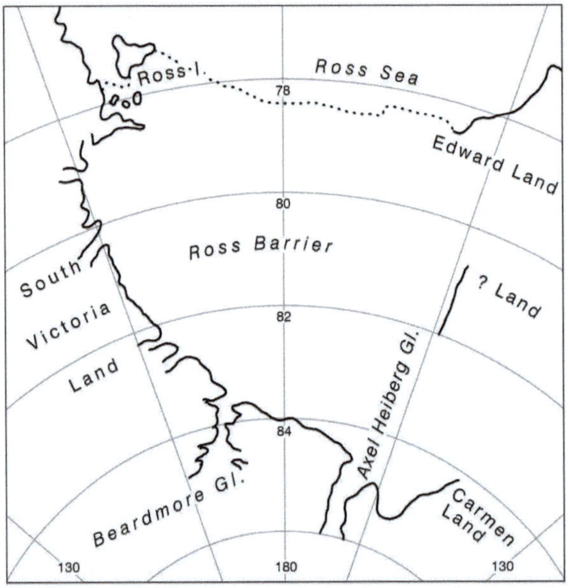


**Figure 13a** Sketch map of the Ross Barrier and the glaciers flowing from the plateau, showing the extent of knowledge of the Ross Sea Sector in the 1920s. Rudmose Brown (1927) page 103; reproduction © British Antarctic Survey, UKRI.

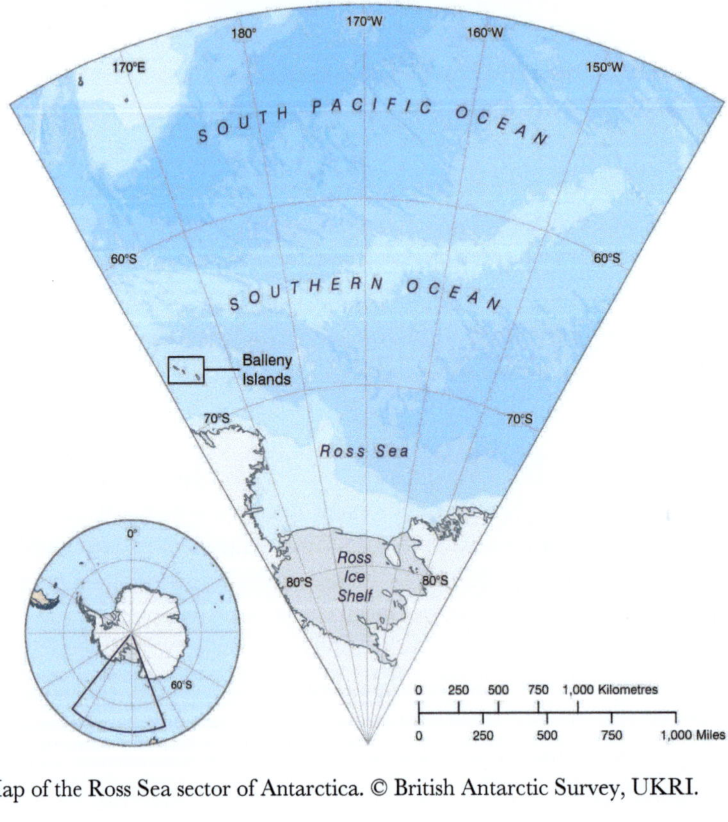


**Figure 13b** Map of the Ross Sea sector of Antarctica. © British Antarctic Survey, UKRI.

> *that part of His Majesty's Dominions in the Antarctic Seas, which comprises all the islands and territories between the coasts of the Ross Sea, with the islands and territories adjacent thereto, between the 160th degree of East Longitude and the 150th degree of West Longitude, which are situated south of the 60th degree of South Latitude, [as] shall be named the Ross Dependency.*(24)

The Governor-General would, as guided by the New Zealand Executive Council, act also as the Governor for the Ross Dependency (Dudeney and Sheail 2019, 134–137).(25)

Rosshavet was granted a 21-year licence, from October 1922, for two floating factories, each with five catchers (Quartermain 1967, 406–407). The Governor-General, Sir Charles Ferguson, refused to grant Salvesen a similar licence until more was known of the whaling potential. Larsen sent only one floating factory that initial season of 1923–24. *Mahronda* had been found lying idle for want of cargo in Liverpool. Larsen had both purchased and refitted the vessel as *Sir James Clark Ross*, its *flytende kokari* being fitted with both a blubber and meat plan, 14 pressure boilers, 10 open cookers, two Hartmann rotary cookers and the most powerful radio aboard any whaling vessel at that time (Figure 14). Having broken through the drift ice belt, the expedition proceeded southward to the Ross Barrier, reaching it on 31 December 1923, spending a particularly cold summer in a narrow embayment where it processed 221 whales (211 blue) that yielded 17,300 barrels of oil, before returning on 7 March. However disappointing 'the pioneer season', the expedition learnt much of both the whaling potential and the 'dreadful climatic conditions' – and made a modest profit. An administrator for the Ross Dependency, George Samuel Hooper, had accompanied the expedition (Villiers 1925; Tønnessen and Johnsen 1982, 348–349; Barr and Watt 2005).

Larsen died on 8 December 1924, at the outset of a second and more successful season, during which 37,000 barrels were obtained from 408 blue and 19 fin whales, caught for the

**Figure 14** The whale factory ship Sir James Clark Ross photographed at 65°S on 15 December 1930. Public Domain, Image created by William E. Howard, held by the University of Newcastle Library's Special Collections. Source was https://livinghistories.newcastle.edu.au/nodes/view/54938.

most part among the ice floes where they mostly congregated. Flensing platforms were erected just above wave height alongside the floating factory. The radio direction-finders, fitted to the five catchers, proved especially valuable for maintaining contact between the different vessels. A record 40,000 barrels of whale oil, obtained from the third season of 1925–26, encouraged Rosshavet to purchase its second floating factory for the fourth season 1926–27. The comparatively new oil-fired tanker, renamed the *C. A. Larsen*, was uniquely refitted with a bow slipway, through which its five catchers delivered the whales. The aggregate catch would have been considerably higher if the vessel had not had to be stationary whenever 'the ship's jaws' were open. Both expeditions spent the latter part of the summer on the previously unexploited grounds around the Balleny Islands (Villiers 1925, 281–285; Liversidge 1963, 121–123 and 128–129; Tønnessen and Johnsen 1982, 351; Basberg 1998, 28–29).

Ominously for both Rosshavet and the New Zealand Marine Department, an unlicensed Norwegian factory vessel equipped with a stern slipway had appeared in the 1926–27 season. Owned by the Norwegian Polaris company and directed from Hobart in Tasmania, the *N.T. Nielsen-Alonso* and its four catchers were rumoured to have obtained a 'fairly profitable' 36,000 barrels of oil. Because the ship kept to the high seas, officials of the Marine Department were powerless to intervene, even to prevent the carcases, once stripped of blubber, from being simply left to float away.[26]

The fifth season of 1927–28 proved sensational. So many blue whales had been encouraged by ice conditions to enter the Ross Sea the three vessels had left fully laden after only two months. The sixth season of 1928–29 proved just as productive but marked a turning point as the companies increasingly focused on the Balleny Islands – not because stocks in the Ross Sea were exhausted but because of the extremely hazardous ice conditions. The last and best summer was 1930–31, when an aggregate of 5,223 whales was caught and 428,211 barrels of oil produced. No catch has ever again attained that magnitude, comprising almost entirely blue whales. Each had yielded an average of 90 barrels of oil, despite the very indifferent use of the carcase (Tønnessen and Johnsen 1982, 351–353).

The refusal of Rosshavet to pay royalties on its whale oil, once it had learnt *Polaris* had not done so, meant the income obtained by the New Zealand government had fallen from £13,962 in the 1928–29 season to £7,871 the next, and thereafter ceased altogether (Quartermain 1967, 414–416). Sir Francis Bell (the leader of the Legislative Council) acknowledged, in a debate after the publication of the Marine Department's *Annual Report*, that the owners of the *N.T. Nielsen-Alonso* had ensured their expedition kept to the high seas. Having discussed the matter with the Governor of the Falkland Islands, they both saw no alternative but to seek an international agreement, citing the precedent of the North Pacific Fur Seals Convention of 1911. Not only had the Ross Sea become, within only six years, an integral part of the global economy, but, as with the Falkland Islands Dependencies, it was a forewarning of the technological changes that would shortly transform the industry. It was a prospect calling for the most urgent attention of both companies and governments.[27]

# PART III

## *Pelagic whaling, 1925–50*

CHAPTER 9

# *Freedom of the seas*

The Governor wrote confidently, in a despatch of 31 December 1925, that the floating factories were obtaining the greatest volume of oil practicable. Bonuses were almost invariably based on output rather than the numbers of whales caught. The Colonial Office was never confident of this continuing. As Darnley minuted, as early as June 1914, everything would depend 'on its being impossible, or at least very difficult for the companies, to operate without using territorial waters', and their regulation therefore by governments.[1]

Companies had continued to investigate the feasibility of becoming wholly pelagic. William Moyes described, as the customs officer aboard the first post-war expedition to the South Orkneys, how, with the pack ice continuing to prevent any entry into coastal waters, the captain of *Tioga* had joined others in using the fine weather to flense the carcases 'while lying hove-to in the open sea'. Moyes watched as both a blue and fin whale were brought under the lee of the ship, where 'the blubber was taken off as expeditiously and as thoroughly as it would have been had we been in a sheltered harbour'. In Moyes's words,

> *it was easy to see that a specially prepared floating factory, not differing greatly from the average up to date floating factory of the present time, could make as much use of the whales at sea, in fine weather, as an ordinary floating factory does nowadays in a harbour.*[2]

The organisational and technical difficulties were obvious, but so too were the commercial rewards from pelagic whaling. No longer would the catchers have to economise on fuel and ensure the whale oil remained of the highest quality, by remaining within a day's steaming of the shore station or floating factory. The factory ship would instead follow the catchers to wherever the whales might be found, making it possible to exploit even the remoter grounds. Any regulation would necessarily be by international agreement rather than by what Peder Roberts calls 'imperial diktat' (Roberts 2011, 38).

There had been numerous attempts to devise ways of hauling the carcases aboard the floating factory on the high seas. Jens Andreas Mørch had patented a slipway as early as 1904 through which the carcases could be hauled aboard, but as Tønnessen and Johnsen wrote, he could find no way of keeping the slipway flush with the sea surface. It was Petter Sørlle of Sandefjord who succeeded in patenting 'a built-in hauling slip' at the stern of the vessel and, just as importantly, was successful in fending off the legal challenges of rivals (Figure 15) (Bonner 1980, 237–238; Tønnessen and Johnsen 1982, 265–267; Tatham 2008, 508–509).

**Figure 15** The innovative built-in stern hauling slipway patented by Petter Sørlle, which allowed whales to be brought on board a factory ship for processing. Courtesy of Jacob Emanuel Ludvigsen—Vestfold Fylkesmuseum Digitalt bildearkiv http://foto.vfm.no/ A-004023, Public Domain, https://commons.wikimedia.org/w/index.php?curid=30238637.

Erik Hamilton circulated a translation of an article from *Norsk Hvalfangst Tidende* (*Norwegian Whaling Gazette*) of October 1924 showing how Sørlle had positioned the ramp in the stern so that it would not interfere with the vessel's steering. The lower edge was both regulated and submerged by a trimming tank. Just as importantly, a claw of sufficient size and strength had been devised to haul the carcases directly onto a flensing plan (Ellis 1992, 359–362). Bjørn Basberg has shown how the take-up was far from linear; further adaptation was required by way of refitting and enlarging both the processing and cooking plant. A threshold had nevertheless been crossed, which was to prove as significant as that made by Svend Foyn in devising an effective way of killing the rorquals. Eighteen vessels had been fitted with slipways by the time of Sørlle's early death in 1933 (Basberg 1998, 25–28).(3)

An article in *Norsk Hvalfangst Tidende* reported that Henrik G. Melsom and Magnus E. Melsom, with Sørlle and others, had ordered a 10,000-ton steamer explicitly for pelagic whaling. A new company, Melsom and Melsom Hvalfangersekskapet Globus of Larvik, had been incorporated for the purpose, its share prospectus laying out the intention to spend the first three months of each season off the West African coast before arriving at the ice-front of the South Orkneys in mid-November. Governor Middleton asked for a ruling of 'the projected factory's' legal status. Darnley could only prevaricate until a letter was received from Globus, in October 1925, explaining that 'our company's catching material is specially fitted out to catch on the free international sea and is able to carry on this independent of land'. *Lancing* had left Norway in June 1924, bunkered at Barry in South Wales and arrived, with its five catchers off the Congo in June (Bonner 1980, 238). A 'comparatively good result' was reported. The 'newly-patented system had proved very efficient' in hauling each whale 'onboard the cookery ship, where it is flensed and utilised'. The 10,000 barrels obtained had compared well with the 3,400 barrels from Tønsberg's floating factory *Hektor* using the same number of catchers. A yield of 30 barrels per humpback was comparable to that of any shore station. Although the French colonial government had protested at the loss of licence revenue and poor use of the carcases, the letter brazenly pointed out the 'well-known fact that the high sea outside the territorial limit is free to anybody'. The catchers had come no closer than eight nautical miles to shore. The purpose of the company's letter was to make crystal clear that it had no intention of purchasing an ordinary licence but was prepared to accommodate 'one or two officials from the Falkland Islands on board, in order that they might at any time satisfy themselves the catch always takes place outside the territorial limits and that our company carries on an entirely rational whaling'. Included with the letter was a copy of the manager's instructions, by which he was instructed (in translation) that 'all hunting and shooting etc must always take place well outside the Territorial limit'. Failure to do so, and consequent confiscation of the catch and gear, would result in both dismissal and his having to recompense the company for its losses.(4)

Hans Borge wrote privately to Darnley in September 1925 that another company, Polaris, had already been formed to operate on the high seas around Australia and in the Ross Sea. It was Borge's contention that regulation had become pointless, in the sense that record numbers of whales had been taken in the previous 15 years without any risk to stocks generally. Borge heavily hinted that it was his intention to follow Globus's example unless granted sufficient licences for five catchers. They were refused.(5)

There was nevertheless a brief transitory period, when instances of both forms of whaling might be encountered on the same grounds. The search for new grounds continued, regardless of the method used. Johan Rasmussen obtained a ten-year licence to operate a floating factory and four catchers off Bouvet Island and other islands within a 300-mile radius in those remote waters to the south of South Africa. The Colonial Office had been of two minds. The grounds were so remote from the Falkland Islands that there would be no practicable way of knowing whether Rasmussen was adhering to the whaling regulations. But with news of 'another Norwegian expedition cruising' in those waters, Grindle was anxious that Rasmussen's licence should be quickly published in *The Times*. An article in the *Evening News* with the title 'Treasure Island' soon followed, recounting that the Norwegian flag had already been planted on Bouvet Island. The Colonial Office learnt later that Lars Christensen had sought deliberately to forestall Rasmussen by despatching *Norvegia* (a sealer and whaler owned by Bryde and Dahl) and being the first to attempt the establishment of a depot there.(6)

Lars Christensen had sent the whalecatcher *Odd* first in 1926–27, and then *Norvegia*, on increasingly ambitious voyages explicitly 'to preserve and expand that industry for the Fatherland'. The crew on *Norvegia* reported Bouvet Island to be 'some 19 square miles of rock and ice, cursed with horrific weather'. Located 1,500 miles to the southwest of the Cape of Good Hope, it had to be 'the most isolated spot in the whole world'. The intention had been for *Norvegia* to continue 'right round the south polar basin', but the vessel had been so badly damaged, while searching for the reputed Thompson Island and others in the vicinity of Bouvetøye, that it was forced to make for South Georgia and repair at Godthul, where its cargo of blubber and skins from 800 seals was transferred to Bryde and Dahl's *Thor I* (Gould 1928, 197–214 and 225–233; Christensen 1935, 28–40; Basberg 2004, 31–32, and 2008, 36; Roberts 2011, 55–59; Dudeney and Sheail 2019, 167–180).[(7)]

Both the Admiralty and the Foreign Office were anxious to avoid any dispute over the sovereignty of Bouvet Island, proposing that the United Kingdom Government should offer to 'sink' any claim in exchange for the Norwegian Government's withdrawal of its claim to the Otto Sverdrup Islands to the north of Canada, named after Otto Sverdrup who had surveyed them from aboard *Fram* between 1898 and 1902. The Norwegian Minister in London, in acceding to the proposal, intimated that his Government was averse to entering into any kind of international agreement to regulate pelagic whaling lest it excite 'other foreign interests to seek a share in the spoils of the industry'. It was to the advantage of both countries that the industry should continue to be 'concentrated entirely in British and Norwegian hands' (Dudeney and Sheail 2019, 176–180).[(8)]

## 9.1 'Ice-front whaling'

Pelagic whaling was propelled not by falling productivity but by the technical opportunities whereby the whaling industry could rise still further in the 'big business league'. The prospect attracted further entrepreneurship, amalgamation and closure, whether by choice, bankruptcy or both. The consequences for governments were similarly far-reaching. Francis D'Arcy Cooper (Chairman of Lever Brothers) left Sir Samuel Wilson (the Permanent Under-Secretary) in no doubt, at a meeting of May 1926, how badly pelagic whaling was affecting licensed operators. The Southern Whaling and Sealing Company had paid £2,750 in rent for its shore station, a further £1,200 in whaling licences and £85,440 in excise duty over the previous seven years. Unless pelagic whaling was prohibited or at least severely curtailed, Lever Brothers would feel 'entitled to indulge' in the high-seas activity. The industry, as D'Arcy Cooper remarked, had never stood still. Floating factories of 4,000 to 6,000 tons had been replaced by vessels twice the size, capable of working up the whole carcase. The Southern Whaling and Sealing Company had purchased a petroleum tanker of 12,400 tons which, when refitted, was intended to process four or five whales at a time. With a capacity of 16,000 tons, there would be no need for a separate vessel to ship the oil to Europe. A further Eagle Oil tanker had been acquired, which, when refitted as the *Southern Princess*, would become the first non-Norwegian pelagic whaler. Sufficient processing plant was being installed to process all the carcases the projected five 250-ton catchers were capable of delivering; they were presently being built at Smiths' Yard at Middlesburgh (Elliot 1998, 24; Jackson 2005, 179–181).[(9)]

Darnley had previously considered that it was for the Norwegian Government to regulate pelagic whaling, as all the vessels were Norwegian. But as British companies soon followed, it became ever more obvious there was no 'really efficacious means' by which the two governments could control pelagic whaling, let alone abolish it. Companies, if threatened with regulation, would simply transfer their vessels to another flag. Pelagic whaling quadrupled over the four austral summers of 1927–1931, as the number of expeditions both increased and extended ever farther across the Southern Ocean (Tønnessen and Johnsen 1982, 385).

The speculative fever, which had gripped Norway, soon spread to London, where it became commonplace for a British holding company to purchase a majority holding in one or more of the existing companies. The first instance was the sale of Tønsberg by its main shareholder, in March 1929, to a new Anglo-Norwegian Holding Company, provoking outrage, on the part of the Norwegian newspapers, at such a humiliating capitulation to speculative foreigners. The onboard management and manning of the expeditions remained almost solidly Norwegian (Tønnessen and Johnson 1982, 373–375; Elliot 1998, 30).[10]

Theodore Salvesen had continued to hold back, fearful of investing further in so hazardous an industry. This, however, was the aspect that made whaling particularly attractive to those with the entrepreneurial skills so conspicuously possessed by his three sons (Harold, Noel and Norman) who had followed him into the family business. Noel and Harold had their mother's academic abilities. Noel had given up what promised to be a brilliant career at the Bar to become a company director in 1920–21. Harold, following military service, had become a student and then a Fellow of New College, Oxford, where he lectured in economics for four years (including a one-year fellowship at Harvard). Gerald Elliot (one of the companies' historians) describes him as both strong minded and formidably intelligent; Harold joined the family business in 1928 and soon established himself as a leading expert on the industry. He went south straightway and continued to do so throughout the 1930s, seeking out further efficiencies in operating both Leith Harbour and the factory ships as well as increasing the British share of the labour force (Vamplew 1975, 141–144; Elliot 1998, 26–27; Tatham 2008, 475).

Whereas Theodore Salvesen had balked at the costs of re-equipping the fleet for pelagic whaling, Harold and Noel believed it to be the key to Salvesen's survival as a whaling company. Ice-front whaling, and therefore the renewal of the entire Salvesen fleet, were essential to achieving this. *Sevilla* was replaced by *Salvestria* and *Sourabaya*. Both were refitted with stern slipways and much greater processing capacity, and 13 ice-strengthened catchers were ordered from the Smith's Dock on the south bank of the river Tees at Middlesburgh for the 1929–30 summer. Harold Salvesen was so appalled by what he held to be the gross inefficiencies of 'the Norwegian pattern of management' that he had already begun to apply the planning and technical expertise befitting 'a modern industrial concern' (Vamplew 1975, 180–181; Tønnessen and Johnsen 1982, 187–188; Elliot 1998, 30–31; Jackson 2005, 181).

However hostile both the imperial and colonial governments might have been to pelagic whaling, there was also acknowledgement of the increasing number of orders being placed with British yards (Tønnessen and Johnsen 1982, 372–373). The Furness Shipbuilding Company, upon being invited to submit a bid for building a pelagic factory ship of 22,500 tons, costing about £290,000, applied to the Department of Overseas Trade for an Export Guarantee, warning that its refusal would lead to the order being placed with a Scandinavian yard. The Board of Trade sought assurance that the grant of the guarantee would not conflict with the endeavour of the United Kingdom Government to suppress pelagic whaling. The Colonial Office, having made due enquiry, intimated there was little prospect of international action against pelagic whaling and that any financial assistance given by the Export Credits Department would not conflict with the wishes of the Secretary of State for foreign affairs.[11]

Anders A. Jahre had come to play an increasingly prominent role as both a legal adviser and board member. He first came to notice for his role in establishing the companies Antarctic and

Pelagos, during which time he had conceived 'an idea which took shape in Kosmos', the name of the company specifically formed to operate a factory/tanker ship of that name, of 17,800 gross tons, designed to be capable of producing 2,500 barrels of oil per day. Kosmos was formed in November 1928; about one-third of the 6 million kroner of the capital invested in it came from British sources. The ship itself was 5,000 gross tons heavier and had a considerably greater oil-storage capacity than any rival vessel. Built by Workman Clark of Belfast, its seven catchers were purchased from the Smith's Dock Company at an estimated cost of 8.96 million kroner (Tønnessen and Johnsen 1982, 375–379). Almost the entire catch obtained from its initial season of 1929–30 was from the previously unexploited grounds northeast of the Balleny Islands. The huge catch of 1,822 whales produced 119,400 barrels of oil. An even greater catch followed the next austral summer, largely along the ice-front off Wilkes Land between 86° and 102° east. As many as 199,190 barrels of Numbers 0/1 oil were obtained from the catch of 2,431 whales. The catchers caught an average of 270 whales at a cost of 33.53 kroner, compared with the Antarctic average at that time of 173 whales for 60.23 kroner. The Kosmos Company had already begun to accrue the capital reserve that was to prove so invaluable in the years of crisis ahead.

Norway's other pioneer of pelagic whaling was Johan Rasmussen. Whereas Jahre's companies remained Norwegian, Rasmussen's Viking Whaling Company was much more a partnership between what Tønnessen and Johnsen (1982, 379) characterise as 'Norwegian know-how and British capital'. The immediate purpose of forming the company had been to finance the projected factory ship *Vikingen* at the Swan Hunter and Wigham Richardson shipyard. By its British registration, the Viking Whaling Company joined the two existing British operators in Antarctic waters. The Kosmos and Viking vessels were not only the largest specifically designed for whaling, but their curved slipways, built into the side of the respective vessels, made them especially distinctive. Their engines and funnels were also placed at the stern, with the bridge brought as far forward as possible. They were in design identical with modern mineral oil tankers but with a factory deck placed above the tanks. They could have been easily adapted for carrying mineral oil had the economic need ever arisen. Each was equipped with a Gipsy Moth aircraft to help locate whales. The aircraft flown from *Kosmos* vanished on 2 December 1929, having made only two or three sorties – it had not been fitted with a radio (Budker 1958, 119; Basberg 1998, 29–33).

The factory ships of over 10,000 tons boasted of 'both adequate working and storage space on which modern efficiency depended'. A giant claw drew the carcase by the tail up the slipway to the shipboard flensing plan where, as the mouths of the larger number of modern digesters had been engineered to be flush with the deck surface, it was possible to achieve an even greater throughput. With the further installation of large-scale water purification plant, there was no longer the need for a system of water-ships. Crew accommodation and recreational facilities were generally better than on other vessels. Whaling continued nevertheless to be hazardous. Wireless communication became ever more critical when a ship was working at the ice-front. There were frequent mishaps as coal and freshwater supplies were transferred and the whale oil was pumped, while still at sea, directly into the attendant tankers for the European market (Figure 16) (Vamplew 1975, 186–190; Jackson 2005, 182–183).

As many as 38 factory ships were working in the 1930–31 austral summer, now operating over some three-quarters of the Southern Ocean. The Colonial Office drew some consolation at the enormity of the catch from the much greater use made of the whale carcase. Although

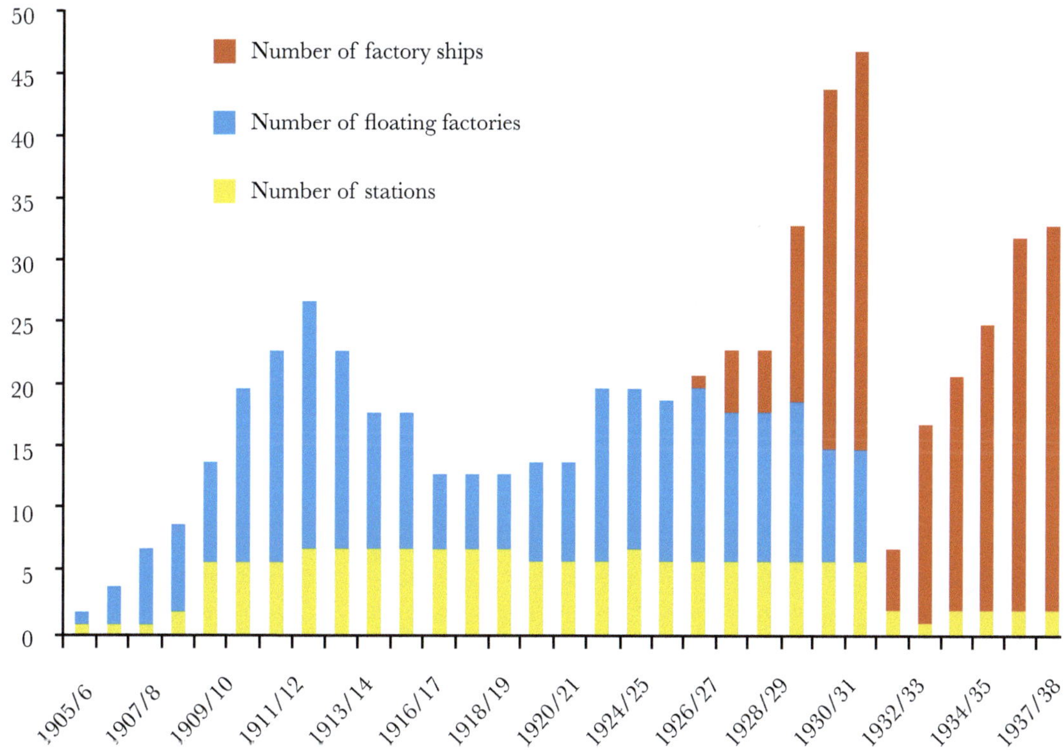


**Figure 16** A time series of the number of shore stations, floating factories (which needed sheltered anchorages and were therefore subject to regulation) and pelagic factory ships (which were not subject to regulation) showing the rapid growth of the pelagic ships, particularly after the so-called Silent Summer of 1931 when most shore stations and floating factories ceased operation. Data source: Headland (2009).

three times as many whales were killed in the 1928–29 summer than a decade previously, almost five times as much oil was produced. Output soared. Over 30,000 whales were killed, and 2,777,100 barrels were obtained in the 1929–1930 season, compared with 11,369 whales and 407,327 barrels, respectively, a decade earlier. More than 40,000 whales were slaughtered (including a record 29,410 blue whales) and 3,250,000 barrels of oil produced in the following summer of 1930–31. The catch of blue whales had already peaked, but it proved to be market capacity rather than any exhaustion of stocks that brought about what the industry and historians have ever after referred to as *Det Stille Ăret* (The Silent Year) (Jenkins 1932, 131–132; Ellis 1992, 347; Basberg 2008, 43).

## 9.2 The diplomatic response

Confronted by so much uncertainty, companies operating from the shore stations and floating factories in the Dependencies sought assurance that their leases and licences would be renewed rather than transferred to another applicant, such as a British company. Pesca was the first to write, in October 1925, pointing out that it had invested £300,000 at Grytviken, purchased a catcher of 'the most modern type' and installed '20 new press boilers'. Further capacity would soon be needed 'for the elaboration of bi-products'. Pesca wanted to purchase the shore station but would accept a further 21-year lease. The Governor objected most strongly to purchase. A further 21-year lease should be granted, subject to the imposition of 'such general restrictions on whaling as may later be found to be necessary'. The separate lease for Jason Harbour should be annulled – the site had never been developed. Brian Freeston (a Principal in the Colonial Office) warned that the renewal of the Pesca lease would be taken by other companies as a precedent. Darnley minuted that every renewal should be for two-and-a-half years as a first step in aligning the leases, four of which were to expire in May 1929, a further three in 1930 and one in 1932.(12)

Johan Hjort impressed upon Darnley the hostility with which Den Norske Hvalfangerforening (the Norwegian Whaling Association) regarded pelagic whaling; the companies wanted it 'repressed'. The renewal of the companies' leases and licences was an essential step in securing the confidence to fight off the unregulated operators. He was accordingly anxious to know

> *what we thought about a suggestion that if it were possible for the Norwegian Government to take effectual steps for the suppression of participation in pelagic whaling by Norwegian industry, companies and subjects, the Secretary of State would be disposed to consider the giving of an assurance relating to the renewal of leases and licences which might perhaps take the form that the British and Norwegian companies should receive equal treatment in the matter during a term of years.*(13)

Darnley acknowledged the Association's hostility toward pelagic whaling but also warned of the ease with which pelagic companies might simply switch their vessels to another flag. As if to confirm this, Sir Francis Lindley (the British Minister in Oslo) relayed, in December 1926, that a Norwegian whaler (recently returned from Hamburg) had reported that a company was being established there 'to fit out an expedition for pelagic whaling in the Antarctic'. Hjort feared 'the whole whaling industry' might be threatened unless urgent measures were 'taken to preserve the stock of whales'. Officials in the Foreign Office were rather taken aback by Henry Maurice's amused response in the Fisheries Department to this 'sudden anxiety' on the part of Hjort for whale stocks. As Maurice explained, Hjort was 'first and foremost . . . a very good Norwegian' and had 'been playing his cards very carefully and well for years past to keep the whaling business in Norwegian hands'. They were of course 'quite satisfactory hands to have them in' as the Norwegians were 'the natural whalers of to-day'. Maurice suspected Hjort's anxiety had arisen in this instance not so much from a concern for whale stocks but from the possibility that the Colonial Office might reduce the number of leases and licences granted as a way of compensating for the impact of pelagic whaling on whale stocks. Maurice personally doubted whether pelagic whaling could ever succeed, but he thought it likely that the Germans should want to 'organise pelagic whaling in the Antarctic on a large scale'. If that proved successful, 'we may be forced to make a move'.(14)

Hjort asked Darnley, in May 1926, for something in writing that he could show the Whalers' Association as evidence of the British Government's hostility toward pelagic whaling. Although highly sceptical of whether the Norwegian Government would ever have the political strength to take effective measures, both Sir Samuel Wilson and Sir Charles Davis (now the Permanent Under-Secretary in the Dominions Office) agreed that Darnley should write semi-officially to Hjort, making it clear that since the only pelagic companies were so far Norwegian, it was for the Norwegian Government to take up the initiative. The Secretary of State would seriously consider any proposals put to him though 'the diplomatic channel'.[15]

Darnley duly reported, on 27 February 1927, that Benjamin Vogt (the Norwegian Minister in London) had approached him on 'the possibility of controlling pelagic whaling'. Darnley presumed that he might be given the authority to discuss with Vogt 'anything but decide nothing', anticipating no difficulties in the discussions except for those inherent in the subject matter. Although the participants were far from hopeful of the result, pelagic whaling had become 'so serious a threat to both British and Norwegian interests that every possibility of controlling it ought to be fully explored'. The Secretary of State, Leo Amery, in minuting his agreement to talks, wrote that the Antarctic explorer, Captain Ernest Mills Joyce, had recently shown him 'photographs of a 17,000 ton Norwegian steamer which hauls the whales on deck bodily, and I gather the destruction of whale life at this moment is appalling'.[16]

Darnley, in the course of discussing the continued growth of unlicensed pelagic whaling with Vogt, on 2 March 1927, duly conveyed the United Kingdom Government's considerable apprehension of the prejudicial effect it would have on both 'our revenues and system of control', and indeed 'the very existence of the whaling industry'. Vogt pressed his Government's desire to 'explore every possible measure for bringing pelagic whaling under control'. There were three possibilities. Both he and Darnley rejected the first: to buy out the pelagic companies, as they would simply be replaced by others. The second was for the Norwegian Government to legislate against pelagic whaling, but Vogt feared this might only 'drive the business away to new headquarters in Hamburg or other foreign ports'. Darnley thought Norwegian citizens might be prohibited from taking part in expeditions organised from outside Norway. Vogt countered that this would be 'extremely repugnant to Norwegian trade and public opinion'. The Norwegian gunners were in any case a very independent class and might simply take Danish or Dutch nationality. As to the third approach – international action – both Governments had reason to fear that every government would want a larger role in controlling the industry. But confronted by pelagic whaling, this might be the lesser evil. Vogt emphasised the keenness of his Government that the two Governments should work together within the limits of what was 'efficacious and politically possible in Norway'. Grindle emphasised that there must be as little interference as possible with 'what goes on the high seas'. Amery agreed.[17]

Hjort continued to emphasise that as 'we have built our industry in the Falkland Islands Dependencies' on the basis of the whaling leases and licences, their continuance must be assured. As Brian Freeston recorded, Hjort had put 'all his cards on the table' with the utmost frankness in providing a confidential account of the Norwegian Government's interest in the negotiations. Darnley countered by insisting that the priority should be to ensure that every lease had a common expiration date. A fresh start could then be made. In doing so, both Governments had to recognise the interests of existing lease and licence holders and the 'British interests' anxious to enter the industry and who were therefore dependent on 'the reversion of certain Norwegian concessions'. Darnley found that, far from wanting to take the lead, Hjort was anxious for the United Kingdom Government to first publish its proposals and thereby enable the Norwegian Government to respond as if spontaneously.[18]

Sir Francis Lindley in Oslo continued to emphasise the need for some kind of intergovernmental initiative. The austral summer of 1927–28 had been so 'unusually favourable' that companies had become ever more confident there was 'no danger to the stock of whales within a measurable future'. Any warning to the contrary was seen as a further attempt to extend British ownership and control of the industry. Grindle himself had meanwhile illustrated the increasing impotence of the Colonial Office by confessing, in a letter of January 1928 to a Norwegian company, that it was 'legally entitled to pursue whaling on the high seas near the Dependencies', despite the prejudicial effect this would have on 'the research work which depends on the revenues obtained from the various leases and licences, along with the export duty paid on the whale oil'. Brian Freeston summarised the dilemma. The summer had been so 'exceptionally successful' that there was plenty of capital for 'the flotation in Norway, London and Germany of new pelagic enterprises'. However much the Norwegian companies would be encouraged by the renewal of their various leases and licences, the effect would be to prevent further British companies from joining the industry. Hjort, for his part, continued to warn of the increasingly real risk that the Norwegian companies would break away from 'a licensing system, with which they had loyally co-operated for the past twenty years'.[19]

Sir Cecil Hurst (the Legal Adviser) minuted in the Foreign Office, on 1 May 1928 that Hjort had found Henry Maurice strongly in favour of an agreement, but it was Darnley who proved to be the problem: 'This gentleman has some difficulty in making up his mind'. Hurst was afraid the Foreign Office would have to intervene, particularly as an agreement with Norway would 'certainly help to strengthen the British position in connection with the Antarctic claims' to the sovereignty of certain territories. Hjort telegraphed the next day from Oslo that Johan Rasmussen, Lars Christensen and he would, on behalf of Den Norske Hvalforeningen, put to its members the United Kingdom's offer of a five-year extension of both leases and licences. The Colonial Office informed the five non-members of the association – two Norwegian companies, the British companies and Pesca – of the offer. Hurst acknowledged in the Foreign Office that 'the position seems satisfactory'.[20]

The draft agreement extended the system of leases and licences and modified the payment of export duty better to accommodate bad seasons. As a further safeguard for those licenced to operate on a particular whaling ground, no other company would be permitted to come within a certain distance of those waters. The separation would be 150 miles in the case of South Georgia, and 100 miles for the South Shetlands, South Orkneys and South Sandwich Islands. Perhaps most critically for the Colonial Office, every company holding a lease or licence undertook 'to conduct their whaling on the high seas within the limits of the Falkland Islands Dependencies Sector as if the regulations in force in the Dependencies were equally applicable to the whole of the area'. Hjort became so pessimistic of Darnley brokering any agreement that he had asked the Admiralty Hydrographer whether Sir Cecil Hurst might intervene. Hurst responded by suggesting that Hjort should 'get Maurice, whom he knows extremely well, to write to the Foreign Office, asking it to interest itself in the question'. This would avoid any impression of the Foreign Office intruding in Colonial Office matters. As Hurst put it, everything should be done 'to prevent all those whalers taking to pelagic whaling'. Representatives of Den Norske Hvalfangerforening accepted a five-year extension. Darnley thought it 'as good as we can hope to get, having regard to the growth of unlicensed whaling'. Leo Amery initialled the agreement straightway, not waiting to confer with the Governor, Arnold Hodson. Although on home leave, Hodson was 'busy upon the Bisley rifle-ranges'.[21]

The annual report of the Norwegian Department of Overseas Trade highlighted the need for further regulation. The very large catch of 801,500 barrels of oil from the 1927–28 summer had made for an aggregate value of £68 million. Increasing numbers of expeditions were being fitted out for pelagic whaling, financed increasingly by the foreign, and more particularly, British, money markets. There was a danger that the industry 'pass into foreign control' at a time when so little was known of 'the actual stock of whales in the Antarctic'. More immediately, there was fear that even a single bad season might bankrupt many companies. Newspapers warned almost every day of the 'the rising wave of speculation', making a 'serious South Sea bubble' ever more likely. A way had to be found to safeguard what had become a highly capitalised industry from the volatility of markets (Tønnessen and Johnsen 1972, 364–365).[(22)]

Lever Brothers produced a memorandum emphasising inter alia the dependence of the smaller Norwegian towns of Larvik, Sandefjord and Tønsberg as well as a good many shipyards on the continued prosperity of the industry. Some 4,000 men were directly employed and thousands worked in the associated shipyards and factories. A considerable portion of the global output of whale oil was now processed by the large refining and hardening plants of De Nordiske Fabrikker (De-No-Fa) at Fredrikstadt and Vera Fabrikker at Sandefjord. Whale oil was increasingly used for a range of medicines, toilet and household soaps and the 'soft soap trade' generally. Further employment opportunities came through the provision of tackle and the wireless telegraphy apparatus required. The government arsenal provided many of the whaling guns, and Niterdalk (the gunpowder- and detonator-maker) manufactured the ingredients for the propellants, bursting charges and fuses. The Norwegian rope factories supplied the forerunners, whale lines and other classes of rope. The purpose of the Lever Brothers' memorandum was therefore to warn of the potential consequences of a shift toward 'high seas whaling' and the consequent risk that, after a brief boom, 'the reliable whaling localities' would have become so depleted that the entire fleet would have to be laid up, resulting in large-scale unemployment within both the whaling and associated industries. The ramifications for the individual companies would be considerable. The Southern Whaling and Sealing Company had made a capital investment of £545,646 in its present 21-year lease for Prince Olaf Harbour, along with a further £196,483 in the floating factory and three catchers employed in the South Shetlands and Graham Land. A total of £89,390 in licence fees had been paid over a seven-year period (1919–20 to 1925–26).[(23)]

Darnley readily acknowledged 'the seriousness of menace to the future of the whaling industry' should high-seas whaling prove to be commercially profitable. It seemed 'unlikely that His Majesty's Government and, as matters stand, the Overseas Governments' were individually in any position to 'place any serious obstacle in the way of such expeditions', where they did not make use of British territorial waters. While acknowledging that an international conference should be considered, 'His Majesty's Government are not prepared at present to take any steps for that purpose'.[(24)]

## 9.3 The erosion of authority

However great the sympathy for those adversely affected by pelagic whaling, there was steadfast opposition from the United Kingdom Government to extending any part of the existing regulations to the high seas. An obvious device would have been to close the ports and waters of the Falkland Islands and Dependencies to pelagic whalers, thereby denying them the provisioning and repair facilities presently offered. An Admiralty spokesman protested that this would run absolutely counter to the long-held principle of 'the freedom of the seas which this country endeavours to maintain'. There would be most undesirable consequences for British mercantile shipping, of which the whaling industry formed a comparatively small part. Nor could it be very effective. The pelagic expeditions would make greater use of the South American ports. Regulation was better achieved by international agreement of the kind that had brought improvement to the seal stocks in the North Pacific.(25)

The whaling industry had once again forged ahead of regulation. It was far from clear how effective the Norwegian legislation or the bi-lateral Agreement with the United Kingdom Governments might prove to be. It would be very difficult to enforce the exclusive whaling areas. Theodore Salvesen had straightway condemned the Agreement for perpetuating what he held to be the Norwegian monopolies, provoking Gilbert Grindle to write that 'his [Salvesen's] firm have been known to us for years as greedy and grasping, and impossible to satisfy'. Although exploiting to the utmost the many advantages that had been given them as a British firm, Salvesen had done nothing to support the United Kingdom Government in its various discussions. Despite all this, Salvesen obtained an interview with Leo Amery the Secretary of State. In briefing Amery, Brian Freestone warned that Salvesen needed 'closer watching and a more stringent control than the majority of their rivals'. There were grounds for suspecting its whaling returns had been falsified and licence conditions breached. William Ormsby-Gore (the Parliamentary Under-Secretary) protested that the Secretary of State must see the evidence, causing Darnley to intervene by affirming that the allegations were true but now were only relevant for emphasising how 'Messrs Salvesens must be watched' much more closely than other companies. Salvesen wrote, thanking Amery for having received both himself and 'his son Harold (Scholar of Oxford)'. At that meeting, Amery had agreed that the Salvesens could have a further catcher for the remainder of the austral summer, a condition Theodore Salvesen now asked to be extended to the forthcoming season. He extracted two further concessions in a later letter but was refused a third: to operate within 50 miles of both the South Orkneys and South Sandwich Islands despite having no licence to do so. On being refused, Salvesens threatened to abandon their South Shetlands' licence and operate their new vessel *Salvestria* as a pelagic whaler. Darnley wrote that this would not only be a clear breach of the bi-lateral Agreement with the Norwegian industry, but the Secretary of State had 'ample hold over Messrs Salvesen'. The lease for the shore station at Leith Harbour would expire in September. Lord Passfield (the recently appointed Secretary of State in the new Labour government) intimated that unless *Salvestria* continued to comply with its licence conditions, he would be 'obliged to consider how far it is consistent with his general policy' to allow 'the enjoyment of other facilities accorded to you in the Falkland Islands Dependencies'. The *Salvestria* was telegraphed fresh instructions.(26)

Noel Salvesen wrote to Dr T. Drummond Shiels (who was both the Parliamentary Under-Secretary and the Member of Parliament for Edinburgh East), protesting that the Salvesens had taken no part in the 'secret diplomacy' with the Norwegians. 'The best legal advice in Scotland' had found no evidence of their having breached their various lease and licence agreements. The letter continued, 'My father would have pitched it in much stronger terms as he considers we have not received fair treatment for the last twenty years'. The 'lightly-veiled threat' which the Colonial Office had made to 'our facilities at South Georgia' was 'so

Machiavellian' that he felt sure 'it could not be justified in the House of Commons'. Darnley countered by briefing Drummond Shiels that 'we have long regarded Messrs Salvesen as the most troublesome of whaling firms, and they are maintaining their reputation'. Whereas generally there was little the Colonial Office could do to prevent the companies from operating without a licence, 'we have ample hold over Messrs Salvesen, who possess leases expiring on 30 September next. The Colonial Office was under no obligation to renew them. Grindle agreed 'we should not let ourselves be defied by a British firm'. As Darnley further wrote, in late December 1928, 'Salvesens are very clever and experienced people'. They knew full well the bi-lateral agreement with Norway was essentially 'a gentleman's agreement' – there had never been any intention of making it a legal document. The zoning system, which had been agreed with the Norwegian licence holders, had been crucial to persuading them to enter the agreement. They would otherwise have become the kind of constant menace that Salvesens were currently exhibiting themselves to be. The bi-lateral agreement was by no means immutable, but 'the present sabotage by Messrs Salvesens' meant it was hardly an opportune time for amending it.[(27)]

Salvesen's obduracy had nevertheless highlighted the loss of authority by governments in the face of pelagic whaling. The first quarter-century of southern whaling had been the heyday of British/Norwegian hegemony, during which the Colonial Government had awarded exclusive rights of access. For the companies, the symbiosis had been driven not by any consideration for whale stocks but the more immediate prospect of bankruptcy. As the Salvesen historian writes, Theodore Salvesen saw the excessive killing as an inevitable consequence of no one owning what they hunted. Should Salvesens catchers choose to spare a school of whales, it would not be left to flourish; other companies would pursue it and kill as many as possible (Vamplew 1975, 192). As the size of the whale catches began to diminish in the Ross Sea, the expeditions did what had happened so many times before. Individually, they showed an even greater rapacity to take what whales remained before moving to another ground. Whereas the blue and fin whales had drawn some protection from their 'oceanic habitat', pelagic whaling meant they were ever more exposed to the pressures that had driven the humpback ever closer to extinction as a commercial species.

The whaling data now collated and disseminated by Sigurd Risting at the Bureau of International Whaling Statistics at Sandefjord helped the individual companies identify where the more remunerative grounds might be found. The Salvesens achieved their highest output to date of 1,101,165 barrels, followed by Pesca with 924,000 barrels. Tønsberg had obtained 892,654 barrels and Sandefjord 864,000 barrels (Hart 2001, 251–252). Their strategies varied, in the sense that Lars Christensen had already begun to follow the long-established precedent of seeking fresh grounds for their immediate bounty and for when the present grounds were exhausted. There was never any doubt that Salvesens would abandon whaling once there were more profitable ventures to pursue. Lever Brothers took a diminishing interest in the Southern Whaling and Sealing Company as the commercial opportunities began to be realised from their vegetable oil plantations spread across the tropical colonies.

Johan Tønnessen wrote that it suited the more speculative companies to project the Colonial Office's concern for whale stocks as 'a devious policy to ruin pelagic whaling' and thereby preserve its revenues from 'shore whaling'. Darnley had acknowledged, as early as March 1926, that pelagic whaling had already begun to undermine British authority in Dependencies' waters, writing despairingly in May 1929 that 'our licences are now of little value to the recipients'. The wonder was that any company was 'still willing to pay our taxes, although these taxes have been much reduced'. As he wrote to Grindle, 'Our former very complete control has now almost disappeared'. The particular exchange of minutes had been prompted by a despatch from the Governor reporting that most of the blue whales killed off South Georgia that season had been immature. The older animals appeared to have been already killed or driven off.[28]

CHAPTER 10

# *Market collapse*

Any inter-war account of 'the golden age of pelagic whaling' will have four parts: the expeditions of 1927–31, the crisis of the early 1930s, the company agreements of 1932–36 and the international agreements of the later 1930s. *Polar Record* published an anonymous piece in January 1932 claiming that the industry had brought about the crisis many had foreseen. Economists had long warned of supply outstripping demand. Scientists predicted the species' extermination (Anonymous 1932, 26–27; Tønnessen and Johnsen 1982, 367–370).

The Falkland Islands Government had become little more than a bystander. Whereas the five shore stations reported in the late 1920s an aggregate output of 11,000 barrels of oil, this was now dwarfed by what the pelagic industry was capable of producing. With average sales of whale oil at £32 a ton and production costs as low as £15 to £18, whaling was as lucrative a form of investment as petroleum. Companies had taken full advantage of the 250,000 tons of laid-up British vessels to invest over £2.5 million in refitting 140,000 tons of that shipping as factory ships and tankers, causing the number of factory ships to increase from 17 to 41. The number of catchers had risen from 80 to 232 (including those attached to shore stations). The total Antarctic catch rocketed from 872,364 barrels (145,394 tons) in 1926–27 to 3,608,348 barrels (601,391 tons) in 1930–31, making for a Norwegian share of 64 per cent and British share of 30 per cent. Although rising from 44,879 tons in 1927–28 to 61,781 in 1929–30 and 120,533 tons in 1930–31, the British market share had nevertheless declined slightly (Jackson 2005, 185).

The differences in the financial underpinning of expeditions became more evident. Whereas the Norwegian companies might require large bank loans even to put to sea, both the despatch and modernisation of the British fleet were financed largely through capital accumulation (Hart 2006, 177). However, the distinction had begun to break down in what amounted to the complete restructuring of Finn Bugge's Hektor whaling company, in a move ostensibly to pre-empt the Colonial Office from transferring its lease and licences for the shore station at Deception Island to a British-registered company. Some two-thirds of Hektor's shares were purchased by the Hector Whaling Company, formed for the purpose of restructuring. The drive had come from Finn Bugge's son-in-law, Rupert de Burgh Trouton. Found unfit for military service, Trouton had joined the Treasury, where he became (as with Harold Salvesen) one of John Maynard Keynes' 'bright young men' as wartime economists. He had thereafter obtained a Double First in economics at Cambridge, joined a stock brokering partnership and become more directly involved with investment in the whale oil trade (Roberts 1966; Tønnessen and Johnsen 1982, 374–375; Jackson 2005, 182; Basberg 2016).

Johan Hjort had meanwhile continued to appeal for greater market discipline, in what was now emerging as essentially a purchasing combine. A first tentative move in that direction, as it affected whaling, was a meeting of Lever Brothers, Margarine Unie and De Nordiske Fabriker (De-No-Far) in early September 1929. Each had begun to fear its over-exposure in marketing. Each pledged to purchase only from suppliers using 'their best endeavours to maintain a regular supply of Whale oil' without risk to whale stocks. Rowland Darnley was asked whether the pledge might be advertised as having the Colonial Office's support. Brian Freeston welcomed what would effectively be 'a purchasing ring' designed specifically 'to freeze out' those 'operating pelagically and wastefully', but he also warned of the 'odium' this would incur from other quarters. Darnley similarly warned of the 'political difficulties of any official backing for what was essentially "a colossal" trust', however welcome its specific purpose of curbing output. The Mercantile Merchant Department of the Board of Trade warned that the United Kingdom Government must not be seen as supporting what would be essentially a trading cartel. Nor should it be too discouraging. It might prove to be the only way of regulating output in the longer run.(1)

The takeover and amalgamation of individual oil refiners and processors gathered pace. Lever Brothers purchased their last two British rivals. The Dutch Companies of Van de Burghs and Jurgens combined to form *Margarine Unie*, and then merged with Hartog (their last Dutch competitor) in January 1929. The further merger of *Margarine Unie* with Lever Brothers, in September 1929, to form Unilever, meant there was now only one significant purchaser of animal and vegetable oils in Europe (Wilson 1970, 301–319; Fieldhouse 1978, 24 and 38; Tønnessen and Johnsen 1982, 369–370). The oil producers Den Norske Hvalfangerforeningen (the Norwegian Whalers' Association), had, for their part, dissolved and reformed the same day (11 May 1929) as Hvalfangerforeningen, combining the previous 25 Norwegian companies with four British, two Danish companies and Pesca. This allowed the Association to act as a sellers' pool, achieving the largest single sale of whale oil ever recorded – 1,590,000 barrels (265,000 tons) – on 29 November 1929. Only four Norwegian companies and Salvesen had remained outside the association. The preparedness of three of the Norwegian companies to settle for £25 per ton largely determined the price at which over 80 per cent of the global output was sold that year (Tønnessen and Johnsen 1982, 364–375) (Figure 17).

However impressive the degree of association achieved among the producers, the marketing advantage had begun to shift from the producers to the oil processors. The turmoil in international markets was to affect both, but the consequences for the oil producers proved more lasting. Whale oil was unusual as a commodity for being mostly sold before it physically arrived in Europe. The forward sales helped the heavily indebted companies and minimised the risk of the oil becoming physically contaminated while awaiting purchase. But as Ian Hart points out, any change in market conditions between the point of purchase and delivery could severely affect one or other of the parties to the sale (Hart 2001, 248–249). As Gordon Jackson wrote, no one appeared to realise that the bonanza could not continue indefinitely. Buoyed by the volumes and profits from sales of margarine and other products, Unilever used its almost complete dominance of markets to offer a guaranteed price of £25 to £30 a ton for as much oil as the producers could deliver in both the 1929–30 and 1930–31 seasons, primarily for the expanding German and East European markets (Jackson 2005, 185–186).

It was the Wall Street Crash of October 1929 and the increasing uncertainties of demand for manufactured products in world markets that brought matters to a head. 'Unilever took

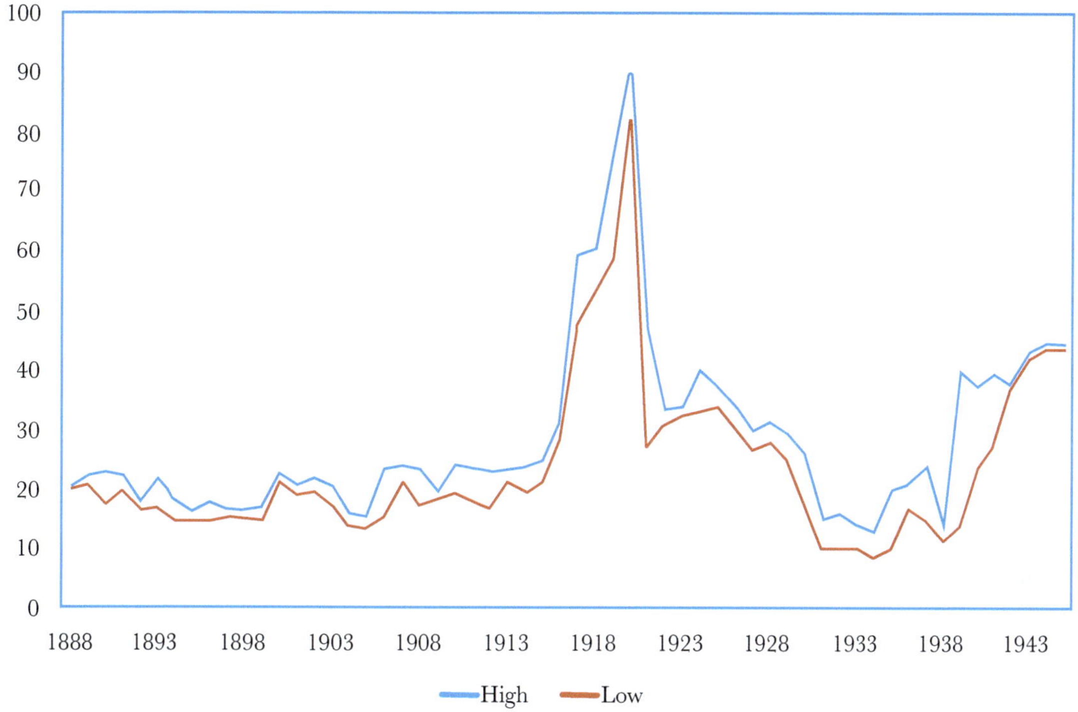


**Figure 17** The annual price per ton of whale oil from 1888 to 1945 with the values adjusted to 1945 prices using the Treasury deflator (annual inflation figures provided from http://inflation.iamkate.com/). The maximum and minimum values for each year are given. The data indicate that the "golden years" for whaling were just after World War I in terms of return on investment whilst the early 1930s were the opposite. Data source: Vamplew (1975).

fright before the season had hardly started', confronted by contracts to purchase all the oil the expeditions could deliver at a price that was now more than double the market value to which the whale oil had fallen. Unilever instigated an immediate suspension of whaling. Rosshavet was the first to order its 'floating cookeries' to leave the whaling grounds immediately. Hval, Tønsberg and Viking followed. Lars Christensen's companies (Bryde and Dahl, Odd and Ornen) and those of Johan Rasmussen soon followed. Hvalfangerforeningen had no alternative but to instruct all its members to lay up their vessels for the 1931–32 season. As many as 33 factory ships, 190 catchers and five transport vessels (350,000 tons of shipping in all) were abandoned. Irvin and Johnson's *Tafelberg* (after only one season) and three newly built factories – *Vestfold* (Vestfold), *Svend Foyn* (Sydhavet) and *Kosmos II* (Kosmos) – went from the builders' yard to join others laid up in the narrow 9-kilometre fjord of Mefjorden, off Sandefjord, in what became known as *Det Stille Ăret* (the Silent Year). Far from compensating the companies, Unilever drew attention to its own losses. The price of number 0/1 oil had plummeted from £25 a ton to £11–£13 (Anonymous 1932, 82–83; Hart 2001, 274 and 2020, 173–174).

Gordon Jackson has argued that the Great Slump of 1929–32 'may have aggravated the difficulties of the whaling trade, but it did not cause them'. Those difficulties had sprung from how recklessly the whaling companies had invested in 'relatively cheap shipping, which had enabled them to produce too much oil and kill too many whales. The whales had been 'slaughtered in unprecedented number less because the additional oil was needed' than to keep their oversized expeditions financially afloat. Hvalfangerforeningen had failed to impose any catch limit by which a degree of viability might have been preserved. It had taken Unilever's refusal to purchase yet more oil to halt the profligacy (Hart 2001, 278; Jackson 2005, 193). The effect had also been to consolidate its market dominance. Henry S. Moss Blundell (Inspector of Fisheries in the Ministry of Agriculture and Fisheries) found, from a meeting with Francis D'Arcy Cooper in January 1931, that the combine now had sufficient stocks of oil until 1933. With more than 2,820,000 barrels (470,000 tons) of whale oil in storage, no further contracts would be placed. Unilever had in essence achieved what the United Kingdom Government had so long wanted – reduced pressure on whale stocks. Whaling by the Norwegian companies was effectively halted (Wilson 1970, vol. 2, 301–308 and 316–319; Tønnessen and Johnsen 1982, 369–371; Hart 2020).(2)

The 1931–32 summer became unique for the absence of a Norwegian fleet. Salvesens were almost exceptional in having no creditors to stave off. The 'good liberal economist' Harold Salvesen had always 'preached that free competition should be allowed to have its beneficial effects' by removing 'surplus' expeditions. Rather than have their fleet idle, Salvesens drew on capital reserves to operate Leith Harbour and to despatch the *Salvestria*, *Saragossa* and *Sourabaya*. Despite the loss of *Saragossa* to fire on the ice-front, the other expeditions obtained an aggregate of 808,560 barrels of oil from 9,527 whales. Much to the annoyance of the Norwegians, Unilever's Southern Whaling Company had similarly had the temerity to despatch *Southern Empress* and *Southern Princess*, with eight catchers assigned to each (Hart 2020, 178–179). Johan Hjort wrote privately to Darnley in May 1931 of his pleasure at seeing 'the stock of whales in the Antarctic [being given] a year's holiday', but he also warned that the breakaway of the British companies had destroyed the kind of collaboration among the whaling companies that both he and Darnley had striven for so long to achieve. Darnley acknowledged the setback but also minuted its 'large foreign element' meant that Unilever was less amenable to any representations made by the United Kingdom Government than might have been the case with an 'entirely British Company' (Anonymous 1932; Vamplew 1975, 197; Elliot 1998, 31–32; Jackson 2005, 187).(3)

Maurice complained that Unilever seemed determined to use 'its controlling power for all they are worth' with the intention possibly of forcing the Norwegian companies to become part of Unilever. Moss Blundell showed much less sympathy for the whaling companies – they had insisted on Unilever meeting its contractual obligations to buy their oil at what had already become 'the exorbitant price of £25 a ton. Having saddled 'Unilever with excessive quantities of oil at a high price', those companies could hardly squeal at how their single customer was now seeking to redress the balance. It was furthermore entirely reasonable that Unilever should use its own whaling company to meet any shortfall. As Moss Blundell warned, 'Unilever would not like our attempting to butt in on their negotiations', nor should the Fisheries Department attempt to do so. It was for the Norwegian whalers to negotiate directly with Unilever, and should their Government be unsatisfied, for the question to be pursued through diplomatic channels. Only then would the Minister have a *locus standi* to interest himself in the matter.(4)

## 10.1 Re-appraisal

How might the now largely pelagic industry sell the largest volume of whale oil possible at a price high enough to secure profitability (Hart 2020, 171)? Both Henry Maurice and Moss Blundell anticipated further restructuring and for the least efficient companies to disappear. For example, Vestfold had abandoned the Stromness shore station and sold its transport ship *Peder Bogen* to Salvesen. Sydhavet's new factory ship *Svend Foyn* was sold to Tønsberg which, with the withdrawal of the floating factory *Orwell* from the South Orkneys, was now an entirely pelagic operator (Hart 2001, 276–278).[5]

Whereas whaling had previously been almost entirely a British/Norwegian matter, the opening of the Ross Sea and further waters of the Southern Ocean had drawn in the Dominion Governments of Australia, New Zealand and South Africa. The Dominions Office had become increasingly involved, with its Polar Committee the obvious venue for interdepartmental discussion (Dudeney and Sheail 2019, 148–149). There was unanimous support for its Chairman, Sir Harry Batterbee, in asserting, at a meeting of January 1931, that it was for the Norwegians to take the initiative; their companies had largely caused the surplus of whale oil. Salvesens, Southern Whaling and the Hector Whaling Company had shown an ability to survive financially, at least in the short term, by stockpiling (Hart 2020, 171). Charles Wingfield (the Minister in Oslo) was instructed to convey both the United Kingdom Government's serious concern and expectancy that the Norwegians (as the 'home of the industry') would initiate 'expert' discussions. Johan L. Mowinckel (the Norwegian Prime Minister) had long complained of too many whaling companies, so there was an obvious pretext for proposing further 'conversations between experts of our two countries'. Prime Minister Mowinckel readily agreed, acknowledging that 'the providential close season occasioned by overproduction' might help whale stocks to recover. But as Wingfield soon discovered, the Norwegian Government had not the least idea what to do beyond agreeing that there should be further informal discussions between 'one or more British and Norwegian experts'.[6]

Lever Brothers had questioned, from as early as 1926–27, the value of retaining its Southern Whaling Company. On balance, as D'Arcy Cooper acknowledged, 'we are bigger buyers than sellers'. Rather than using its ownership of Southern Whaling to drive down prices through competition with other suppliers, the greater productivity that had come by pelagic whaling had achieved that goal. The blindspot, as Ian Hart writes, was that Unilever had failed to anticipate how pelagic whaling might also produce so great a glut of whale oil as to outstrip market demand. By the close of the 1929–30 season, Southern Whaling's two factory ships had achieved an aggregate output almost half as much again as before those vessels were commissioned. (Hart 2020, 172).

John Borley was highly sceptical of whether Unilever's purchasing power would ever be sufficient to provide the 'final solution' to 'conserving the future of whaling'. A licencing system remained the most direct and practicable way of achieving this, particularly where underscored by knowledge of what might constitute an optimal stock of whales. Whereas it had been pretty hopeless to imagine the companies would readily cooperate when their profits were so high, now, with more than half the previous season's production still unsold, there might be the political strength to force them do so. The long-desired relaxation in the pressure on whale stocks could therefore come from the most unlikely of quarters: the whaling companies themselves. A range of obvious measures could be adopted, albeit the whale's behaviour between any two summers was so erratic there could never be certainty in how effective any strategy would be. Borley doubted whether further curtailment of the season was enough. The rotational closure of the whaling grounds was no solution as it would increase pressure on those that remained open. Although measures were needed to halt the notoriously large proportion of immature whales killed off southwest Africa, this would never be enough to secure global whale stocks. The present, and what might prove momentary, reduction in whaling must nevertheless be seized – the first step being to agree with Norway an effective way of achieving it.(7)

Alexander Clutterbuck (Secretary to the Polar Committee) struck a similarly cautious note in the Dominions Office, warning that any stringent regulation would be effective only if every national government participated. This would require patient negotiation, a preparedness to legislate and 'continuous propaganda calculated to influence public opinion in all parts of the world'. Even if all of this could be achieved, there would still be the obvious difficulty of determining the optimal size of the individual expeditions, the output per vessel/expedition and how such regulation might be enforced. It would be completely unrealistic to imagine a consensus being achieved on strictly scientific lines. There was no alternative but for the United Kingdom Government to press for an immediate and all-encompassing licensing system, however imperfect this might prove to be. To delay would risk becoming 'doctors wrangling over a patient only to find, upon agreement, the patient was dead'. The best hope continued to be through inter-company agreement. The recent laying up of the Norwegian fleet had shown how effective this could be. Even so, it would not be easy to apportion 'the spoils' of such prudence between the diverse parties.(8)

Moss Blundell had found Unilever's directors both shrewd and farsighted, potentially far more effective than the whaling companies, whose circumstances and interests were often so divergent. Tighter market discipline, as imposed through Unilever, might make companies more receptive to the kind of regulation required, but, as Maurice cautioned, Unilever had done only what might have been expected of business judgement. Wingfield later reported from Oslo in mid-May 1931, that Unilever's use of the Southern Whaling Company to meet any shortfall had incurred considerable hostility. Banks were hardly encouraged to advance the monies so desperately needed by the Norwegian companies. The Whale Harpooners' Association threatened to retaliate with a boycott of Unilever's factory ships *Southern Empress* and *Southern Princess*. With 900,000 barrels of oil still unsold and companies facing bankruptcy, a boycott of foreign vessels seemed the only way of saving the Norwegian industry. It was within that bitter and confusing context that Wingfield confirmed the Norwegian Government's preparedness to enter into 'expert' conversations, despite the Secretary-General at the Ministry of Foreign Affairs confessing that he was 'really at a loss' to know what to propose beyond informal talks between 'experts' to identify a basis for more formal negotiations. There was also a sense of impotency on the British side. James O'Grady, upon becoming Governor, had pressed the position that every holder of a whaling lease and licence should be compelled to employ a specified number of Falkland Islanders. Borley acknowledged that this should have been done long ago, but it was now far too late. Nothing must be done to further discourage companies from renewing their leases and licences.(9)

The Colonial Office was confronted by a debt of £10,761, accrued by the Hector Company through its failure to pay any rent for the shore station at Deception Island. The station had remained closed since the Norwegian fleet had been laid up (Basberg 2016). A letter of December 1932 explained that the company's predicament since 'the heavy slump in the value of whale oil' had become so serious that its creditors had given it 'a breathing space of at least two years'. Although this enabled the company to operate through the 1933–34 summer, there was no way that even its secured loans and interest payments of more than £400,000 could be met before 1936. It would be foolish of the Falkland Islands Government (as an unsecured creditor) to press for the outstanding rental payments and thereby force the company into liquidation. The signatories of the letter saw the only hope to be renewing the station lease as a first step to recouping the rental monies. While the Colonial Office agreed to this, it was obvious that even if marketing conditions improved, the cost of the lease and licences would have to be considerably reduced. Whereas no more than 8,000 barrels per catcher had ever been produced in the Falkland waters, the pelagic companies might obtain 25,000 barrels per season.(10)

No longer did the whaling companies compete for the Colonial Government's leases and licenses. Even the few that still showed an interest drove the hardest of bargains. Vestfold, which had made no use of Stromness for two years, demanded both substantial reductions in rent and more licences for catchers. Although Pesca continued to operate Grytviken, its London agents demanded the easiest of terms, including its use as a base for pelagic whaling. An obvious course was for the shore stations to focus on the manufacture of byproducts from the whale carcase. They had the ground space for both the extra machinery required and sheds for storing the guano and meal. Such diversification brought little profit. Tønsberg closed Husvik Harbour upon becoming wholly pelagic. Leith Harbour was essentially the operational base for the Salvesen pelagic fleet. Grytviken managed to survive by capitalising on its sealing license as sales of seal oil provided in some years over a quarter of Pesca's income (Basberg 1993). As George A. Jones (a Principal in the Colonial Office) warned, any further closure of the shore stations would risk political 'difficulties with the Argentine Government' in its claim to the sovereignty of the Dependencies. By the time the five-year whaling agreement with Norway had expired in 1934, there was little to distinguish those whaling grounds from the remainder of the Southern Ocean.(11)

CHAPTER II

# *Market recovery*

However much the 1931–32 season proved to be a watershed in Antarctic whaling, market forces continued to decide how much whale oil was produced (Hart 2020, 180). Sales were significantly helped by the discovery in the late 1920s of a process to produce hydrogenated whale oil at a melting point that enabled margarine to be produced almost entirely from whale oil. The breakthrough and continued endeavour to produce as much whale oil as possible meant a further record global catch of 54,835 whales was attained immediately before the Second World War. 46,039 of those whales (84 per cent) were caught by a fleet of 34 factory ships and 270 catchers. But ominously, only 2.7 million more barrels of oil were obtained – an increase of only 11 per cent. The industry could well have gone into further deep recession had it not been for the stockpiling of whale oil by governments as war threatened and then broke out (Hunt 1969, 63–64; Hart 2001, 280–283 and 2006, 326; Jackson 2005, 201).

It was the pelagic rather than the longer established companies that led the recovery in the early 1930s; the key to their growth was access to the required capital, together with the entrepreneurship, innovation and organisational skills that helped them succeed in devising a quota system through inter-company negotiation. The whaling fleet was transformed. Although 21 of the 40 vessels operating in the 1930–31 season were scrapped, the aggregate capacity of those built between 1931 and 1938 was almost three times greater than the capacity in 1930–31 (Basberg 1985, 103–106). Governments became essentially by-standers. Charles Wingfield reported from Oslo, in February 1932, an increasing anxiety that the Norwegian fleet might have to remain tied-up for a second summer, causing even greater suffering among the whaling communities of Vestfold. It was the decision of Procter and Gamble (the American corporation at Cincinnati) to purchase Thomas Hedley of Newcastle and thereby enter the European retail market that enabled the Norwegian expeditions to go south (Jackson 2005, 200).[(1)]

Unilever, rather than invest further in the Southern Whaling Company, found it more cost-effective to purchase whale oil from the cheapest source to the extent of investing in other companies (Tønnessen and Johnsen 1982, 333–334). Sir Sidney Harmer wrote privately to Francis D'Arcy Cooper (the President of Unilever) stressing the timeliness of a fundamental re-appraisal of the entire industry and claiming that Unilever was in 'the exceptionally strong position' to assess what was required. Meeting Harmer in March 1932, D'Arcy Cooper acknowledged Unilever's failure to have anticipated the scale of overproduction. Unilever had lost so heavily that it was under no obligation to the producers. As Henry Maurice remarked in the Fisheries Department, having read Harmer's account of the meeting, the Norwegians had clearly profited from Unilever's improvidence, so they had little cause to complain now that the boot was 'on the other leg'. As the world's largest consumer of whale oil, Unilever wanted 'a steady supply of oil at a steady price and a conservation policy to maintain the whale population' (Hart 2020, 177).[2]

It was within the parameters set by Unilever that the whaling companies, driven by fear of further lay-ups, devised a series of annual production agreements extending over the four austral summers of 1932–33 to 1935–36. The agreements drew heavily on the statistical summaries of the annual catches prepared by Sigurd Risting (Secretary of Den Norske Hvalfangerforening) using the returns that members of the Association had submitted since 1919. A synthesis, published by *Nature* in October 1929, showed an aggregate 7,750 blue and 6,919 fin whales to have been slaughtered and 1,054 foetuses encountered, in the ten years for which data had been available. There could no longer be any doubt of the regularity of the annual migrations, with the main body of blue whales moving slightly ahead of the fin. It seemed from the foetal records that pairing could occur at any time of the year but was much more likely in the austral winter months of June, July and August, at which time the whales were in temperate waters. Both species appeared to attain puberty in their second year; the period of gestation was roughly a year. Females might not give birth every year (Anonymous 1929).

The Producers' Agreement of 9 June 1932, based on this gathering body of knowledge, became a milestone in the history of whaling. Charles Wingfield (the Minister in Oslo) had first learnt of the negotiations from a conversation in May 1932 with Svend Foyn Bruun (a whaling entrepreneur and representative for Vestfold in the Storting). Carl J. Hambro (the Storting President) had asked Bruun to seek Wingfield's support in persuading the British companies to join with Hvalfangerforeningen in developing a strategy based on both 'a barrel quota' and 'a whale quota'; he also requested a further shortening of the whaling season to 20 October to 30 April. The overriding intention was to reduce oil output by some 38 per cent of the 1930–31 production figure and thereby achieve a stable output of around 2.5 million barrels annually. In achieving their prescribed quotas, companies would be forbidden to take any right whales, calves or an adult female accompanied by a calf. It was hoped to limit output through these improved efficiencies to between 202 and 205 million barrels per season – an amount the global oil and fats market could absorb (Hart 2020, 174–175).

Every company signed the agreement except Laboremus and Southern Whaling. Very significantly, Salvesen had agreed to join, but whereas Lars Christensen, Anders Jahre and Johan Rasmussen (all managers of pelagic companies) had devised a simple barrel quota, Harold Salvesen insisted, as a condition for joining, that there should also be a whale quota, both to ensure better use of the carcase and to benefit the larger and more efficient expeditions. Salvesen had similarly set great store by the older and smaller companies being encouraged to dissolve themselves, as they would probably have been compelled to do by a free market. Companies should similarly be encouraged to acquire and transfer quotas as a way of further reducing the number of older vessels to achieve higher productivity. The bonuses paid to expedition leaders and crews should be based on oil yields rather than production alone (Vamplew 1975, 198–200; Elliot 1998, 36; Hart 2001, 278).[3]

A committee of five was appointed to supervise the quota agreement, with Salvesen representing the foreign member companies. As Tønnessen and Johnsen (1982, 402–403) explained, every company was allocated a portion of the aggregate 'whale quota', that is, the equivalent of 'a barrel quota'. By ensuring that every factory ship and shore station kept to its whale quota, there would be every encouragement for companies to produce at least 110 barrels per blue whale unit, and so meet their respective barrel quota. The concept of a 'blue whale unit' (BWU), as it had begun to evolve in the 1920s, equated one blue whale in terms of oil output with two fin whales, three humpbacks or sperm whales and five sei whales. By adopting this formula, whereby bonuses were based on the number of barrels filled, those aboard the factory ships would be encouraged to squeeze every possible drop of oil from each whale. An excess of up to 10 per cent should be permitted, where a company exceeded even this level of high productivity (Anonymous 1932).

The term 'recovery' must be used with care. Even the prescribed output for 1932–33 was little more than half of the peak production achieved in 1930–31 (Basberg 1985, 90). It was nevertheless a watershed portending 'a new era of consolidation and control', whereby stability and profit came from inter-company agreement. It presaged what Elliot (1998, 33) adjudges to have been a period unique in whaling, whereby the Norwegian and British companies wielded 'effective control of catching' free from any direct government intervention. Whereas 35 factory ships and four shore stations had been expected to operate in the 1932–33 season, only 15 factory ships and one shore station did so. Output was only one-third of the 1930–31 season. Every company had been given a quota of the aggregate 2,031,455 barrels, as divided by 110, or an aggregate of 18,584 BWU (Tønnessen and Johnsen 1982, 402–403; Hart 2020, 175).

The outcome exceeded all expectations. The 16 whaling expeditions achieved an aggregate 2,090,825 barrels of oil from 126,985 BWU, or 116.4 barrels per BWU – a level of productivity well above the 110 barrels anticipated. The average production per day per factory ship rose from 488 to 970 barrels, and per whale catcher from 100 to 140, as compared with 1930–31. There was now an obvious incentive for the companies to retain only their best gunners and crewmen aboard the most powerful catchers. As many as 21 floating factories were replaced by 11 newly-built vessels and two converted vessels. Their aggregate capacity was three times greater than that of the old factories. Tønnessen and Johnsen (1982, 403–404) warned that the benefit to whale stocks was far less clear. Gunners still tried to kill the largest, and more particularly female, whales, for their greater ratio of oil to size. But whatever the limitations, 'the British-Norwegian agreement based on a catching quota' proved to be 'the best and most successful' ever devised. An average seasonal catch of some 29,550 whales and production level of 2.48 million barrels were attained between 1932 and 1936 (Hart 2020, 175). Kirkpatrick Dorsey wrote that 'the Norwegians had created a clever system'. Whereas aggregate output might have been expected to fall by at least one-third, the more efficient use of the carcase meant it fell by only 15 per cent, to a level of 2,089,000 barrels (if the 360,000 barrels produced by Unilever's Southern Whaling Company are included). Efficiency had increased. Sixty thousand more barrels were obtained than expected from the prescribed limit of 16,000 BWU. It had prevented the slaughter of the equivalent of some 2,500 BWU (Hart 2001, 278–279; Heazle 2006, 209; Dorsey 2013, 53–55).

The Whaling Committee, appointed by the ICES, both welcomed the agreement and urged every company to adhere to it. The aggregate output might still be higher than what whale stocks could withstand in the longer term, but John Borley acknowledged that the proposed reduction was 'drastic', and Johan Hjort was probably correct in saying that no government could have achieved so much in a single step. It proved, however, to be no more than a valuable *interim* measure; it was no substitute for government action.[(4)]

## 11.1 Quota restrictions

Richard Ellis reflects in his book *Men and Whales* that 'concerned men' might have shown a greater generosity in more prosperous times, but the early 1930s were 'not a good time for altruism'. There was deep recession worldwide. Whaling companies collapsed, stations closed and the older fleets were scrapped. The largest fats- and oil-hardening plant in Norway De-No-Fa (De Nordiske Fabriker) lost £792,000 on its oil stocks in 1931, and Unilever lost £114,000. Hector was forced by its bankers to sell its oil at a loss. Salvesen's profits were the lowest since 1920–21. Lars Christensen tried, but failed, to sell out. It proved difficult to secure inter-company agreements for the 1932–33 season. It was even harder the next year (Budker 1958, 159; Tønnessen and Johnson 1982, 405; Jackson 2005, 197–199).

Charles Wingfield reported, in mid-August 1933, that Hvalfangerforeningen was keen to renew the quota agreement, subject to an increase of productivity, to '115 barrels per standard blue whale with the right to increase production by averaging up to 120 and allow 1% toleration'. It would encompass Jahre, Bruun and Lippe (*Pelagos, Kosmos I* and *Kosmos II*) 350, 590 barrels; Rasmussen (*Sir James Clark Ross* and *Vestfold*) 290,871 barrels; Lars Christensen (*Thorshammer, Ole Wegger* and *Solglimt*) 265,311 barrels; Tønsberg Hval (*Svend Foyn*) 188,258 barrels; Hektor (*Hektoria*) 131,131 barrels; Melsom (*Lancing*) 100,000 barrels, and the *Skytteren* (79,261 barrels). Three more signatories were obtained: Salvesen (*Salvestria, Sourabaya, New Sevilla*) and Leith Harbour 348,397 barrels; Irvin and Johnson (South Africa) *Tafelberg* 128,000 barrels, and Pesca (Grytviken) 52,951 barrels. An aggregate output was set of 1,934,770 barrels: namely 17,074 BWU, which would rise to 2,051,224 barrels when companies achieved an average of 120 barrels per BWU.[(5)]

The agreements reached for the successive seasons of 1932–33 and 1933–34 limited the catch to about two-thirds of the 1930–31 catch (Tønnessen and Johnsen 1982, 406–407). The market price had remained depressed. With some 300,000 tons of oil still unsold, Christensen thought it economic madness to put to sea without a guarantee that Unilever would buy the oil. It would be better for the fleet to remain laid up for a further season. Moss Blundell found the companies' rejection of Christensen's proposal 'quite inexplicable'. For Anders Jahre, it was not disregard for unsold stocks but the foolishness of laying up the fleet when British companies put to sea. The Norwegian proportion of the catch had already fallen from over two-thirds to less than a half. But as Wingfield also reported, on 22 May 1934, Christensen had entered into an agreement whereby Unilever chartered his entire fleet to supply 660,000 barrels of oil at a price determined by production costs and 'development of the market'. The other companies and Norwegian public generally were reportedly furious at what would amount to a doubling of the Christensen Group's previous quota to the obvious detriment of other companies. For *Aftenposten* and the other Oslo newspapers, the only hope of salvaging something for the Norwegian industry was for Christensen to enter into some kind of sharing arrangement with them. The bitter animosity would otherwise inflict immense harm (Vamplew 1975, 202; Tønnessen and Johnsen 1982, 407–408).

By with what Ian Hart describes as 'a conservation measure', the Norwegian Government intervened by imposing a statutory whaling season of only four months: from 1 December to 30 March for expeditions south of 50° south latitude. No more whales were to be delivered to a factory ship than could be processed within 36 hours of their being shot. Companies were prohibited from transferring any vessel in their possession to another flag. Cecil Dormer (Wingfield's replacement as the British Minister at Oslo) wrote that the Storting's obvious intention was to prevent Christensen's group from rupturing Norwegian solidarity. The effect was to make the 1934–35 season what Ian Hart describes as a watershed. By a further measure of June 1935, the season was limited to 1 December to 15 March in waters south of 40° south. An incidental effect of the Norwegian legislation was to thwart any ambition that Unilever might have harboured for dominating both the marketing and production sides of the industry. The shorter statutory whaling season, along with the weather and comparative sparsity of whales, meant that it was impossible for the Christensen Group's fleet to obtain even half the prescribed catch. The shortfall, plus the proportional increase in production costs, meant an aggregate loss of some £154,000, which Unilever agreed to offset by only £45,000. Whereas Salvesen and most of the other companies had kept to 'the Norwegian season', the Southern Whaling Company had operated without limitation (Tønnessen and Johnsen 1982, 407–410; Hart 2020, 182).[(6)]

Henry Maurice had meanwhile learnt more of Unilever's attitude from 'a confidential source', minuting in July 1934 that 'we have been making a mistake' by failing to grasp how the Ministry's interests in whaling were far closer to those of the whaling companies than to Unilever. Whereas the producers would always want plentiful stocks, Unilever would be hardly affected by the whale's demise. Unilever's priority was to bring to maturity its long-standing investment in the West African plantations and those on a much smaller scale in the Pacific. At a meeting of April 1935, Maurice secured the agreement of all the British companies to the shorter whaling season. The Unilever representative thought it pointless – companies would simply expand their expeditions to maximise their catches in the permitted weeks. Borley acknowledged there could be no final solution until the catching power of the individual expeditions was similarly reduced, but, as Maurice admitted to Johan Hjort, the company's consent to the shorter season had only been secured by pledging that there would be no further restrictions.[(7)]

Hvalfangerforeningen had responded to pressures from the Norwegian Government by announcing a further production agreement on 6 July 1935, in which every expedition, operating to the south of 40° south latitude in the 1935–36 season was limited by a quota based on an aggregate of 1,478,500 barrels. Johan Rasmussen (who had led the negotiations) was allotted 118,750 barrels for *Sir James Clark Ross* and *Vestfold*, and 95,000 barrels for *Vikingen*. The Jahre-Bruun and von der Lippe Group was allotted 118,750 barrels each for *Kosmos I* and *Kosmos II*, and 95,000 barrels for *Pelagos*. The Tønsberg group was granted 161,500 barrels for *Svend Foyn*; the Melsom Group 76,000 barrels for the *T. Neilsen-Alonso* and 74,000 barrels for *Lancing*; Skytteren 95,000 barrels for *Skytteren*; the Salvesen Group 95,000 barrels each on *Sylvestria*, *Sourabaya* and *New Sevilla*, and 60,000 barrels on the shore station at Leith Harbour and a similar amount to Pesca at Grytviken. The agreement was contingent on the Norwegian Government restricting Christensen's fleet to 95,000 barrels on *Thorshammer*, *Solglimt* and *Ole Wegger*, and 66,500 and 57,000 barrels on the Norge factory ships *Sydis* and *Suderøy*, respectively. Harold Salvesen made his adherence conditional on the Southern Whaling Company's observing the prescribed season. A further integral part of the agreement was to lay up the factory ships *Frango*, *Fraternitas*, *C. A. Larsen* and *Anglo-Norse*, their owners being compensated from a fund of £145,000, to which the other companies contributed in proportion to their quotas.[(8)]

Henry Maurice welcomed the quota agreements, but he opposed any move whereby governments became responsible for their imposition and enforcement. Maurice himself chaired a meeting of the British companies in March 1936 at which the representative of the Southern Whaling Company continued to oppose even voluntary quotas. Maurice and company representatives nevertheless attended a meeting in Oslo, where the Norwegian Minister for Commerce sought a reduction of the total catch to 1.8 million barrels in the 1936–37 season, Rasmussen won considerable support for a limit at 2.25 million barrels. Maurice contended that delegates should first agree on 'what the existing stock of whales could stand without prejudice to future operations' and then to anticipate what the oil market could absorb. Hjort protested that far too little was known of the reproductive rates of the different whale species for this to be possible. Maurice countered that the British inspectors aboard the factory ships had found unmistakable evidence of a decrease in the general size of the blue whales caught.(9)

A joint committee of the Norwegian companies and the seamen's unions agreed with the Norwegian Government, in May 1936, to a global limit of 2.1 million barrels. Anders Jahre thereupon proposed an aggregate limit of 2,203,000 barrels of oil, with the Norwegian quota of 980,000 barrels divided between the three class A expeditions – *Kosmos*, *Kosmos II* and *Sir J.C. Ross* (95,000 barrels each); the class B expeditions – *Pelagos*, *C.A. Larsen*, *Thorshammer*, *Ole Wegger*, *Solglimt* and *Skytteren* (76,000 barrels each); and class C expeditions – *N.T. Nielsen-Alonso*, *Lancing* and *Sydis* (62,000 barrels), and *Suderøy* (53,000 barrels). The foreign quota of 1,233,000 barrels should be divided between the class A expeditions – *Terje Viken*, *Svend Foyn*, *Tafelberg* and *Vestfold* (95,000 barrels each); class B – *Vikingen*, the new ship *Jan Wellem*, *Southern Princess*, *Southern Empress*, *New Sevilla*, *Sourabaya*, *Salvestria* and *Hektoria* (76,000 barrels); class C – *Anglo-Norse* (62,000 barrels); and class D, *Frango* (53,000 barrels). The land stations of Grytviken and Leith Harbour should be allotted 60,000 barrels each. Should the three factory ships *Fraternas*, *Polar Chief* and *Strombus* also participate, they should be allocated 62,000 barrels, 53,000 barrels, and 36,000 barrels, respectively, and a proportionate reduction made to the quotas of the other companies, so as to retain a limit of 2,265,000 barrels in total production.(10)

Maurice met the representatives of all but one of the British companies on 24 June 1936, their number now comprising Anglo-Norse, Falkland Whaling, Hektor Whaling, Polar Whaling, St Helier Shipowners, Sevilla, South Georgia, Salvesen, Southern Whaling, Star Whaling and the United Whaling Company. Not only were the individual allocations condemned as unfair, but there was anger at the involvement of the Norwegian trade unions in their negotiation. Norman C. Watt (manager of the Southern Whaling Company) contended that the most effective way of saving the whale would be to let the market regulate itself (Hart 2020, 180). Maurice adjudged that any move to impose quotas would cause companies to transfer to another flag. On 30 July 1936, the Norwegian Government announced its support for every factory ship being limited to 90,000 barrels, that is, an amount sufficient to achieve profitability and an 'equilibrium' in stocks. A meeting of the British companies at the Ministry on 7 August resolved that it was sufficient for the 14 expeditions to be limited to an aggregate of 900,000 barrels on the premise that the Norwegian companies would keep within an aggregate of 1,499,000 barrels, and the South Georgia shore stations to 65,000 barrels each.(11)

Tønnessen and Johnsen (1982, 438–439) describe the British companies' insistence on so large a proportion of the catch as 'a stain on the history of whaling'. Gordon Jackson perceives any attempt on the part of historians to follow the negotiations and detail of the annual agreements as pointless 'for the simple reason that they did no good'. However much a quota system, when agreed to, might be taken as an 'earnest of good intent', each was in practice accepted only 'when necessary and broken when possible'. It might even have encouraged overproduction by implying that all was well provided companies kept within their contrived limits (Jackson 2005, 198). There was a political dimension too. While Borley was impressed by the give-and-take shown by the British companies, his forebodings about the increasing influence of the trade unions proved all too prescient. Not only did the newly elected Labour Government in Norway insist that every expedition be limited to 90,000 barrels, but it also showed increasing sympathy for the seamen union's demand that every whaling vessel should be wholly Norwegian manned (Vamplew 1975, 204).(12)

## 11.2 The Norwegian boycott

Tønnessen and Johnsen declared that 'volumes could be written' on the dramatic falling-out of the Norwegian and British companies. Whereas the Norwegian companies had borne the brunt of the decision to lay up the fleet, the British share of the global oil output rose from one-third in 1925 to over one-half in 1932–33. The newly elected minority Labour Government (Det Norske Arbeiderpartiet) prioritised the protection of jobs in the whaling industry and appointed a Whaling Council under the chairmanship of Birger Bergersen (a zoologist and loyal Labour Party man), comprising both company owners and the seamens' groups (Tønnessen and Johnsen 1982, 433; Walløe 2009; Roberts 2011, 71–72). The Foreign Office had previously assigned whaling matters to its Western Department, but the increasing antipathy between the two governments caused responsibility to be transferred to the Northern Department, which had otherwise been responsible for Norway. Laurence Collier (Counsellor and Head of the Department) detailed Bernard A. B. Burrows (a Third Secretary) to prepare a brief. Burrows attributed the increasing belligerency of the Norwegian seamen's unions to what they saw as the increasing threat to their historic near monopoly of jobs in the industry.(13)

Harold Salvesen resorted to every device he could find to publicise the boycott by the seamen's unions (Norsk Sjömannsforbund, Norsk Styrmannsforening and Norsk Maskinistforbund) of *Sourabaya* and its eight catchers to prevent them from leaving Tønsberg. The action had been taken in retaliation to the Salvesens' intended replacement of some Norwegian seamen by British citizens. Harold Salvesen emphasised, in the course of 'a personal interview' with the President of the Board of Trade, that the *Sourabaya* and its catchers did not dare to put to sea without 'the strongest official support' for their position. Collier minuted in 15 June 1936 that the obvious way to make the Norwegians see reason would be for Unilever to stop purchasing their oil. The Mercantile Marine Department of the Board of Trade protested that the unions' demands were so monstrous that 'not even the present Norwegian Government could countenance' them, causing Laurence Collier to retort that the Norwegians were 'capable *de tout!*' Cecil Dormer (the British Minister) delivered, as instructed, a Note of 25 June 1936, warning that if there was further obduracy on the part of the unions, the British companies would cease employing any Norwegian crew. The Norwegian Ministry for Foreign Affairs countered by protesting its inability to intervene.(14)

Meanwhile, the dispute had widened considerably to involve every British company. Henry Maurice warned the Foreign Office that with the whaling season so imminent, the Fisheries Department must soon notify the British companies of the season's prescribed length. Should the British companies be forced to employ an entirely British labour force, the Norwegian Government should be warned that all restrictions on whaling would be removed. Unilever responded to the picketing of the stores and fuel intended for the Southern Whaling and Sealing Company's vessels by sending a party of 200 to remove both *Southern Princess* and *Southern Express*, and their respective catchers, now that they had been 'entirely deserted by their Norwegian crews'. Berg, on being summoned to the Foreign Office, protested that his Government had no practical means of intervening. Dormer formally presented a *Note Verbale* seeking the Norwegian Government's protection of both the vessels and replacement crews. Berg again protested that his Government had neither foreknowledge of the unions' ultimatum nor the legislative powers to force its nationals to work aboard British ships.(15)

The newspaper *Arbeiderblad* speculated that the unions might try to win the support of 'their English comrades'. John Thomson (the Fisheries Department's Chief Inspector of Fisheries) took the precaution of explaining the position fully to a representative of the Transport and General Workers' Union in London. Dormer had meanwhile learnt from 'a reliable source' that the Norwegian Government was furious with the unions but at a loss to know how to end the boycott. Nor did Burrows and Thomson find the companies in any mood to agree to a face-saving compromise; they were increasingly confident that they could recruit sufficient British labour. Unilever positively relished the Southern Whaling Company's being free of Norwegian labour and associated legislative restriction. The company had already taken the precaution of hiring a tug from the German port of Cuxhaven to remove its vessels from Sandefjord. A replacement crew of about 120 was due to land at Bergen; another 18 were being sent through Oslo. The Under-Secretary of State at the Norwegian Ministry for Foreign Affairs acknowledged, at a meeting of 25 August, 'a risk of attack by excited mobs', but he assured Dormer that the replacement crew would have adequate protection. There were reports the next morning of union members using motor boats to 'swarm' aboard the Southern Whaling Company's factory ship, their purpose being to 'let down one of the anchors which had been raised, thus preventing the vessel from moving'. The engine room had been locked and the key thrown overboard. The Ministry of Justice had responded by sending 30 state police to Sandefjord. The unions had been instructed to allow the Unilever parties aboard.(16)

Dormer recounted much later that the replacement crews had arrived at Bergen and travelled to Hönefoss by rail; the intention was then to bus them to Sandefjord or a small port from where they could be fetched by boat. The Ministry of Justice had instructed the state police to provide protection. Union representatives had met the men at Hönefoss but had failed to persuade any of them 'to change sides'. The hired motor-coaches, however, had been halted after some distance, where some aboard had left with the union members for Oslo. The rest had spent the night aboard the coaches and in the morning the state police escorted them to Sandefjord. Eight of the ten men sent to Oslo had also gone over to the unions. Dormer was highly critical of the lack of police protection. The Minister of Justice had even told the chief of police at Hönefoss that he might take the day off. Dormer suspected he had assumed the unions would have won over the replacement crews long before they had needed personal protection. Clement Davies informed the Foreign Office, much more confidently on 27 August, that there were now 'sufficient officered-crewmen' for the *Southern Princess* and six of the catchers to be towed out to sea by the Dutch and German tugs sent for the purpose. The remaining vessels would leave as soon as further tugs arrived.

With the release of the Unilever and Salvesen vessels, 'the burning question' was how to devise a common whaling policy. Dormer had learnt in confidence that the British companies were now prepared to accept a maximum British catch of 900,000 barrels as part of a further quota arrangement. As Dormer cabled in late August, the Norwegian crews had begun to realise how much they stood to lose through their intransigence, causing George G. M. Vereker to minute in the Foreign Office that 'the Norwegian Trade Unions are receiving a lesson from which I hope ours, too, will profit'. But while Clement Davies blamed Norwegian intimidation for the recalcitrance shown by some of Unilever's workforce in siding with the unions, the Fisheries Department later obtained a copy of a notice signed by Ernest Bevin (Secretary-General of the Transport and General Workers' Union), instructing members to meet their contracts but otherwise not to board any ship until the Norwegian boycott was over.(17)

A further Norwegian note again called for a closed season, a production quota and this time a limitation on the number of catchers employed. Collier said that if a reasonable overall catch limit could be agreed on and the companies awarded a fair quota, the British side was prepared to 'let bygones be bygones'. However, the blockade must first be lifted to enable negotiations to proceed on equal terms. Berg seemed genuinely relieved that the United Kingdom Government was prepared to treat the boycott itself as 'a secondary consideration', despite the extreme anger of the British companies. He undertook for his part to persuade the Norwegian Government to obtain an agreement as quickly as possible. On 1 September 1936, the unions offered to call off the boycott provided Salvesens engaged no more than the present 100 'Scotchmen at Leith'. Salvesen agreed and Dormer confirmed the next day the boycott had ended. Dormer thought 'the Oslo authorities' had become only too anxious to arrive at a reasonable settlement. Collier passed on to Berg a few days later the preparedness of the British-registered factory ships *Terje Viken* and *Svend Foyn* to use only seven catchers, *Salvestria*, *Sourabaya* and *Hectoria* to use six and the *Polar Chief* five. The Southern Whaling Company was prepared to restrict *Sea Empress* and *Southern Princess* to 250,000 barrels – 10,000 barrels fewer than the previous season. Salvesen's *New Sevilla* would not exceed the previous two years' production of 110,000 barrels. Although acknowledging the fears of the Norwegian unions that the training of wholly British crews would displace their Norwegian members, there was no statutory way the United Kingdom Government could force companies to employ foreign labour.(18)

Collier acknowledged, in a long Note to Colban of 9 September 1936, that both Governments were endeavouring to protect stocks and ensure 'an equitable distribution of opportunities' for their national fleets. The United Kingdom Government had accordingly accepted the principle of restrictive quotas but continued to oppose the way the Norwegian authorities sought to allocate them. Thomson, in the Fisheries Department, thought Colban's response almost a 'complete "climb-down"'. The Norwegian Government agreed that the overall limit, previously set at 2.1 million barrels, should be raised to 2,265,000 barrels and that the season should be extended to three-and-a-half months. The two companies directly affected by the boycott would be exempted from any limit on their catchers. Collier pledged the immediate implementation of what had been agreed to. The Oslo newspapers largely welcomed the settlement. As the newspaper *Aftenposten* remarked, the Norwegians should have chosen their counsellors more carefully. Dormer was confident the dispute would soon be forgotten. Burrows in the Foreign Office thought that much of the Norwegian anxiety had arisen from the prospect of their national fleets being joined by German and Japanese expeditions and of the need therefore to have in place 'the severe general limitations' needed to safeguard whale stocks.(19)

## 11.3 The Japanese and German whaling fleets

There are many turning points in the history of whaling. One was obviously the mid-1930s when market conditions compelled companies to adopt, albeit temporarily, a quota system. As Gordon Jackson writes, it was overproduction rather than a concern for whale stocks that had made some form of rationing unavoidable if the industry was to survive. Johan Tønnessen recalls the inter-company agreements as 'undoubtedly the best form of regulating whaling that there has ever been' – they evolved from within the industry itself. It was therefore particularly demoralising for the participating companies to witness a Japanese expedition in 1934–35 and then a German fleet some two years later freely operating on those same grounds (Tønnessen 1970, 287; Bonner 1980, 240; Jackson 2005, 196).

The Japanese had pursued coastal whaling for centuries, principally for whale meat. The combination of pelagic whaling and the lucrative European market for whale oil soon began to attract their interest in Antarctic waters. Sir Cecil Dormer reported from Oslo, in November 1934, that the Nippon Hagel Kabushiki Riesha had evaded the Norwegian Fisheries Department by purchasing the redundant factory ship *Antarctic* and five old catchers. Their purchase had been timed to occur between the expiration of the latest quota agreement and the imposition of a Norwegian law prohibiting such purchases. Renamed *Tonan Maru*, the factory ship left straightway for the South African port of Walvis Bay, where it was boarded by a crew of 166 Japanese, with a Norwegian manager, engineers, gunners and flensers. Although having a target of 40,000 barrels, the overriding purpose was to gain experience and see how others operated in those waters. George E. Boyd Shannon (an Assistant Principal in the Dominions Office) forewarned the Foreign Office that Japanese operating costs would be significantly lower and that the Japanese Government would do everything possible to avoid joining an international agreement until compelled to do so. Borley similarly warned of an immense assault on whale stocks, judging by the strides the Japanese had already made in high-seas fishing.[20]

The British ambassador at Tokyo, Sir Robert Clive, confirmed in early July 1935 the intention of the Japanese to send a second expedition, despite the first having caught only 213 whales, from which as few as 77 barrels per BWU were obtained. The few Norwegians still aboard refused to sign up for another Japanese venture. As Japanese soap manufacturers used fish oil (of which there was a glut in Japan), the obvious intention was to compete with the British and Norwegian companies for the European market. Rather than enter a joint approach, which risked 'putting up the backs of the Japanese from the start', the Norwegian and United Kingdom Governments appealed separately to the Japanese Government that the *Tonan Maru* should adhere strictly to the regulations being followed by the other national expeditions. A spokesman for the Japanese Ministry for Foreign Affairs pledged the fullest use of every carcase.[21]

There was soon a report that the Mitsubishi Combine had ordered a factory ship from the Kawasaki Dockyard at Kobe. Although the Furness Shipbuilding Company had failed to win the contract for building the vessel, largely because it would need 15 months to complete the order, the vessel's construction continued to hinge on Mitsubishi persuading the Furness yard to sell the plans used previously for building the *Sir James Clark Ross*. Harold Salvesen urged Henry Maurice to do everything possible to prevent the yard from doing so. Maurice responded that a government department could hardly intervene in what was a purely commercial transaction. It was subsequently learnt that the Furness yard had sold the plans reportedly for £7,000. The *Tonan Maru*'s second expedition had met with greater success, obtaining 44,145 barrels of oil, of which 30,000 barrels were almost immediately shipped to Europe. It was noted that there were women among what was now an entirely Japanese crew. Japan's second factory ship *Nisshin Maru* was launched within only 157 days and caught as many as 1,116 whales between 13 November and 17 March 1937. Most of the oil was sold to the German market (Tønnessen and Johnsen 1982, 415–422).[22]

Germany imported almost half the global output of whale oil for both its domestic market and Central Europe, so any move to develop a German whaling fleet was bound to have major ramifications, particularly for Norway. The Weimar Government had designated a Reichsstelle för Öle und Fette (Oil and Fats Directorate) as the sole purchaser of oils and fats for Germany, thereby supplanting Unilever as the principal buyer for the German and East European markets. Any trading profits made from Unilever's German subsidiary had to be spent within Germany, causing Unilever to become very quickly one of the shipbuilding industry's biggest customers. Unilever was by no means alone. Svenska Handelsbanken was similarly compelled to use 'blocked' Reichsmarks to finance the completion of the factory ship *Tjerke Viken* in a Bremen yard for its owners United Whalers, the vessel becoming the largest to fly the British flag (Scholl 1991, 47–49 and 54–55; Hart 2020, 183–190). Walter Rau Walfang (the German margarine, vegetable and animal fats conglomerate) had long argued that Germany should have a whaling fleet of its own. For August Dierks (a member of the chamber of commerce at Wesermünde), building a German fleet would go far in alleviating local unemployment. Although never on the scale that Dierks and others had anticipated, an explicitly German whaling fleet was made part of the National Socialist government's second Four Year Plan of 1937–1940, whereby both Walter Rau and Henkel (the soapmaker of Dusseldorf) secured the means to build fleets of their own along with assurances from the Fats Directorate that it would purchase any unsold oil. They were joined by two additional companies, enabling as many as 61,000 barrels to be obtained from the first season's catch of 920 whales. The crews, except for the gunners, were entirely German. As many as seven German factory ships and 56 catchers had joined with six Japanese factory ships and almost 50 catchers operating in the austral summer of 1938–39 (Tønnessen and Johnsen 1982, 422–432; Scholl 1991, 49–55).

The German 'fats gap' in the domestic market was never closed, even though its fleet produced 86,500 tons of whale oil in the third and last season before the Second World War. The United Kingdom Government had acquired four times as large a reserve as Germany by the time war broke out (Tønnessen and Johnsen 1982, 454; Scholl 1991, 61–62). Any joint strategy between Britain and Norway, which might have discouraged the German and Japanese industries, had never got off the ground. Not until March 1937 had the two Governments discussed any kind of joint arrangement to meet the German need for whale oil from their respective fleets. Bernard Burrows in the Foreign Office had doubted from the very beginning whether any talks to that end were worth pursuing – for Germany at least the driving force was as much 'a symptom of *Drang nach Antarctic*'. The more realistic course, as both governments soon came to realise, was for them to become exemplars of how both whale stocks and markets should be managed.[23]

# PART IV

# *The Discovery Investigations*

CHAPTER 12

# *Establishing the Discovery Investigations*

There was an ever-widening chasm between the whaling companies and those alarmed by what they perceived to be the industry's impact on whale stocks. Far from acknowledging the need for caution, the whalers ridiculed all notion that their exploitation posed any kind of threat to whales. However, a very small but vociferous lobby had begun to press the United Kingdom Government for the most urgent investigations. It was in response to that concern that the rudiments of a survey and research programme had begun to emerge as an integral part of the post-war appraisal of the Falkland Islands and, more particularly, of the Dependencies. Thus, more than a decade before the International Whaling Convention was signed at Geneva, it had become the deliberate policy of the United Kingdom Government to acquire extensive knowledge of the individual cetacean species.

There had generally been a positive response to the *Report of the Interdepartmental Committee on the Research and Development in the Dependencies of the Falkland Islands* of April 1920. Henry Maurice formally welcomed the Report on behalf of what had become the Ministry of Agriculture and Fisheries. The most eagerly awaited response was from the Admiralty and more particularly whether it would provide financial support for what was proposed. Although agreeing that the Report should be published, the Admiralty was at pains to emphasise that this did not imply support for all the Committee's recommendations. Costs had risen steeply. The estimate of £85,000 for building both research vessels was already inadequate; the consulting naval architect, acting for the Crown Agents for the Colonies, estimated the current cost to be three to four times higher than estimated. Sir Frederick Learmonth (the Admiralty Hydrographer) finally indicated, in October 1920, that the severe economies being imposed on every government department meant the Navy Votes could not possibly meet the cost of one, let alone the two, vessels proposed by the report.[(1)]

The Colonial Office appointed a Research Ship Committee 'to consider what action is now practicable, or likely to become practicable in the near future, as regards the proposed employment of two special vessels to carry out the researches recommended'. The Admiralty nominated its Assistant Director of Naval Construction and an engineer from the Chief Engineer's Department. They were joined by a member of the Crown Agents for the Colonies and their Consulting Naval Architect (Sir Fortescue Flannery), John Borley from the Fisheries Department, Sidney Harmer (British Museum (Natural History)) and James Wordie (Royal Geographical Society). The DSIR, whose representative had chaired the interdepartmental committee, asked to be kept abreast of the committee's activities.[(2)]

A report by Sir Fortescue Flannery's company (Messrs Flannery, Baggallay and Johnson) recommended, in June 1922, the purchase of *Discovery*, which the Dundee Shipbuilding Company had built for Robert Falcon Scott's 'Antarctic Exploration work' in 1901 (Figure 18). The vessel had since been employed by the Hudson's Bay Company as a cargo vessel and was now 'lying afloat at the buoys in the West India Docks' of London. An initial offer of £3,500 was rejected. Winston Churchill (Secretary of State for the Colonies) approved further negotiations up to a figure of £10,000, as well as £750 to cover the costs of survey and refurbishment once in dry dock. The vessel's woodwork (except for the bow) was found to be so rotten that both costs and time schedule of nine months were far exceeded. The opportunity was also taken to remedy faults that Scott himself had complained of a quarter-century before (Darnley 1923, 727–728).[3]

Adrian Howkins writes of a neat circularity, whereby the taxes levied on the whaling industry were used to finance the scientific underpinning required to regulate the largely Norwegian industry. Mindful that 'an international free-for-all had created a tragedy-of-the-commons in the case of sealing', Howkins (2016, 70) writes that both British government officials and whale scientists strove to formulate 'the conditions for a sustainable whaling industry in the Southern Ocean'. The rationale for the funding of the Discovery Investigations was far more prosaic. Darnley knew there was not the slightest chance of obtaining monies from the Imperial Treasury. The only hope of the investigations proceeding was the imposition of a steep increase

**Figure 18** RRS *Discovery* at Melchior Harbour, Palmer Archipelago in 1927. Courtesy of National Oceanographic Library, Archives, National Oceanography Centre, Southampton, Record DI7012.

in the oil duty the whaling companies were required to pay. First levied in 1912, this duty had already been increased from 3½d to 18d a barrel in October 1919, with the intention of covering the Colonial Government's administrative costs for the Dependencies, with any residual amount set aside for a research and development fund. The Interdepartmental Committee had proposed this being raised to 5 shillings (60d) a barrel for 'research into the life history of the whale'. Grindle thought the increase 'stiff' but acknowledged that there was so great a demand for whaling licences that 'if the money is earmarked for research' of benefit to the industry, 'I do not suppose there will be great opposition'. Leo Amery (the Parliamentary Under-Secretary) agreed, minuting that

> *we must push ahead with the research ships. When the tax has covered those required in the Falklands area we might keep it on to provide a ship for the other regions where we wish to assert our territorial claims.*(4)

The whaling companies were formally notified of the increase in May 1920.

All too predictably, Theodore Salvesen protested, claiming 'the majority of the objects of the proposed expedition' would be of no interest to the present companies. Any discovery of new grounds would require new companies to be formed. There was in any case 'an exaggerated idea' of how much taxation the industry could bear. What with 'the enormous depreciation in the value of whale oil' since the end of the war and much higher coal and labour costs, the Salvesen South Georgia Company had recorded a deficit of £79,183 in 1918–19. The Southern Whaling and Sealing Company responded with what Henry Allen described as 'a most reasonable letter, and a pleasant contrast to Salvesen's wails'. Johan Rasmussen wrote on behalf of Den Norske Hvalfangerforeningen (the Norwegian Whaling Association). Although acknowledging the need for research, the Association asked that the proposed increase should be postponed until there was greater stability in the market. Gilbert Grindle both acknowledged that there was undoubtedly a 'general depression' and noted that the Norwegian Government had imposed a considerably more modest duty of only 19d per barrel on the oil produced in the waters off the Norwegian coast. A compromise of 30d per barrel was eventually agreed on for the 1922–23 season.(5)

Both Darnley and Grindle acknowledged the impossibility of setting any definitive rate of tax until the costs of operating the intended research vessels were known. A further complication was the keenness of Den Norske Hvalfangerforening for the whale oil duty to be apportioned according to the quality of the oil – a move that Charles Watt (managing director of the Southern Whaling and Sealing Company) strongly opposed on the grounds of unfairness. As he pointed out in a letter of January 1923, 'we, a British Whaling Company . . . were making every endeavour, at considerable expense in plant and by the services of competent chemists, to improve the quality of oil produced'. Any differential rate of taxation, would be to his company's detriment. Erik Hamilton was even more alarmed by how any differential rate would encourage fraud – companies would deliberately under-estimate their volume of better-quality oils. As Hamilton warned, 'The returns made by the Whalers themselves are the only source of information available' to the Colonial Government. There was no other practical way the magistrate and whaling inspectors could adjudge the aggregate, let alone composition, of the oil aboard any single vessel. The optimal course was to adopt a sliding scale based on the average price of Number 1 grade oil, whereby 24d a barrel would be refunded when the average market price fell below £35.(6)

John Middleton arrived as the new Falkland Islands Governor on Christmas Day 1920 and was immediately confronted by the post-war slump in wool prices. Over and above the general post-war depression in commodity prices, there had been a long-standing decline in the quality of the colony's grasslands and consequent fall in sheep numbers. One of his first moves was to appoint Hugh Munro (a senior stock inspector from the New Zealand Department of Agriculture), whose findings quickly led to the establishment of an experimental farm. As well as the appointment of a geologist and forester, Middleton was anxious to obtain a much closer management of the whaling industry, most obviously through the employment of a patrol vessel to keep closer watch on the individual expeditions. Although recognising 'the vast scientific importance' of the Interdepartmental Committee's recommendations, the greater priority was to obtain much closer supervision of the expeditions on the various grounds. In what Allen acknowledged to be 'a long reasoned despatch' of 29 November 1921, Middleton argued that it should be for the Colonial Government to decide how the revenue raised from the oil duty should be used. Only when those needs had been met should any residue be set aside for a research fund (Tatham 2008).(7)

Middleton went on to recount that the Colony had a total population of only 2,100, of whom 712 were under 15 years of age. Most (1,698) lived on East Falkland, with 897 in Stanley. The Falkland Islanders, Middleton wrote, were 'a hard-working and thrifty community' of British descent, who presently had 'none of the ordinary amenities of life'. They had neither 'the opportunity nor prospect of advancement'. The main industry, sheep farming, was so depressed that there had been only the most minimal investment in the capital works that the Colony so desperately needed. Housing was both inadequate and unsatisfactory. Even Stanley had no main water supply, drainage or lighting system. There was only one primary school and no secondary school; many adults were said to be illiterate. Children living on the sheep farms might have only a few days of schooling a year, when an unqualified travelling teacher visited them. There was a strong feeling among the Colony's residents that their needs had been ignored. Middleton therefore sought compromise, whereby 40 per cent of the whaling revenues should be used to administer both the Colony and the Dependencies. That would still leave a significant sum for scientific research.(8)

Allen dismissed all notion of the Colony's needs being ignored. The industry's need for both 'large capital and much technical knowledge' had meant the Falkland Islanders could play little part, but, as Darnley pointed out, they had pocketed both the lease and licence revenues. The initial duty of 3½d paid on each barrel of whale oil was also to cover the administrative costs incurred in administering the Dependencies. There had already accrued a sum of £50,000 in that account. The further increases in duty had been intended explicitly for research and development, and therefore the proposed investigations. The Governor had to realise that although directly responsible for managing the whaling industry, his present knowledge of how this might be achieved was no better than 'a captain without chart or compass'. The proposed investigations would, if successful, identify ways in which 'the industry can be expanded with safety'.(9)

Far from persuaded, Middleton wrote, in a further despatch, that it would be the Colonial Government's intention to use the entire income from the oil duty to the mutual advantage of colony and dependencies. An obvious instance would be the provision of a slipway at Stanley for overhauling the whale catchers rather than sending them to Montevideo as was presently done. Darnley was unmoved, minuting in August 1922 that 'we shall never see an end to these attempts to snaffle the funds of the Dependencies' without their being separated entirely from those of the Colony. It would in any case take 'several years before we can say with accuracy what the running expenses [of the proposed Investigations] are'. Grindle agreed in the sense that 'we impose heavy taxation on the whaling industry and are bound to use the proceeds for its benefit, and not for the relief of the Falkland Islands sheep farmers who, in ordinary times, do very well'.(10)

## 12.1 The Discovery Committee

A Colonial Office press notice of March 1923 headed *Whaling in the Falkland Islands Dependencies. A Research Expedition* announced the purchase of *Discovery*, the purpose of which was to study the whaling industries of both South Georgia and the South Shetlands, where

> *the present amount of scientific knowledge regarding the numbers and habits of the whales is insufficient to enable the industry to be controlled in such a way as to afford security against the depletion of the stock.*(11)

The *Discovery*'s immediate task would be 'to ascertain the geographical limits of the stock of whales, to trace their migrations, and to form some idea of the numbers and the rates of reproduction'. There would also be opportunity for studies in oceanography, meteorology and magnetism.

The venture would far exceed anything the Challenger Expedition had attempted, most obviously in extending over five or more years. As Grindle minuted, 'The business is a big one'. Sir James Masterton-Smith (in commending the 'Whaling Research Expedition Executive Committee' to the Secretary of State) recalled that Darnley 'had made this question his own' and how much 'we are indebted for the care and attention that – with the assistance of Mr Allen – he has given it'. Both William Ormsby-Gore (the Parliamentary Under-Secretary) and the Duke of Devonshire readily concurred with Darnley's appointment as chairman, with Sidney Harmer as vice-chairman and chairman of a scientific sub-committee. The title 'The Discovery Committee' was adopted at its third meeting of 8 June 1923. Members included the Admiralty Hydrographer, Rear-Admiral Henry P. Douglas, Henry Maurice (the Fisheries Secretary) and James Wordie (of the Royal Geographical Society). Sir Fortescue Flannery was retained for his specialist advice about the vessels. The priorities were to appoint the requisite scientists and their despatch aboard *Discovery*.(12)

Darnley drafted the advertisement for the post of director of research at a salary of £1,000 for two-and-a-half years in the first instance. Three particular qualities were sought: a research record in biology, marine experience and an ability 'to coordinate and control the work of a scientific staff'. Three advertisements were placed in *The Times* and two in *Nature*. Copies were sent to the Royal Societies of London and Edinburgh, and to the Zoological Society of London. Attention was particularly drawn to Darnley's article in *Nineteenth Century and After*. At Darnley's prompting, the Assistant Private Secretary responsible for appointments to the Colonial Office was asked to obtain from his 'correspondents at Oxford and Dublin Universities' details of three possible candidates: Julian S. Huxley and John B. S. Haldane at Oxford and James Bronte Gatenby at Dublin. In doing so, Darnley warned that even if 'disposed to become candidates and their biological qualifications were excellent, they might be ruled out for lack of experience at sea or in the organisation of research'.(13)

With only one suitable applicant identified, the post was re-advertised in July 1923. Separate approaches were made to Alister C. Hardy (Oxford), L. Harrison Matthews (Cambridge) and James H. Orton (an assistant naturalist at the Plymouth Laboratory of the Marine Biological Association). Of ten applicants considered in September, three stood out: Michael Graham at Cambridge, Alister Hardy (now at the Fisheries Laboratory at Lowestoft) and Stanley W. Kemp, who happened to be on home leave from Calcutta and was invited to apply. Kemp had begun his career as an assistant naturalist in the Fisheries Branch of the Irish Department of Agriculture, investigating the fisheries off the west coast of Ireland. His supervisor, Ernest W. L. Holt, had particularly encouraged him to specialise in crustacean taxonomy. Kemp had moved to the Indian Museum, Calcutta, in 1910 and later became Superintendent of the Zoological Survey of India (Hardy 2004; Wilkins 2020).(14)

Alister Hardy applied for both the directorship and, should that fail, a 'subordinate post of zoologist at a salary of not less than £700'. He had matriculated at Exeter College, Oxford, at Michaelmas 1914 and spent the greater part of the war with 'the HQXIII Army Corps in France'. He had thereafter taken the special course for demobilized officers, graduated with distinction and was awarded the Welch Scholarship in Biology as 'the candidate who showed the greatest aptitude for original research'. After six months at the Biological Research Station at Naples, he had been appointed a junior naturalist at the Fisheries Laboratory at Lowestoft in August 1921. Concerned that his interview had focused too closely on his development of a 'torpedo plankton-collector', Hardy emphasised in a follow-up letter his strong belief that the proposed investigations must be 'as economic as an expedition to conquer tropical diseases'. They should identify the best way of preventing the extinction of any whale species, and therefore of securing the maximum number of all types of whale so as to provide the 'maximum *continuous* yield of oil and products'. There must be sufficient data to enable some elastic system of restriction to be enforced, as would include the times and routes of migration; times and localities of breeding and foetus development; the determination of any sub-species found; and 'selection of the statistics that are going to be of the greatest value in building such an elastic system'. Hydrographical and planktonological factors must be observed, and the 'boundaries of the main plankton areas' must be mapped as they are the key to interpreting both the feeding habits and migratory behaviour of the different whale species.(15)

Hardy and Kemp were shortlisted for the director's post. Hardy was twenty-eight and Kemp forty-one years of age. John Borley and Professor Ernest W. MacBride (Imperial College of Science) preferred 'Hardy on his merits'. Harmer and Edgar J. Allen (head of the Marine Biology Association at Plymouth) strongly recommended Kemp. With "expert opinion" so finely balanced, Harold Allen (as the Financial Member) recalled how the post had been advertised at £1,000 per annum. Kemp wanted £1,500 and the retention of the superannuation rights he had accrued over fourteen years with the Indian Government. He would be twice as expensive to employ as Hardy. Although 'happily free from Treasury control and from effective criticism in the Colonial legislature', there was nevertheless a need for economy. A further meeting was arranged to consider Hardy and Kemp, together with another candidate, James Orton. Three of those present preferred Orton; two Hardy and one Kemp. Allen asked, speaking as the Financial Member, that the Secretary of State be explicitly told that his 'decided preference' was for Hardy. The proposed investigations were 'really economic with a view to the preservation of the Whaling Industry', and Hardy's research on herring stocks was much 'more analogous to what was wanted' on whales. Should that not be the case, Allen wrote, he would have grave doubts of whether 'the special taxation' could be justified 'to provide for the heavy expenditure involved'.[16]

At the Discovery Committee meeting of February 1924, Allen proposed that Hardy should be appointed at the advertised salary of £1,000 a year - Kemp's appointment would cost nearly twice that amount. His motion was neither seconded nor put to the vote. Sir Fortescue Flannery suggested that the post should be re-advertised at the salary required to obtain Kemp's acceptance. Others acknowledged the point to be 'a very serious one' but thought any further delay would be impracticable. Any criticism of Kemp's appointment 'must be faced'. Harmer thereupon moved that the Secretary of State should be recommended to appoint Stanley Kemp at a salary of £1,500 a year. With the acceptance by the Government of India of his pension settlement, Kemp was able to return to his post in Calcutta in the knowledge that he would be appointed Director of Research of the Discovery Investigations in early June.[17]

The other eight staff appointed to the Discovery Investigations were recent graduates in their early to mid-twenties. All were placed on short-term contracts without pension provision, their salaries being met by the Research and Development Fund. While some stayed, the combination of poor remuneration and lack of tenure meant that others soon left (Marsden 2001, 71). Hardy was unanimously appointed in May 1924, and was followed in seniority by Neil A. Mackintosh (aged twenty-three), a London graduate and demonstrator at the Royal College of Science. As the senior zoologist appointed specifically to take charge of the proposed marine station for King Edward Cove, Darnley wrote of Mackintosh, 'We are all satisfied that he is a first-rate candidate'.[18]

Of the other biologists appointed, Erik Hamilton was seconded to the Investigations in May 1924. E. Rolfe Gunther was still a Cambridge undergraduate when interviewed and appointed as a junior zoologist for three years (Marsden 2002, 187). Darnley described him as 'a specially suitable young man', whose appointment was unanimously recommended and supported by Kemp. The last to be appointed was L. Harrison Matthews. He had just returned from a Cambridge University expedition, describing himself as in excellent health, 'a good sailor' and extremely keen on zoology, especially marine biology. He had attended zoology classes at the Marine Biological Association's Plymouth Laboratory and had spent vacations working aboard fishing vessels. Darnley recorded his appointment as completing 'the trained scientific staff of eight persons, besides the Director' (Tatham 2008, 358–359).[19] Darnley perceived the Investigations as 'the first serious attempt to place a great industry, that of whaling, upon a scientific basis'. In the same way as forestry science 'ascertained the natural rate of increase of various kinds of trees and the best method and proper amount of cutting', so 'we are also attempting', in Darnley's words, 'the more difficult task of ascertaining the natural rate of increase of the economically valuable species of whales, and the extent to which they are to be taken without putting the supply in danger'. Among the secondary objectives of the proposed Investigations, hydrographic survey was perhaps the most important – far too many casualties were caused by the lack of proper charts. Allen saw the 'hydrographic value of the work' as the main justification for demanding 'large sums from the industry for the purpose of the proposed investigations'. An official letter might be sent 'to all the whaling interests' asking where 'the most important charting work' should 'be undertaken from their point of view'. Allen particularly recommended 'running surveys, deep sea soundings and plans of useful anchorages and harbour refuges' in the South Shetlands and Graham Land. Grindle accordingly wrote to the companies in February 1925 that every opportunity would be taken, as far 'as the primary objects of the expedition may allow', to pursue this very necessary hydrographic work for which the Admiralty would nominate a fully trained surveyor.[20]

## 12.2 The despatch of *Discovery*

A press statement announced that the Falkland Islands Government had commissioned 'the well-known Antarctic ship *Discovery*' to undertake 'research on the whales of the Southern Ocean'. The Colonial Office had long felt the need for a much closer knowledge of the life history and habits of the larger whales to secure the greatest possible annual production by 'some system of farming the whales'. Researchers needed to know much more about their breeding and growth rates and the 'migration and local movements which they make'. As this is not 'a simple question', it would require comprehensive surveys of the temperature, salinity and so forth of the waters in which the whales were found. Another question of great importance would be whether the whales caught in Dependencies' waters were part of the herd also 'being fished in other places besides the Southern Ocean'. A way had to be found to mark the whales without injuring them, as this would go far in providing information about their movement. As the press statement emphasised, the whole question was being approached without 'any preconceived ideas, but with the sole intention of finding out the truth of things as they really are'. Every opportunity would also be taken to correct and considerably improve the charts presently used by those navigating dependencies' waters.[21]

Discovery House, the name given to the Marine Biological Station at King Edward Point, was opened in May 1925 – long before *Discovery* was ready to sail. Designed in the style of an Edwardian bungalow, the building comprised laboratories and accommodation for 'four officers and three men' (Figure 19). The initial party comprised Mackintosh as officer-in-charge, L. Harrison Matthews, John F. G. Wheeler (zoologists) and A. J. Clowes (hydrologist), with a laboratory assistant and steward. Although accessible by rough track from Grytviken, the kilometre-long journey to King Edward Point was usually made by boat, along with equipment and samples taken for laboratory analysis.[22]

*Discovery*'s complement would be a master, three deck officers, three engineer officers, two cadets, five petty officers and three stewards, with 13 seamen and firemen. It soon became apparent, as Darnley minuted Grindle in November 1924, that 'the best of the polar ice captains' had been appointed ship's master. Joseph R. Stenhouse had both 'an intimate acquaintance with sailing ships, their rigging and fitting out', and experience of the Antarctic and Russian Karelia. Allen protested at his annual salary of only £700, compared with the Director's salary of £1,500 and indeed that of £500 to £600 paid to junior scientists. Although acknowledging, as Allen phrased it, service pay in the Royal Navy 'does not compare well with that attached to Scientific posts', Ernest Shackleton had paid Frank Worsley £1,000 a year, as navigation officer aboard *Quest.* Worsley had applied for Stenhouse's post. The Discovery Committee voted unanimously that Stenhouse's salary should be raised to £1,000. Although 'the jump to £1,000 is a long one', Grindle agreed.[23]

Everything continued to hinge on how soon *Discovery* could put to sea. On the assumption that most of the sea passages would be 'under sail only', major improvements were made to 'the sailing qualities' of which Scott had complained. But even when these were made, it was

**Figure 19** Discovery House Marine Station at King Edward Point South Georgia. Courtesy of National Oceanographic Library, Archives, National Oceanography Centre, Southampton, Record DI7250.

obvious that 'so heavy a vessel' could never be 'made into a fast sailer'. The ship's main engine and boilers were also overhauled and the main and mizzen masts renewed. But when opened in Messrs Vosper's dry dock, the ship's woodwork was also found to be so badly affected by dry rot as to require its entire renewal. Not only was it difficult to obtain and fit 'the very special material' required, but, as Darnley wrote, the art of 'wooden shipbuilding' had become 'nearly extinct'. The reports from the Consulting Naval Architects and Messrs Vosper became ever more discouraging, as they warned of a further six-month delay. As Allen minuted Darnley in December 1924, 'We have been very patient hitherto, but I think the time has come to put on the screw'. Darnley agreed it was 'time to apply some ginger'. The scientific accommodation, as eventually fitted out, comprised two laboratories (one for biology and the other for chemistry), a dark room, two holds and a locker, as well as a main winch, three smaller deck winches and a sounding machine. The Admiralty's Department of Scientific Research had used the ship's passage from London to Portsmouth to test the 'echo-sounding gear' installed primarily for hydrographic work.[24]

The Research and Development Fund was never able to support the two larger vessels envisaged by the Departmental Report. It was however essential, as Darnley noted in December 1924, the Investigations should have a whale catcher of its own, but one that was larger and fitted out with two fuel bunkers sufficient for the much longer voyages envisaged. Although primarily a 'whale-marking vessel', *William Scoresby* would undertake 'the economic survey of the fishing grounds between the Falkland Islands and the Continent of South America' (Figure 20). A commercial-scale otter trawl, two deck winches and a trawl winch were fitted for the purpose. As with *Discovery*, the *Scoresby*'s departure was considerably delayed by the need to install 'so much scientific and other gear' in so confined a space.(25)

It was important 'for political reasons' that the ships were registered with the Falkland Islands Government at Stanley rather than in London. The Admiralty readily agreed to this but objected to *Discovery* being given the prefix 'His Majesty's Research Ship' (HMRS) as the vessel would not display the White Ensign. By the time King George V had consented to the simple prefix 'Royal Research Ship' (RRS), the question had arisen of whether the *William Scoresby* should have the same prefix. Sir Samuel Wilson (the Permanent Under-Secretary in the Colonial Office) ruled that 'we must be content with the *Discovery* being styled 'the Royal Research Ship *Discovery*'. *Scoresby* was accordingly described as the 'Research Ship' on its first commission. With the King's consent in 1929 to *Discovery II* also having the prefix 'Royal Research Ship', the *Scoresby* followed (Hardy 1967, 13). Indeed, every British research vessel in Antarctic waters has since carried the prefix.(26)

RRS *Discovery* finally left in October 1925, bound for South Africa, where more coal was taken aboard, and then to Stanley and South Georgia. Darnley had always acknowledged that the research programme must be subject to 'modification as circumstances dictate', particularly as the priority was the study not only of the whales themselves but also 'their feed and causes which influence it'. Kemp personally led the initial expeditions of 1925–26 and 1926–27, with *Scoresby* often accompanying them. Alister Hardy wrote much later, 'The old *Discovery*, splendid for working in ice', was soon found to have neither the power nor speed

**Figure 20** The research whaler RRS *William Scoresby* launched in 1925 to support the Discovery Investigations at Grytviken in 1929. © British Antarctic Survey, UKRI.

required. The combination of a square rig and slab-sided hull made her especially awkward to operate as an oceanographic vessel. Contrary winds and shortage of coal often meant cutting the number of sampling stations to the bare minimum (Matthews 1931, 150–151; Hardy 1967, 15 and 263).[27]

The achievements of those pioneering expeditions were all the more impressive for what had proved to be a lamentable choice of vessel. Although the Crown Commissioners for the Colonies had paid only £5,000 for what had essentially been a wooden cargo vessel, the extensive repairs required to the ship's structure, along with the complete refitting for scientific work, had brought the eventual cost to almost £140,000. The protracted delays and inadequacy of *Discovery* had made for further costly disruption of the intended programme. Kemp himself paid generous tribute to Stenhouse and his officers, who, through their perseverance, had enabled so much of the scientific work to be accomplished. On the premise that the Investigations had always been envisaged as extending to wherever whales might be encountered, *Discovery*'s purchase had been a costly blunder, made before any of the scientific staff had been appointed. As Anne Savours has phrased it, there was no one available at the right time who might 'have recognised that a speedy, widespread network of observations was beyond the capability of a square-rigged sailing ship'. It was not a polar ship that was needed but an oceangoing one (Kemp 1926 and 1928; Hardy 1928, 213 and 1967, 14–15; Savours 2001, 178).

Margaret Deacon writes more positively of Kemp as an excellent administrator and inspiring leader, greatly respected for both leading from the front and for obtaining a replacement vessel so remarkably quickly. Built to Kemp's specifications, *Discovery II* incorporated the much-needed improvements discovered in the initial expeditions. Commissioned in 1929, *Discovery II* was a faster, purpose-built steel research vessel, fitted with triple expansion oil burning steam engines intended for the high seas and light Antarctic ice, capable of a range of 10,000 miles at a cruising speed of 13.5 knots (Figure 21). Built for the comparatively modest sum of £60,000 as the first purpose-built research ship to have a global reach, *Discovery II* was acknowledged by general consent to be 'the finest research ship to have been built at that time' (Ardley and Mackintosh 1936; Deacon 2010, 15–23). Alongside the logistical achievement of taking personal charge of the expeditions of 1929 and 1931, Kemp had insisted on the immediate writing up of the research results, which became a series of monographs published by Cambridge University Press for the Colonial Office. Entitled quite simply the *Discovery Reports*, they became what Hardy later described as a fitting tribute to Kemp's 'vision and energetic leadership' (Matthews 1931, 150; Calman 1947, 469; Coleman-Cooke 1963, 53–54; Savours 2001, 178; Hardy 2004, 179–180).

Although the Investigations were nominally financed from the fiscal revenues of the Falkland Islands Government, Sir Gilbert Grindle (now a Deputy Under-Secretary in the Colonial Office) noted on 9 December 1924 that it would have been 'quite impossible to give the Falklands Legislature any real control' over the Expedition – the Legislature was in any case 'a small and insignificant body'. But now that the preparatory period was drawing to a close, the time had come to assimilate the financial procedures into 'the Colonial system as closely as circumstances permit'. For Grindle, it had been 'a record of good work' and an example of 'what can be done when the Colonial Office has money to spend untrammelled by a local legislature or a Home Department'. Leo Amery, now the Secretary of State, agreed, minuting that 'Mr Darnley and Mr Allen are to be congratulated on their work'. Allen was able to write, in September 1925, that the whole enterprise had now been put 'on a constitutional basis by

**Figure 21** The custom-built oceanographic research vessel RRS *Discovery II* built for the Discovery Investigations, seen here in Deception Island on Christmas Day 1930. Courtesy of National Oceanographic Library, Archives, National Oceanography Centre, Southampton, DI4580.

local laws and the finances of the expedition . . . regulated by approved instructions'. The Research and Development Fund was sufficient to meet capital expenditure, and provided there was no 'radical change in the whaling position', there should be sufficient income 'to cover maintenance expenses in the future'.[28]

## 12.3 The initial Investigations

The Investigations were by no means the first to gather data on the industry. Sigurd Risting had been appointed the founder-secretary of Den Norske Hvalfangerforening (Norwegian Whalers Association) in 1919 and had begun straightway to analyse the statistical data derived from schedules completed by the Norwegian companies of the catches made and species caught. The statistical evidence suggested the blue migrated farther north. Rather than the previous isolated measurements, Risting's figures, 'systematically drawn from the large number of animals of a number of years', had provided 'the first really accurate information' on the average length of the male and female blue in the southern hemisphere: 23.68 metres for the male and and 24.72 for the female. Animals of over 31.38 metres (100 feet) were rarely caught. As to oil production, Risting calculated the average yield of one blue to be the equivalent of two fin whales; two-and-a-half humpbacks and six seis in the same region with the blue yielding an average 84.87 barrels of oil per animal. From the analyses of whale foetuses, Risting concluded that pairings occurred throughout the year, with most in the winter months of June, July and August, spent in temperate waters. Both blue and fin appeared to reach puberty in their second year. With a gestation period of about one year, the females did 'not as a rule produce young every year' (Anonymous 1929).

Harmer had meanwhile continued at the British Museum (Natural History) to analyse data the Governor had forwarded to the Colonial Office, typically finding that the evidence appeared to confirm 'the hypothesis that whales undertake extensive migrations from south to north and vice versa, according to the season' – observations that had at times been 'received with unnecessary scepticism'. Any sound policy for whaling must therefore be based on 'the assumption that the whales frequenting the Dependencies of the Falkland Islands constitute a single stock equally subject to depletion whether attacked at one spot only or in several'. The obvious goal for the Discovery Expedition should be to investigate the seasonal differences in the condition of the whales; the predominance of males, which had been recorded in some whaling returns; the nature of the small blue whales, known to whalers as 'Marsh Bears', and the disproportion in the numbers of fin to blue whales at South Georgia and South Shetlands.[29]

As the Development Commission remarked in March 1920 fisheries research had not existed some 20 years previously. Now as the principal funding body of that line of investigation 'nothing less was required than a complete knowledge of the sea and the life in it, and especially the interdependence between fish problems and all the conditions of fish life'. Whereas 'free research' should continue to fall to the universities, the Development Commission saw the priority of government departments to be what the commission described as 'economic' research, albeit a rigid demarcation should never be drawn between the two. The scientists employed should never be completely cut off from 'free' research. The same sentiments might well have been expressed of cetacean research (Development Commission 1920; Lee 1992, 91–92). There was obvious guidance to be drawn from the methods being developed by both the Fishery Laboratory at Lowestoft and the Marine Biological Association at Plymouth. As an acknowledgement of the help the Director, Edgar J. Allen, had already given to establishing the Investigations, Harold Allen thought it 'a graceful act' to contribute fifty guineas toward its appeal for a new laboratory. Darnley agreed its advice would assuredly be required again.[30]

Fisheries research had been considerably strengthened by the re-opening of a fishery research laboratory after the war. As Henry Maurice wrote, in his Ministry's first post-war report,

> *The first and last concern of a Fisheries Department must be to foster for the benefit of the fishermen of its own country the rational exploitation of the sea, and the first essential of rational exploitation is knowledge of that which it is proposed to exploit. Without such knowledge not only is intelligent administration impossible but the practice of fishing is almost inevitably wasteful of time and energy or of fish, if not of all three. (Ministry of Agriculture and Fisheries 1925)*

The same sentiment might equally have been expressed of the Discovery Investigations for their potential relevance to public policy. The surveys and research at Lowestoft before the war had already begun to reveal through tagging experiments the migratory and feeding behaviour of plaice and some indication of mortality rates. As with whaling, the post-war landings of fish had been both significantly greater and the proportions of smaller fish noticeably higher (Lee 1992, 99–100 and 108–109).

Both Kemp and Hardy had begun to publish brief reports in *Nature* and the *Geographical Journal* from October 1926 onwards (Kemp 1926, 1928, 1929 and 1931). Hardy gave the first substantial account of what was being achieved in a lecture to the Royal Geographical Society in April 1928. He began by recalling the expedition led by Captain James Cook to the Southern Ocean, during which Cook had claimed South Georgia for the British Crown. Johann and Georg Forster (the naturalist aboard) had famously speculated that 'if the northern ocean should ever be cleared of whales by our annual fisheries, we might then visit the other hemisphere'. Hardy said that time had come and the Southern Ocean risked sharing a similar fate. Although believing, as did Kemp, that modern whaling would have collapsed long before there was risk of the rorquals being exterminated, the industry's demise would deprive the world of a valuable animal product. But for any restrictive legislation to be of value, there must first be scientific research of the most relevant kind (Hardy 1928, 212).

The President of the Royal Geographical Society had, in introducing Hardy to his audience, recalled that Harmer had pressed for 'a thorough scientific investigation', to avert any risk of 'killing the goose which was laying the golden eggs'. Kemp confessed, in opening the discussion following Hardy's lecture, that the Investigations did not have 'the same glamour' and popular appeal as the earlier Antarctic expeditions. Many seasons must elapse before sufficient data were gathered 'to produce really satisfactory and reliable results'. Harmer paid tribute, in the ensuing discussion, to Darnley's foresight in establishing a Research and Development Fund paid for by the industry. The veteran polar commentator, Hugh Robert Mill, warned with considerable prescience of the risk of the Discovery Investigations meeting the same fate as had so many previous polar ventures: of being 'prematurely abandoned' leaving 'an efficient staff' to drift into other occupations (Hardy 1928, 227–231). As it was, Alister Hardy had already been appointed professor of zoology at the new University College of Hull. It was not until the tenth volume of the *Discovery Reports* in 1936 that he found time to publish a substantive account of what he had achieved in the Southern Ocean (Tatham 2008).

The Investigations (apart from hydrography) fell organisationally into two parts, the first being on the whales themselves and the second on their environment. Neil Mackintosh and John Wheeler had, on the opening of the Marine Biological Laboratory at King Edward Point, continued to accomplish what Barrett-Hamilton had begun some ten years earlier. Their findings constituted what became the first Discovery Report of 1929; a summary of their examination, with Harrison Matthews, of as many as 1,636 carcases over two-and-a-half summers. Harrison Matthews would later provide a firsthand account, *South Georgia: The British Empire sub-Antarctic Outpost,* of the seasonal rhythm of the transport ship arriving in later September with the labour required for the forthcoming summer at Grytviken, departing with a cargo of oil before Christmas, and then returning with further coal supplies before March. It finally left in early May with oil, guano and most of the workforce aboard, leaving only a maintenance gang of fewer than 50 (Matthews 1931, 129). Mackintosh's group had also left, but in their case it was for Irvin and Johnson's shore station at Saldanha Bay, some 50 miles northwest of Cape Town.

Hardy wrote many years later that both he and Kemp had greatly admired Johan Hjort's work (Hardy 1967, 53–54). Following his resignation from the Norwegian Government as

Fisheries Director in 1916, Hjort had moved first to Copenhagen and then to Cambridge, where his personal standing was acknowledged by his election to the Royal Society as a foreign member. He had returned to Norway and a personal chair at the University of Christiania in 1921, both as the professor of marine biology and as director of a laboratory which the Ministry of Commerce and Den Norske Hvalfangerforening had jointly established and the association had endowed. There he had focused on whale population dynamics, seeking to establish 'the optimal catch' the industry might take annually. In doing so, Hjort gathered once again a highly talented group around him (Andersson 1949, 4–7; Hardy 1950, 173–174; Smith 1994, 214; Schwach 2002, 40–41).

It was so obviously impossible to brand and release whales that an alternative approach was developed, inspired by the discovery of United States bomb-lances embedded in blue whales caught off Finnmark – and similar instances on the Chilean coast (Mørch 1911, 669–670; Murray and Hjort 1912, 713–714). Harmer had used his 'bookish knowledge' of the whale to craft a rough type of arrow in April 1921, which members of the Royal Toxophilite Society thought might be practicable. Accounts exist of Harmer, dressed in his morning coat, striped trousers and bowler hat, supervising the firing of arrows at a large oil cloth model of a whale erected in the grounds of the Natural History Museum (Burnett 2012, 156–157). Whale-marking is similar to the practice of bird-ringing as a means of discovering their migration routes (Dorsey 2013, 38). The Naval Ordnance Department produced, at the Colonial Office's expense, small numbers of barbs sufficient to adhere to the whale carcase. There was soon suspicion that the whales ejected them in much the same way they rid themselves of skin parasites. A lead-tipped, stainless steel dart was developed, which was fired from a modified shotgun, capable of penetrating the blubber sufficiently to remain in place until retrieved on the flensing plan.(31)

Hardy and Macintosh had accompanied Hjort aboard *Michael Sars* on a cruise in the North Atlantic in 1924, during which they had further developed a standardised method for marking the whales. Mackintosh drew heavily on it in conducting trials three years later from *William Scoresby*, using a dart resembling a large steel drawing pin (about three inches long) fired from a specially made gun. Light, robust and cheap, the pin was inscribed 'Reward for return to the Discovery Committee, Colonial Office, London.' Illustrated leaflets were distributed among the whaling companies describing the dart's purpose, together with a form to complete giving the mark number; whaling station, date and position of capture; species, sex and length from tip of upper jaw to notch in the middle of the tail (measured in a straight line). The form would have the 'name and address to whom reward is to be sent' and would be signed by the station manager (Hardy 1950, 176).(32)

The outcome was little short of a fiasco. As with *Discovery*, *William Scoresby* was far from adequate for its purpose. Accommodation was 'cramped', *Scoresby* was too big, awkward and under-powered to pursue whales, and hardly a suitable base from which they could be 'shot'. Only eighteen of the seventy-six markers fired had struck home, causing Kemp to write, in April 1930, 'I am afraid whaling-marking by present methods is rather a wash out' (Marsden 2002, 188–189). Worse still, few of those that had struck the whale were ever retrieved. They had either missed, fallen out or been overlooked on the flensing plan. A more successful marker was developed following further trials by Hardy, Hjort and a London gun-maker. Comprising a stainless steel tube with a leaden head, shaped like a truncated cone of about ten inches in length and three-quarters of an inch in diameter, it was designed to penetrate the blubber and become lodged within the underlying muscle. More than 5,000 had been fired by the Second World War, mostly from catchers hired for the purpose. Fewer than 300 were retrieved. The sample, however small, was thought to confirm the whale stocks of the Northern and Southern Hemispheres were discrete, with each moving seasonally and therefore out of phase with each other between the subtropical breeding grounds and sub-polar summer-feeding grounds (Rayner 1939; Budker 1958, 53–55; Mackintosh 1965, 81; Hardy 1967, 51–53; Ommanney 1971, 165–171).

It was not until the 1929–30 summer that Hjort first visited the Antarctic whaling grounds, an experience that caused him to analyse the statistical data previously obtained from the now depleted Icelandic grounds. What particularly impressed Hjort and his co-authors was the manner in which the Icelandic catches had dropped off steadily and then with increasing rapidity.

> *Whether the whaling ground is large or small, and whether the stock of whales is greater or less in the same waters, we invariably find that the whaling activities ultimately reach a point where progress is out of the question and it is even impossible to maintain the industry at the same level. (Hjort 1932, 29; Smith 1994, 221)*

Hjort wrote a brief resumé of the various investigations for *Polar Record* in 1932. The blue whale and fin whales seemed to calve every second year: blue whale calves were some 7 metres in length at birth; calves of the fin whales were 6.5 metres. Suckled for six to seven months, the blue whale might grow to 16 metres, and the fin to 12 metres in length. By sexual maturity at two years, the blue might be 23 metres and the fin 20 metres in length, achieved, as Hjort phrased it, by the ingestion of 'countless masses of crustacea in the ocean', and, in doing so, 'created boundless wealth for mankind'. In a similar way to how the technologies developed for catching whales had won 'our profound admiration', so the biochemical sciences, Hjort wrote, were now busily discovering ways of using every part of the whale. Whaling had become ever more lucrative, now employing some 10,000 people; from whaling, Norway obtained some 100 and 200 million kroner a year (Hjort 1932, 31–32).

As intended, the Discovery Investigations had begun to provide, alongside their main purpose, the more detailed charts needed for navigating the whaling grounds of the Southern Ocean. Drawing heavily on the guidance and resources of the Admiralty Hydrographer, considerable use was made of the echo-sounding apparatus the Admiralty had provided. Detailed surveys were undertaken, whenever the opportunity afforded, of both harbours and almost the entire length of the South Georgia coast. A running survey was completed of the South Orkneys and South Sandwich Islands, with further surveys of the South Shetlands and Palmer Archipelago. A further dimension was the discovery of how the Southern Ocean functioned as 'a single and unified whole'. George E. R. Deacon had joined the second commission of *William Scoresby* and taken part in three of the commissions of *Discovery II*, with the object of determining the temperature and chemical composition of the waters through which the vessels passed. It was through this detailed record of the water movement in the Southern Ocean (both horizontally and vertically) that Deacon was able to identify, in parallel with surveys made by the German survey vessel *Meteor*, the existence of an Antarctic Convergence Zone (now called the Antarctic Polar Front), where cold polar waters meet the warmer temperate waters to the north and, being denser, sink beneath (Charnock 1985).[33]

Part V, The Scientific Overview, will appraise the significance of what was accomplished during the inter-war years in the context of the further advances made by the Investigations in the decades immediately following the Second World War.

## 12.4 Challenged by the Governor

The Discovery Investigations were never without challenge. The whaling companies and Colonial Government saw them as largely irrelevant – a stigma from which the Discovery Committee never broke free (Marsden 2002, 186). Even those sympathetic to the purpose of the Investigations never agreed where the balance should be struck between studies of an immediate value to the industry and those designed to provide the knowledge needed of whale biology and the environmental conditions in which stocks might flourish. However much the Investigations required their 'sympathy and active cooperation', H. Henniker Heaton (the Colonial Secretary) found the companies 'sceptical to say the least'. Governor Middleton wanted far greater priority to be given for hydrographic surveys of the coast and harbours – nothing would obtain their active cooperation more.[34]

It was extremely fortunate that the Secretary of State for the Colonies, Leo Amery, recognised the need for specialist advisers. As he wrote much later, it was right that the Colonial Office should be organised along geographical lines and therefore able to address the general needs of each territory, but there was also need for specialist advisers in agriculture, forestry and fisheries. He particularly recalls, in his autobiography, the appointment of 'a Fishery Adviser for the Discovery Antarctic fishery work' (Amery 1953, 338). Harold Allen (the Financial Member) had endeavoured from the beginning to counter what he saw as the Discovery Committee's inclination to focus on the more academic aspects. Strongly supported by John Borley (the Fisheries Department representative), he pressed, at a specially convened meeting of March 1928, for the appointment of an additional member explicitly to develop the economic dimension of the Investigations (Marsden 2001, 74).

Rosalind Marsden writes that Allen thought Borley to be the obvious person for filling that role. Kemp threatened to resign. Henry Maurice encouraged him to do so; James Wordie thought Kemp indispensable. Harmer and Darnley together persuaded him to stay (Marsden 2001, 74–77). Grindle had meanwhile compromised by writing to the Ministry of Agriculture and Fisheries, in March 1928, laying out his Secretary of State's need for a scientist of high standing to advise him more generally on the fisheries of the empire. Borley appeared 'admirably fitted' for the post. He had shown 'ample evidence' on the Discovery Committee of his 'breadth of view and sound and cautious judgment'. Borley, close to retirement, accepted Amery's offer of a five-year appointment, during which time he would also serve as an ex officio member of the Discovery Committee. His salary of £1,500 a year would be met by the Research and Development Fund.[35]

Kemp had meanwhile become increasingly frustrated by the adjustments made by the Colonial Office to his programme of research – never more so than when he received an overtly political instruction to assist the pioneer aviator, Sir Hubert Wilkins, in making what was intended to be the first flight from the Antarctic Peninsula. Not only were two months of research lost from the beginning of the 1929–30 summer, but, as Kemp protested from aboard *Discovery*, Wilkins had flown only a few brief sorties before abruptly abandoning the mission and returning to Deception Island (Marsden 2001, 76–77; Dudeney and Sheail 2019, 161–165).

There was meanwhile frustration of an altogether different kind. As was the practice, Alexander E. Beattie (the Colonial Secretary) had acted as the officer-in-charge during what proved to be an unusually long interval between the departure of Governor Middleton and the arrival of Arnold Hodson. Beattie had meanwhile further encouraged the development of the experimental sheep farm, which had been opened in February 1926 to bring much-needed improvements to the colony's sheep industry. Hodson, after only the briefest of tours, closed the farm and abolished the post of stock inspector. As was minuted within the Colonial Office, the Secretary of State would be compelled to close 'every Agricultural Department in the Empire' should he have chosen to accept Hodson's line of reasoning. It did not bode well for the Discovery Investigations (Tatham 2008, 65–66). Beattie had meanwhile similarly warned of the hostility felt by the Executive Council toward the Discovery Investigations. The local manager of the Falkland Islands Company and the Unofficial Member of the Legislative Council, H. H. R. Gresham, had protested, in a letter to Beattie of July 1927, at 'the enormous sum of money' being spent 'on researches of doubtful utility'. Had the colony's legislature been given charge of those monies, the colony's 'financial position would have been [put] on a fine basis for many years to come'. Gresham demanded the repeal of the Whaling Ordinance of 1924, by which the Dependencies' Research and Development Fund was established. However much 'the movement of plankton and the collection of specimens' might interest scientists, it was wrong in principle that the revenues obtained from the whale oil duty should pass directly 'into the hands of a scientific or any other body'. If the whales chose to absent themselves, 'no amount of scientific endeavour will entice them to go there'.(36)

Hodson wrote, within a month of his arrival, of his high opinion of Gresham and that the annual revenues from the Research and Development Fund provided 'a unique opportunity . . . to finance through investment all the essential services of the Colony for all time'. The claims of the colony by way of education, housing, roads, and the medical and sanitary services, should come before scientific knowledge of an 'almost entirely academic' kind. Sir Samuel Wilson acknowledged that his draft reply to the Governor might 'appear at first sight to be "putting it across" too much' but, as he further explained, 'I don't see how we can avoid it, if he [the Governor] takes it upon himself' after less than six weeks in the colony, to reverse previous policy. It should have been obvious to Hodson, had he taken the trouble to read his predecessor's correspondence, 'the principles governing the appropriation of the revenues of the Falkland Islands Dependencies were laid down after the most exhaustive discussion'. Nothing had since changed. The Letters Patent of 1908 and 1917 were never intended to make the Dependencies 'an integral portion of the Falkland Islands Colony'. Once the Colony's costs for the administration of the Dependencies were met, any further income was for 'the benefit of those Dependencies'. Any 'failure of the whaling industry' would render 'the Dependencies practically uninhabitable'. 'Success in securing the maximum supply of whales' could only come through 'a full knowledge of the habits and life history of the animals'. Use was being made of 'all known methods of enquiry', both statistical and anatomical.(37)

William Ormsby-Gore (the Parliamentary Under-Secretary) signed the formal Colonial Office response of January 1928, in which he emphasised that the Secretary of State was responsible for the entire expenditure incurred by the Investigations; the Discovery Committee dealt with only the year-on-year costs. Ormsby-Gore was confident therefore that Hodson would realise on reconsideration the impracticality of the Investigations being conducted from Stanley. There would be an inordinate delay in obtaining 'many of the data required to frame the estimates'. There was also an international dimension. The New Zealand Government was responsible for the governance of the Ross Sea. As to Hodson's complaint of the salaries paid to the Investigations' staff being higher than those of his own officers, this reflected the scarcity of officers of the calibre required for the Investigations, the risks and discomforts incidental to working at sea and the fact that these appointments were both temporary and non-pensionable.

Far from mollified, Hodson fired off a further long despatch of 8 May 1928, describing the Discovery Investigations as 'a grave public scandal' and that he was responsible for the collecting and management of revenues in the Dependencies. The Ordinance of 1924 should be repealed, and an end put to such

> *abrogation of my powers and the powers granted under the constitution to the Executive and Legislative Councils, composed, as they are, of experienced and responsible men with full knowledge of local conditions, in favour of a Committee occasionally convened eight thousand miles and more away from the scene of the operation which it is empowered to direct.*[(38)]

Hodson went on to recall that the purchase of *Discovery* had been a scandal of the first order. There had been no need for *William Scoresby* – the whaling companies could have provided a catcher. Reports of drunkenness in the streets of Stanley and the mishaps to *Scoresby* at both Grytviken and Stromness showed, when taken together, how poorly the Investigations were being conducted from London by a committee entirely devoid of 'practical experience and free from responsibility'. The same might have been expected of a committee appointed at Stanley to study 'the cause of decline of the ostrich industry in South Africa'. The obvious course was for the Colonial Government to take direct charge of the Investigations from the marine station at King Edward Point. Kemp might remain Director of Research but answerable to 'the administrative supervision of the Colonial Government'. Free from 'that quality of avarice which I am unable to dissociate from the authorities of the British Museum', Hodson wrote that his sole purpose would be to ensure that the revenues raised in the dependencies were used 'in the manner best calculated to secure the moral and material prosperity of this section of His Majesty's subjects and not to acquire the doubtful advantage of a more intimate insight into the generative organs of the whale'.

Grindle found it 'rather disturbing to think that the character and fortunes of the Government officials in the Falklands should be at the mercy of a man with so little sense of responsibility'. A twofold response was called for. 'The scientific and professional members of the Discovery Committee' should receive 'some definite expression of the Secretary of State's continued confidence in them', and second, 'some plain language should be used' in telling the Governor that 'he will not be allowed to substitute himself for the eminent experts on the Committee'. The Governor had in effect challenged the wisdom of the Secretary of State in his delegating responsibility to the Discovery Committee. Sir Henry Wilson thought Hodson should

> *now be given a warning by you that unless in the course of his next tour in the Falkland Islands he shows more judgment than he has done in the past, you will have to consider whether you should advise His Majesty to allow him to continue in his appointment or not'.*[(39)]

The *Daily Mail* had meanwhile published an article headed *£130,000 Spent on £5,000 Ship. Waste of Public Money. Vessel Unsafe for Antarctic Work.* The Investigations had not only cost 'the little colony of the Falkland Islands' no less than £317,500 in three years, but they had been humiliating for how little had been achieved. Where, demanded the newspaper article, were

> *the successors to Scott, Shackleton, and other heroic figures who enriched the Empire by their discoveries and scientific contributions to research? We have not now the men. We cannot, judging from the fiasco of the* Discovery *find the suitable ships for research. We have only the money – to waste.*(40)

Darnley thought the Governor was unlikely to have instigated the article, but he was 'substantially responsible' for it.

It had already been arranged for an afternoon session of the annual meeting of the British Association for the Advancement of Science to be devoted to the Discovery Investigations. Kemp, in introducing the papers, defined the purpose of the Investigations as one of assisting, 'if possible by scientific inquiry, an industry that was possibly in grave danger'. The marine station at South Georgia had already examined a far greater number of whale carcases than the aggregate of all previous investigations. The findings would be published within six months. Work on the *Discovery* and *William Scoresby* had focused on the natural conditions in which whales lived; a further year's patient analysis was required for identifying the microscopic organisms and minute crustacea on which the whales fed. A third line of investigation was to determine the routes and dates by which the whales migrated – information of fundamental importance both for the industry and any protective measures that might be necessary. The data obtained would be 'of immediate economic value and of a valuable scientific character' (British Association 1928).(41)

Kemp's introductory statement was followed by three papers. The first given by Neil Mackintosh used the morphological data collected from both the Grytviken and Saldanha shore stations to show the disproportionately high number of immature whales being caught off the South African coast. Analyses at the marine station at King Edward Cove had shown the gestation period of the whale species studied to be almost 12 months, with the calf and mother inseparable for a further seven months. In the second paper, Rolfe Gunther reported that the blue whale was almost entirely dependent upon a single species of *Euphausia* – Antarctic krill *Euphausia superba* – and more particularly the euphausid 'nurseries' commonly found in association with upwellings of water rich in phosphate characteristic of the more sheltered waters around the Dependencies. Alister Hardy gave the last paper, highlighting the potential of his continuous plankton recorder for mapping the distribution of those 'nurseries'. Transects made over some 2,400 miles of ocean had revealed a striking patchiness in their distribution (British Association 1928).

Walter Garstang (the chairman) concluded the Session by remarking that the interim reports, given by members of the Discovery expedition, had shown

> *the scientific work of the expedition has been well planned and well executed, and unanimously records unabated confidence in the satisfactory progress of the expedition towards valuable and well-founded results.* (42)

His motion was passed unanimously.

The Colonial Office's press officer had secured coverage of the particular session. The *Times* led with the headline *Valuable Work of the* Discovery; the *Morning Post* and *Daily Telegraph* carried reports, the *Telegraph* under the headline *Expedition Vindicated.* The *Daily Mail* by contrast reported that representatives from most of the Colony's sheep farmers had recently launched a petition in London protesting that 'the cost of the research work should fall entirely upon the Falkland Islands'. Grindle minuted that 'the Falklands would no doubt like to get the money', but the whole cost of the Investigations was borne by the whaling industry.[(43)]

As Peder Roberts writes, Hodson 'for all his blustery ignorance' had identified an important issue. An exercise in imperial development was being organised and executed from London almost despite the wishes of the colony, which it was charged with advancing (Roberts 2011, 38–41). Far from ignoring the point, Darnley was acutely aware of the constitutional position, minuting the whaling industry had developed as an enterprise 'emanating, not from the Colony, but from Great Britain and Norway'. The Discovery Committee had meanwhile prepared a 34-page rebuttal of Hodson's charges of mismanagement by recalling that *Discovery* had been purchased without a dissenting vote, and refitted for a sum of £115,500 at a time of high post-war prices. While the Committee was now giving urgent attention to a replacement vessel more suited for making 'ocean traverses in the rapid execution of hydrological and biological surveys', there would remain the need for 'the strongest vessel in existence for work in pack ice'. *William Scoresby* was similarly required for so wide a range of tasks that none of the whaling companies' catchers was suited to the task.[(44)]

The Discovery Committee, in its rebuttal of Hodson's criticisms, recounted that the Investigations had found 'a profound divergence of opinion among Norwegian and other whalers as to whether dangerous depletion was going on'. There was 'no reliable basis' upon which to determine the extent to which whaling might be 'acting to its own detriment and bringing about its ultimate extinction'. The Committee had responded by following the 'widely established practice with other economic enquiries', that is, mounting an ambitious scientific study. As to the suggestion that the Investigations would be better organised from Stanley, not only was 'a properly equipped laboratory' needed but any investigation of 'the fluctuations in the catches observed from year to year' required 'wisely planned investigations into the hydrographical and biological character of the regions affected, carried out on an adequate scale over a sufficient area and through a sufficient period by properly equipped research vessels'. The inception of pelagic whaling had made that need all the more pressing before a further 'very powerful vested interest' became established and called an international convention. As some 60 per cent of the global whale oil now came from the Dependencies, the Colony had most to gain from a sane management policy and most to lose through neglect and ill-informed regulation. The Discovery Committee employed 'the largest, best equipped and most experienced staff engaged in whaling research'. The Norwegian companies and their Association, far from being ill-disposed toward the Investigations, had closely collaborated with them, providing valuable information that could not possibly have been obtained from any other source. With 'the widening spread of whaling' about the Antarctic, and 'the intimate effect' of the destruction of stocks in one area on another, the Discovery Committee urged that every effort be made to secure the participation of 'the Dominions lying about the Antarctic', the Norwegians and the French in respect of the West African industry.[(45)]

Grindle brought the correspondence to a close in December 1928 by ruling that the Secretary of State considered the Discovery Committee to have 'completely and satisfactorily answered the very serious charges' that Hodson had brought. Its work should not only continue but, Grindle wrote, the Secretary of State was 'not prepared to consider any proposals for transferring the control of their operations to you in your capacity as Governor of the Falkland Islands'.[(46)]

## 12.5 Challenged by the Secretary of State

It was ironical that as whale stocks came under unprecedented pressure, the next Secretary of State should himself want to disband the Discovery Investigations. It hardly helped that the key figure administratively, Rowland Darnley, should have been retired prematurely. Darnley had continued to be Chairman of the Discovery Committee in his capacity as the Assistant Secretary responsible for the Colonial Office department covering the South Atlantic colonies. There are hints of an increasing frustration with Darnley's tardiness and indecision generally. Sir Samuel Wilson (the Permanent Under-Secretary) tried to secure Darnley's retirement in 1930, citing his long absences for health reasons. Lord Passfield (the Secretary of State) personally gave him 'a serious warning of the need to pull his weight'.[(47)]

The transfer of ministerial responsibility for parts of the former Ottoman Empire from the Colonial to the Foreign Office meant a post of Assistant Secretary had become redundant. The Treasury agreed that Darnley was the obvious person to retire on the grounds of 'his inadequate contribution to the discharge of the duties of the office'. Sir Philip Cunliffe-Lister (the new Secretary of State for the Colonies in the National Government) met Darnley on 29 November 1932, with both Darnley's counsel and solicitor present. He said that he had found Darnley definitely inferior in efficiency to the other assistant secretaries and had no choice but to select him for retirement. Despite the intervention of the Association of First Division Civil Servants, both Cunliffe-Lister and Leslie Hore-Belisha (the Financial Secretary to the Treasury) confirmed Darnley's retirement three years short of his sixtieth birthday. His pension and honorarium (as the chairman of the Discovery Committee) would approximate to his previous income.[(48)]

The more fundamental threat to the Investigations came from Cunliffe-Lister's determination to use what remained of the Dependencies Research and Development Fund for other purposes. As he informed his officials, he had also become increasingly sceptical of whether 'the whaling research investigations, apart from their scientific value, were serving any really useful purpose'. The monies would be far better invested in, say, the Imperial College of Tropical Agriculture in Trinidad.[(49)]

A Written Parliamentary Question had, some six years earlier, raised the possibility of using monies from the Fund for other purposes – in that earlier instance for protecting game animals. Leo Amery had on that occasion explained that the Fund was intended exclusively for 'the benefit of the Dependencies [from] whence it is derived'. However, the Slump and ensuing Depression were forcing every Department to look for economies and the Discovery Investigations were, for Cunliffe-Lister, an obvious target. His biographer describes him as a voracious worker, reading and commenting frankly, with a forte for executive action rather than working out longer-term policy. Impetuous and reluctant to change his mind, he was not too concerned about procedural niceties (Parliamentary Debates, Commons, vol. 197, 23 June 1926, 375–376; Cross 1982, 118–121).

The Discovery Committee responded with a spirited written defence, arguing that whaling had ceased to be 'the problem of a single sector of the Antarctic' but now encompassed the entire Southern Ocean. The global output of whale oil had risen by over 200 per cent: as many as 42,874 whales had been killed the previous season. Only scientifically directed, international regulation could prevent the industry's extinction and economic loss to the British Empire. Any foreclosure of the Investigations would both waste the monies already spent and cause the United Kingdom Government to lose its predominant position in effecting any kind of international regulation – a place 'which she has fully earned by her consistent efforts at control'. Not only was the British Empire responsible for about one-third of the global whale oil but those companies and holdings in foreign companies now represented a total investment of £3,815,000. A significant contribution had similarly been made to British shipbuilding. Almost every factory ship was built in the British Isles – the largest at Belfast in 1931. As many as 99 of the catchers employed in the 1930–31 summer were similarly British-built. The Discovery Committee's Note to Cunliffe-Lister emphasised the remarkable progress made by the Investigations, particularly when compared with fisheries research generally. The Research and Development Fund now stood at £413,000. Fiscal and licence revenues had fallen from £194,153 in 1928 to £122,814 in 1929, £84,546 in 1930 and now to £48,133 in 1931. The estimated revenue for 1932 was only £36,695 and £20,245 in 1933 (Dudeney and Sheail 2019, 84–89).[(50)]

The Discovery Committee had responded by reducing the forthcoming year's Estimate to £47,038, a budgetary reduction of £10,835. The Marine Biological Station would close after seven seasons. Faced with the choice of re commissioning *Discovery II*, substituting *William Scoresby*, and even laying up both vessels for a season, the committee had recommended that *Discovery II* should be employed for at least two further seasons and for a whale catcher to be used for whale-marking purposes. Monies would continue to be put aside for writing up and publishing what had been accomplished.[(51)]

The Committee's submission concluded by emphasising the wider relevance of the Investigations. Marine life was so integral to the economy of the Dependencies that, should the Investigations be abandoned, they would have to be resumed at some later date. They had a political purpose too. More than half the sub-Antarctic islands and three-fifths of the Antarctic continent (including the more valuable parts), were part of the British Empire. International law required 'constant action and proof of interest' for sovereignty to be retained.

> *The retention of a ship in commission in Antarctic waters would not only enable economic research and surveying to be carried out, but would be of great service in securing the observance of economic regulations and would materially strengthen the basis of British sovereignty in the Antarctic.*[(52)]

Cunliffe-Lister met the Discovery Committee, on 30 May 1933 to inform members that sufficient information had been obtained to meet the requirement of an international whaling convention and convince whaling interests of its need. He intended therefore to ask the law officers of the Crown for an opinion of whether the Research and Development Fund could be diverted to tropical agricultural research. Among those present, Henry Maurice emphasised that the Investigations had been conducted in the most economical manner. The Admiralty Hydrographer said the Admiralty would greatly regret any closure of the Investigations. Kemp was fortunately in London and made a separate submission, further emphasising 'the extreme "richness" of Antarctic waters' for marine life and generally stressing that if 'we throw in our hand', other countries, and more particularly the Argentines, 'would probably be quick to take our place'. Sidney Harmer wrote privately to Stanley Baldwin (the Lord Privy Seal and a Trustee of the British Museum (Natural History)). The Colonial Office assured Baldwin's office that the Secretary of State was fully conversant with all the points made by Harmer in his letter. The British whaling companies were an obvious absentee from this lobbying effort. Harold Salvesen had remained particularly critical, claiming 'the whole business' to be 'conducted most inefficiently and at criminal costs' (Elliot 1998, 23; Dudeney and Sheail and 2019, 84–89; Burnett 2012, 179–80).[53]

The Colonial Office submission to the Law Officers of the Crown stressed the urgency of raising funds for research, more particularly for the Imperial College of Tropical Agriculture, Trinidad, and the East African Agricultural Research Station at Amani. The Secretary of State was accordingly seeking their advice on whether part of the balance of £380,000, now standing to the credit of the Fund, might be spent 'on purposes not *eiusdem generis* as those specified' by the Falkland Islands Ordinance of 1924. The Law Officers – the Attorney-General (Sir Thomas Inskip) and the Solicitor-General (F. Boyd Merriman) – concluded, in July 1933 that it was not permissible for any part of the balance to be used 'for any purpose unconnected with the Dependencies'. Cunliffe-Lister responded by instructing the Colonial Office 'to work out among ourselves a figure that might possibly be offered to the Discovery Committee as the maximum amount available for winding up their affairs'. Henry Maurice intimated, in a phone call to the Colonial Office, that his Minister, Major Walter Elliot, had spoken to Cunliffe-Lister and obtained the impression that *Discovery II* might make a further cruise. Cunliffe-Lister approved the cruise four days later, minuting 'I have always contemplated that one such cruise might be necessary'.[54]

Despite the personal intervention of the Minister of Agriculture and Fisheries, Cunliffe-Lister persisted in taking the issue to the Cabinet, signing a memorandum on 11 December 1933, which Sir John Shuckburgh (a Deputy Under-Secretary) was instructed to prepare. Cunliffe-Lister acknowledged that with the spread of pelagic whaling, the 'easy and simple course' was 'to let sleeping dogs lie' and for 'the whole of this large fund' to be exhausted. He thought that would be wrong. There was no certainty that even if sufficient information was known of 'the birth, life and death of whales', it would be any easier to draft, let alone enforce, 'an international convention to control whaling in the Southern seas'. Some part of the Fund should be diverted to research of greater importance to the Empire – 'eggs' might be transferred 'to more promising scientific nests'. A Treasury official summarised Cunliffe-Lister's purpose as one of using £125,000 to wind up the Investigations, a further £135,000 to meet future deficits in the Dependencies' budget, and the remaining balance of £100,000 to be spent at his own discretion. Sir Ernest Strohmenger (a Treasury Under-Secretary) thought it 'a breach of trust', and that the Treasury 'should strongly object to the diversion of any part of the Fund for a purpose entirely unconnected with the Dependencies'.[55]

The Cabinet, with the Prime Minister, J. Ramsay MacDonald, in the Chair, considered the memorandum on 20 December 1933. The Cabinet conclusions summarised Cunliffe-Lister's introductory statement, and note that £125,000 would be set aside for completing the Investigations. Further discussion was postponed until Cunliffe-Lister's return from an overseas tour. The formal conclusions of the resumed discussion, of 25 April 1934, record that the Ministers decided only moderate support should be given for further whaling research.

*It was recognised that the whaling industry exercised injurious economic effects on some of the tropical Colonies by producing a competing raw material for the manufacture of margarine. There was some comment on the somewhat barbarous methods of the whaling industry.*[56]

The Conclusions recounted that Cunliffe-Lister admitted 'the truth of the Chancellor of the Exchequer's contention that the £100,000' for a new Colonial Research Fund 'was legally a breach of trust'. As a Treasury official later phrased it, ministers agreed after considerable discussion that 'the Secretary of State for the Colonies and the Chancellor of the Exchequer (Neville Chamberlain) should consult one another on 'the allocation of the £235,000 referred to'. A further official, John A. N. Barlow (an Under-Secretary) was able to minute on 11 October that there was 'no need to trouble the Chancellor' further. Cunliffe-Lister had reportedly abandoned his Colonial Research Fund (Dudeney and Sheail 2019, 84–89).[57]

There remained a final 'side-show': Darnley's dismissal as chairman of the Discovery Committee. Shuckburgh informed him, on 18 October 1933, with *Discovery II* re-commissioned, that the Secretary of State had decided 'to make other arrangements for the Chairmanship'. Darnley protested vigorously. Although 'reasonably sympathetic towards Mr Darnley's feelings over his elimination', Shuckburgh insisted there could be no 'possibility of re-opening the question'. Darnley responded by taking what Shuckburgh considered the most extraordinary step of protesting to the Prime Minister. Cunliffe-Lister indicated that the Prime Minister's Private Secretary should be told 'not to have any correspondence with Mr Darnley'. Henry Maurice had meanwhile moved a resolution upon the Discovery Committee acknowledging Darnley as the initiator of the Investigations, his 'effective services to the control of whaling in the Falkland Islands Dependencies' and 'his efforts towards securing the international regulation of whaling' (Burnett 2012, 180).[58]

The *Daily Herald* speculated on the reasons why Darnley had been retired to his 'small villa in a Surrey village'. The *Daily Mail* expected there to be Parliamentary Questions. The *Morning Post* published both the greater part of Darnley's letter of dismissal and its rebuttal. The leader writer feared the appointment of Lord Plymouth (the Parliamentary Under-Secretary) as chairman presaged a further attempt to close down the Discovery Investigations. The *Western Mail and South Wales News* responded by defending Lord Plymouth's appointment. The Colonial Office's Establishments Officer and one of Cunliffe-Lister's Private Secretaries had previously met a newspaper representative and emphasised to him that 'the normal course of administration' was for the Parliamentary Under-Secretary to take personal charge of an important committee. The Discovery Committee was simply being brought into line (Dudeney and Sheail 2019, 84–89).

Among the personal papers of Bjarne Aagaard (the Norwegian historian and propagandist) is a letter from Darnley, of 28 December 1933, confident that the intention of Lord Plymouth's 'extraordinary appointment' was 'to divert the Committee's funds to other uses' As Darnley had anticipated in his letter to Aagaard, Lord Olivier put down a Parliamentary Question in the House of Lords, extending over five columns of Hansard. Keeping closely to the text prepared for him by officials, Lord Plymouth responded by informing the House that *Discovery II* would be re-commissioned for a further cruise of 12 months. Viscount Elibank (the opposition spokesman) welcomed the assurances given as to the future of the Investigations and that Darnley had not been dismissed. He accordingly suggested that Darnley might be 're-attached' to the committee. Lord Plymouth, having spoken privately to one or two members, minuted it had become quite clear that 'Mr Darnley's re-appointment to the Committee as an ordinary member would not be acceptable to the majority of the members of the Committee' (Parliamentary Debates (PD), Lords, 91, 978–88).[59]

## 12.6 The near-demise of the Investigations

Ralph Pugh wrote, in his history of the Colonial Office, that it was the practice of the Crown Colonies Division to appoint committees and bureaus to tackle technical issues. Some were temporary – others more permanent. All were intended to improve the health of human, animal or plant populations through scientific study (Pugh 1959, 761). A distinction can be drawn, however, between the Discovery Investigations and those intended to function close to, and indeed employ staff from, the respective colony. The Discovery Investigations were directed from London and pursued, from the Falkland Islander's perspective, far from their shores.

The Investigations continued to make little use of the officials employed by the Falkland Islands Government. There was no doubting the ability of Erik Hamilton (the Government Naturalist). It had seemed an inspired move to second him to the Investigations for his unique knowledge and experience of those waters. His obvious cynicism of the manner in which the Investigations were being conducted and Kemp's refusal to appoint him to the post vacated by Alister Hardy meant that he soon returned to his duties as the Falkland Islands Naturalist (Tatham 2008, 262). The Investigations similarly ignored such figures as Edward Binnie who, as 'a native "kelper"', customs officer and magistrate, had witnessed at first hand 'the tremendous growth of the industry'. As Ian Hart writes of him, in *Antarctic Magistrate*, Binnie became increasingly critical of 'the policy of permitting foreigners to develop the potential of the Colony' when so 'many Falkland Islanders were more than willing to participate'. Binnie himself resigned from the Colonial Government in October 1927 and moved to Norway where the whaling and sealing companies made full use of his administrative and linguistic abilities (Hart 2009, 224).

But beyond the Dependencies, there continued to be no Empire-wide approach to fishing policy. John Borley's appointment as the Fisheries Adviser to the Secretary of State could have provided that opportunity to improve the productivity of the colonial fisheries – not least in helping to assuage concerns at the lack of protein in the colonial diet. Christian Jennings is wrong in blaming the Discovery Investigations for denying Borley the funds to develop, as Fisheries Adviser to the Secretary of State, an Empire-wide fisheries investigation, but he rightly perceives Borley as being ideally suited for the task. Borley had begun to craft an Empire-wide fisheries programme, whereby both research vessels and trained staff would be deployed as the fisheries potential of the respective colonies was developed. A Standing Committee on Colonial Fisheries was appointed in 1930 modelled on the Discovery Investigations – but with an essential difference. The costs would have to be met by the Colonial Development Fund and respective colonies. The Slump and then the Depression of the early 1930s destroyed all possibility of that happening. Borley's post was abolished following his eventual retirement in 1937 (Maurice 1939; Jennings 2011, 253–260). The Discovery Investigations came perilously close to sharing a similar fate.

The political priority for the United Kingdom Government in Antarctica had continued to be the requirements of international law, and more particularly, the legal precedents set in other remote areas of the world. The International Court of Justice at The Hague had ruled in an arbitration award of 1928 that discovery alone was not enough to claim sovereignty; there had to be evidence of continued physical occupation. As the technologies for polar exploration improved, governments appraised those ice-bound environments increasingly for their economic potential. A further precedent was set by the International Court in 1933, in adjudging whether there had been sufficient acts of governance for the Danish government to claim the sovereignty of East Greenland.

Peder Roberts writes that the primary purpose of the Discovery Investigations had become one of preserving, if not strengthening, the stance of the United Kingdom Government in negotiating the various claims to Antarctica (Roberts 2011, 41). Adrian Howkins details, in his environmental history of the Antarctic Peninsula, the ambition of Leo Amery (as the Secretary of State) 'to paint the Antarctic red', describing his stance at the Imperial Conference of 1926 as 'the apogee of high modernist desires to control space and nature and a refutation of imperial decline' (Howkins 2017, 44). But in focusing upon the acquisitiveness of such figures as Amery, there is also the risk of overlooking the decisive role played by the Foreign Secretary and the Cabinet as a whole. It was the Foreign Office and Treasury that ultimately determined overseas, and therefore colonial, policy (McKercher 2006, 95).

It was Sir Cecil Hurst (the Legal Adviser to the Foreign Office) who ultimately determined both the extent and timing of Antarctic claims in his ruling that there must be a three-stage process, beginning with the modest announcement of an intention to acquire the territory. The second was to establish as large a physical presence as conditions permitted within the respective territory, which, once achieved, might lead to a third and final stage of formal annexation. Even then, there must be a sufficiency of presence to avoid suspicion of the claim having lapsed (Dudeney and Sheail 2019, 129–147). It was to provide that demonstrable evidence of occupation that the two-year British Australian New Zealand Antarctic Research Expedition (BANZARE) went ahead as the means of fulfilling that second stage in the Australian government's claim to the Australian Antarctic Territory. *Discovery* was lent for the purpose of physically demonstrating interest in that coast and an assessment of its economic potential (Dudeney and Sheail 2019, 184–189). There was no better way of achieving this in the Falkland Islands Dependencies than through the scientific activities of the Discovery Investigations. But even as Hurst's second stage in confirming territorial claim went forward, the imperial fervour was fading rapidly. There was every prospect of *Discovery II* being either sold or scrapped, and the expertise and experience obtained by its scientists and crew abandoned (Dudeney and Sheail 2019, 225–230).[(60)]

Within the larger issue of the ability and desire of the United Kingdom Government to demonstrate an imperial presence in those territories, the question also arises of whether the individuals charged with the Discovery Investigations did enough to secure and retain the political interest and investment needed. William T. Calman, in his obituary notice of Stanley Kemp, praised the Investigations for being the most comprehensive oceanographical venture up to that time. A better knowledge of the hydrology and plankton life of the Southern Ocean was obtained than for any other water body, except for the North Sea and North Atlantic. Compared with other mammal species, an unprecedented amount was learnt of the breeding, offspring, growth rate, life duration and general ecology of cetaceans.

Yet, as Calman also acknowledges, the scale of that achievement was scarcely noticed even in academic circles. It could be argued that the publicity given to the achievements of the turn of the century polar expeditions had greatly exaggerated their scientific importance, but the Discovery Investigations were the reverse. For all Stanley Kemp's strengths, 'his unaffected modesty and intense dislike of anything approaching self-aggrandisement' were a liability. It was not until Francis (Dick) Ommanney published a personal account of his experiences that there was any inkling in print of 'the grim realities underlying the restrained language of the official reports' (Ommanney 1938; Calman 1947; Coleman-Cooke 1963, 29; Marsden 2001, 64–70).

Kemp resigned as Director in 1936. Increasingly frustrated by what had become an essentially desk-bound job in London, he was greatly attracted by the prospect of becoming director of the Marine Biological Association's Laboratory at Plymouth (Calman 1947). Mackintosh later paid fulsome tribute to how, through Kemp's far-sighted leadership, the Investigations had become the well-equipped and many-sided organisation required for the rational development of whaling and the other natural resources of the Falkland Islands (Mackintosh 1945). Now as a member of the Discovery Committee, Kemp lobbied on its behalf, typically emphasising to Anthony Bevir (a Principal in the Colonial Office), in March 1938, both the political and the economic and scientific necessity of continuing with the Investigations. As the only government vessels in those waters, the Discovery Committee's vessels also served an important political purpose. Although the Argentines had continued to claim the South Orkneys by their occupancy of the Laurie Island observatory, this had lost much of its political force since *Discovery II* had made a hydrographical survey of the entire archipelago. Full surveys were similarly being made and published for the rest of the Dependencies. Kemp warned that should the royal research ships be sold and their officers and men (with their unique knowledge of ice navigation and Antarctic conditions) be dismissed, any political hold over some three-fifths of the Antarctic would become increasingly tenuous.[61]

Mackintosh had, as Kemp's successor, been immediately required to reduce and make savings, but whereas Alister Hardy might have possessed the personality to raise their public profile, had he remained with the Investigations, Mackintosh shared Kemp's intense dislike of publicity (Anonymous 1974; Marsden 2001). By largely limiting publication of the research findings to the *Discovery Reports*, which no ordinary library could possibly afford to purchase, their significance, and indeed their very existence were all too easily overlooked (Roberts 1975). Harmer eventually resigned as chairman of the Scientific Sub-Committee in 1944, some 17 years after he had retired as Director of the British Museum (Natural History). Beyond his earlier Presidential Lectures to the Linnean Society, he had made no further attempt to publicise the research findings. It was indeed only 15 years after his own retirement in 1950 that Mackintosh finally published a summary paper of what the Investigations had achieved (Mackintosh 1965; Roberts 1975; Marsden 2005, 171).

Even if those leading the Investigations had been more vigorous in publicising their achievement, it is doubtful whether the closure of the Investigations in the late 1930s could have been avoided. Colonial Office officials had always despaired of attracting Treasury monies. The Financial Crisis of 1931 had made Cabinet ministers even more determined to impose the strictest financial orthodoxy – solidifying their commitment to both low taxation and public spending. Even when the economy began to improve after 1932, there remained a reluctance to invest even in a rearmaments programme (Young 1997, 99). The Prime Minister, Neville

Chamberlain, warned that the increasing costs of defence could prove as fatal to national survival as an all-out enemy attack from another country. It would further sap the nation's economic vitality (McKercher 2013, 167). By such reckoning, and however judged by their value to the whaling industry or for 'showing the flag', the Discovery Investigations were of comparatively little significance.

The Investigations continued to be reduced. *William Scoresby* was laid up, and by the time *Discovery II* had completed its cruise in May 1939, the income to the Research and Development Fund had fallen to the point that the residue, £150,000, was needed for writing up what had been achieved. The ICES had intervened to the extent of nominating *Discovery II* as uniquely equipped to cooperate with a German vessel in making a survey of the Faroes-Shetland Channel of the North Sea, scheduled for 1940–41. Anthony Bevir was deputed in the Colonial Office to draft an application to the Treasury, in January 1939, for a grant of £20,000 from the Development Fund to enable this to take place. In doing so, he warned that 'we should be open to criticism if we did not make every effort that we can to keep the ship and crew in being'. Both the Permanent Under-Secretary of the Ministry of Agriculture and Fisheries and the Parliamentary Under-Secretary of the Colonial Office supported the application, which was soon lost in the turmoil of events leading up to the Second World War. Both royal research ships were immediately requisitioned for military use.[62]

CHAPTER 13

# *International regulation*

Malgosia Fitzmaurice writes, in her *Whaling and International Law*, that whaling was 'a complex, conflict-ridden and divisive field', exemplifying 'the close link between, and the dynamic interplay, of science and the law' (Fitzmaurice 2015, 1–3). North America had shown what diplomats could achieve, where the scientific evidence and public sentiment combined, most obviously in the case of species whose life cycles crossed national boundaries. The consequences were mixed. The Inland Fisheries Treaty of 1908 brought little improvement to fish stocks in the Great Lakes and Puget Sound. The Migratory Birds Treaty of 1916 continues to provide protection for waterfowl crossing the Canadian/United States border. The seal colonies of the Alaskan seaboard would almost certainly have been wiped out but for the North Pacific Fur Seal Convention of 1911. Pelagic sealing had already destroyed more than two-thirds of the population. Scientists had warned that both the colonies and industry would soon be lost. There was public revulsion at how the bodies of a large proportion of those shot in the water were never recovered, their pups being left to perish. The eventual Convention, as agreed by the Canadian, Japanese, Russian and United States governments, imposed an immediate seven-year moratorium to enable the colonies to recover (Dorsey 1998).

Whaling already had a sufficient history for writers to warn of its disappearance without some form of international regulation. Travis Jenkins (the Superintendent of the Lancashire and Western Sea Fisheries) warned, in a further edition of his *History of the Whale Fisheries*, that 'a most interesting group of marine animals' was being hunted to the verge of extinction and 'a great national asset rendered worthless to enrich a small group of speculators and capitalists'. Only those directly benefitting financially could possibly condone 'the present methods of unrestricted slaughter'. The ambition should rather be to secure 'a reasonably productive industry without risk to any particular species. Should the two clash, the latter must prevail (Jenkins 1932, 5–6). Keble Chatterton also warned of the whale becoming 'banished as the prehistoric animals are to us to-day'. The challenge was to take the 'protective measures [required] to control hunting' without crippling 'the industry as a dividend earner'. While the obvious course would be to ban whaling for as long as it took stocks to recover, no company could afford 'to let its plant and shipping remain in harbour idle and deteriorating'. What would have become of the skilled gunners and flensers? These were just 'some of the problems which would have to be faced sooner or later and we cannot possibly ignore them too long' (Chatterton 1925, 248).

Whaling historians have particularly focused on Sidney Harmer's Presidential Lectures to the Linnean Society of London, both of 1928 and 1930, calling for greater regulation of the whaling grounds. It had become more incumbent than ever that both producers and purchasers of whale oil should ensure that the industry was properly regulated. As Harmer calculated, in his earlier lecture, at least 30,000 whales had probably perished during the 1925–26 season, if account was taken of the many killings that went unrecorded. Not only were stocks already threatened, but an even higher catch was likely to be achieved as more up-to-date methods of capture, "flensing", and extraction of the oil were perfected. There was, in Harmer's words, a 'grave danger those species would escape extinction only with difficulty'. It was the purpose of the Discovery Investigations to establish whether 'this danger is a real one'. Should that prove to be the case, it was hoped 'whalers will unite with naturalists in seeking some method by which such an untoward result may be averted' (Anonymous 1929; Harmer 1928 and 1930).

Alistair Hardy recalled many years later that the Discovery scientists were particularly despondent at how powerless the Colonial Government was beyond the three-mile territorial limit. Not only were the companies now operating for the most part beyond that limit but unashamedly making as much money as possible while the going was good, on the assumption that they would eventually move to other grounds or leave the industry altogether for more remunerative investment. Only a small minority, which included the Discovery scientists, thought it folly to destroy what would otherwise have been a permanent source of wealth for 'a short period of dazzling prizes' (Hardy 1967, 442–444). Precedents of relevance might be found in the measures taken earlier to prevent the over-hunting of game animals. Their need had been highlighted by the Eighth International Zoological Conference at Graz in 1910 and by the International Congress for the Protection of Wild Life, held in Paris in 1923. The Swiss naturalist and ethnographer, Paul Sarasin, had famously denounced the 'short-sighted plundering' of the world's oceans, warning that

> *posterity will strongly condemn them and be not a little astonished that while we fill our museums at great expense and with every care with the fossilized remains of extinct mammals, we are so little concerned with the total disappearance of the no less interesting species existing to-day, whose existence is of far greater importance than the palaeontological relics. (Jenkins 1932, 47)*

The difficulties and disagreements had arisen when advocacy moved from the general to the specific, particularly in those cases where so little was known scientifically of the particular species (Dorsey 2013, 4 and 11).

The dilemma, as Charles W. Dixon (a Principal in the Dominions Office) wrote, was the knowledge that any move, on the part of an individual government, to impose stricter measures would encourage companies to register their vessels under another flag. While this might be prevented through international agreements, those participatory governments would expect a considerably closer involvement in the conduct of what had remained an essentially British/Norwegian-based industry. It had been hard enough for the two Governments to devise a common policy affecting their respective industries.(1)

Sidney Harmer, in a letter to the Colonial Office in July 1923, had pressed the urgency of international action, causing Darnley to protest the inexpediency of doing so. Harmer acknowledged it was 'probably undesirable to complicate the matter by involving other governments', but the prospect was already beginning to excite interest. The Argentine scholar and lawyer, Dr José Leŏn Suárez, had approached the British Legation in Buenos Aires, in July 1924, seeking information and advice as to how the Argentine Ministry of Agriculture and Fisheries might contribute to the protection of 'the marine life in the Southern Seas'. Suárez, citing the seal stocks of 'the North and Bering Seas', was convinced the question of whaling would have to be similarly addressed. However, Argentine-British relations were so bad (as a consequence of the Argentine claim to the Falkland Islands) that Henry Allen saw 'no reason why we should go out of our way to supply the Argentine Government with special information'.(2)

Adrian Howkins, in his environmental history of the Antarctic Peninsula, perceives a further, more deep-seated reason for fending off international regulation, writing that the industry had been built on a 'declensionist narrative', whereby 'a single responsible power was better for the conservation of the environment than an international free-for-all'. It was, as Howkins (2017, 25) writes, to demonstrate the British political claim to those waters that both scientists and officials had embarked on 'an ambitious programme of oceanographic research' that had become the Discovery Investigations. It might be remarked perhaps that the Norwegians, altogether free of the trappings of empire, had long investigated those waters. The Colonial Office resisted lobbying from the Union Government of South Africa, on the premise that international discussions would be premature, even for the African coast, until at least the preliminary findings of the Discovery Investigations became available. Suárez was similarly informed.

It was by chance that the League of Nations became involved. The League was never intended to play a significant part in wildlife protection (Boardman 1981, 30). Its Assembly, however, instructed the League Council, on 22 September 1924, 'to convene a Committee of Experts . . . to prepare a provisional list of the subjects of international law the regulation of which by international agreement would seem to be most desirable and realisable at the present moment'. From the responses received, a further seven questionnaires were distributed, one of which raised the possibility of international agreements affecting 'the exploitation of the products of the sea'. From the further replies received, member governments were found to be 'generally in favour of some action as regards whales'. The League's Economic Committee was accordingly instructed, in September 1927, to study in collaboration with the International Council at Copenhagen and any other organisations specially interested, 'what species and in what areas, international protection of marine fauna could be established'. From these discussions an Expert Committee was formed to consider whether 'the problem can be solved to any extent by a Convention' and second, to suggest the form it should take.(3)

The United Kingdom Government strongly opposed the move. George R. Warner (a Counsellor and Head of the Foreign Office's Treaty Department) thought the League was hardly equipped to tackle so complex a question. However deplorable the 'merciless and uneconomic slaughter', nothing could be done before the Discovery Investigations had published its preliminary findings. The Admiralty Hydrographer agreed. The Fisheries Department warned that the League's intervention might set a precedent for the fisheries of the North Sea and North Atlantic. The Foreign Office formally notified the Secretary-General of the League, in November 1926, that 'we consider that this is not a subject the regulation of which by general international agreement is feasible'. It was a technical rather than a legal matter. Foreign Office officials became nevertheless increasingly frustrated by so negative a stance. Sir Cecil Hurst (the Legal Adviser to the Foreign Office) had been a key figure in the founding of the League of Nations and was a member of the Expert Committee. He pointed out that there was far more to be gained from inter-governmental discussions than by 'putting difficulties' in the way of devising some scheme for 'the preservation of the whales in the Antarctic'. Why further risk the particular species becoming extinct? (Went 1972, 77).(4)

Henry Maurice played something of a dual role as the Fisheries Secretary and as President of the ICES; he was particularly well placed to alert the Foreign Office to how José Leŏn Suárez had continued to lobby both the League of Nations and individual governments, securing support, for example, of the French Academy of Sciences. Jean-Baptiste Charcot (the Antarctic explorer and a leading figure of the Academy) had pressed the French Colonial Department for support of a range of protective measures, which might include the establishment of whale sanctuaries, protection of juvenile whales and greater use of the whale carcase. ICES responded by forming a Whaling Committee. To the embarrassment of both Hjort and Maurice, the French representative circulated among members at its first meeting in Paris in April 1927, a draft convention *Projet de Convention réglementant la capture des cétacés hors des eaux territoriales*. Hjort ruled from the chair that it was 'a political proposal' and therefore outside the Council's terms of reference.(5)

Even so, there was a growing realization that pelagic whaling required some form of international regulation. Maurice conceded, in July 1927, that there was 'a good deal to be said for framing a Convention at once to prohibit pelagic whaling'. Not only had stocks already fallen to a level at which steps should be taken to prevent wasteful exploitation, but 'we have a lot to learn about the stock of whales and the stages of their life history at which protection is most desirable and can most effectively be given'. As Maurice and others quickly found, once discussion turned to the practical application of restrictive measures every government became protective of its 'vested interests or potential interests'. A meeting convened at the Dominions Office, in June 1928, with both Maurice and New Zealand's Minister of Marine present, saw little prospect of international action. As one official from the Colonial Office minuted, the whaling companies would transfer their fleets to 'the ports of non-adhering countries', some governments would cite their involvement as the pretext for entering the industry, and the Argentine Government would probably obstruct any 'engagement in which Great Britain might enter'.(6)

## 13.1 The International Whaling Convention

The League's Economic Committee had nominated Sir Sydney Chapman (Chief Economic Adviser to the United Kingdom Government and a former Permanent Under-Secretary of the Board of Trade) to become a member of a committee formed to appraise the whaling industry. For the Foreign Office, this was another reason for Sir Cecil Hurst to attend the interdepartmental 'whaling conference' convened by the Colonial Office in October 1927. Hurst had often spoken of how 'the Arctic and Antarctic fauna of the deep seas are constantly mentioned in the discussions at Geneva' as being particularly suitable for codification under international law. Sir William Tyrrell (the Permanent Under-Secretary) wrote that 'we shall be in Sir C. Hurst's debt if he will attend', not least for the French enthusiasm for the subject, despite having no whaling fleet.(7)

Darnley chaired the interdepartmental meeting and warned of the danger of the British and Norwegian industries being put 'at the mercy of those who do no whaling at all'. There was no 'adequate basis on which an international convention could be framed'. Kemp spoke nevertheless of the increasing rarity of the humpback and how the blue and fin whale might similarly be 'in danger of extinction'. For the industry to survive, the annual catch must not exceed the natural annual increase. The Chairman of the Fishery Board for Scotland recalled how hesitant the ICES was in deciding whether the threat was sufficient to warrant an international convention. Sir Cecil Hurst countered that there had been enough evidence to convince the majority of the League Assembly that action was needed. It was incumbent upon the United Kingdom Government to support the League's Economic Committee. The purpose of the interdepartmental meeting should be to identify 'whether and on what terms, for what species and in what areas, international protection of marine fauna should be established', and that ICES should do everything possible to assist and, by so doing, prevent 'a less-qualified League body' from being appointed for the purpose. The Dominions Office representative emphasised the need to include the Dominion Governments in any discussions – the Union Government of South Africa was already trying to obtain 'some measure of protection' along the entire African coastline.(8)

Darnley's interdepartmental meeting concluded that however premature the French draft resolution might have been, 'such action would probably be ultimately necessary'. The Board of Trade representative warned of the difficulty of obtaining any priority in the parliamentary timetable for the legislation to ratify any convention agreed on. There could well be parliamentary opposition to the enforcement of what might be internationally agreed to. The obvious response to any infraction was the withdrawal of the company's licence, but this might be too extreme for minor infringements of any internationally agreed-to regulations. Darnley's meeting went no further than to welcome the establishment of some kind of regulatory body, which drew upon the expert advice of ICES.(9)

Those in the Fisheries Department were hardly enthusiastic, with John Borley noting that the continued focus on the whale's extinction betrayed ignorance of the resilience of marine populations. Even the southern fur seal had begun to reappear in the Dependencies despite having been hunted almost to extinction. However important 'the humane aspect' and regrettable that any species should become extinct, Borley felt strongly that the League's Economic Committee should confine itself to the economic aspects, where paradoxically there was most comfort to be drawn because whaling was so expensive. The industry would have been long abandoned before most species could possibly be threatened with extinction. Maurice accordingly advised Sir Sydney Chapman that the Expert Committee should focus on 'rational exploitation', of 'getting the most you can out of the sea without injuring future prospects'. Marine conservation was so complex that without the background knowledge of the particular species, any move to formulate a global convention would end almost certainly in embarrassment. Henry Moss Blundell (Chief Inspector of Fisheries) was even blunter in writing to Richard Casey (the Australian government's liaison officer) that 'nobody really knows what regulations or restrictions' should be applied 'to conserve the stock of whales and yet allow the whaling industry to continue'. No one had the least idea how such regulations might be effected. Knowledge of the habits, migrations and 'everything else connected with whales is totally inadequate'. It would take years to discern what regulations were needed.(10)

ICES called on governments, at its annual meeting in 1928, to consider a range of measures. These included more comprehensive collection of data on the individual whaling grounds; a ban on hunting young, immature and female whales with calves; the designation of sanctuaries; and the complete protection of particular species (Jenkins 1932, 56–57). The Council's Whaling Committee went even further at its meeting in April 1929 by appealing for an immediate reduction in the scale of whaling pending further scientific investigations. The Committee strongly believed

> *the enormous expansion of the whaling industry in recent years constitutes a real menace to the maintenance of stocks of whales, and that if the expansion continues at the present rate there is a real risk of these stocks being so reduced as to cause serious detriment to the industry. (Jenkins 1932, 58–59)*

While acknowledging that there was insufficient knowledge to impose permanent measures, immediate steps should be taken to obtain the fullest use of the carcase and that the remuneration of gunners and crews should no longer rely on 'the number of whales killed'. There had to be a uniformity in the legislation enacted by the different whaling nations (Went 1972, 78). Travis Jenkins, in commenting on the various ICES recommendations, acknowledged that the costs of modern whaling were now so high that they already afforded a considerable degree of protection. Whaling was bound to end whenever and wherever the individual catches became too small to be profitable. This might preserve those species still relatively abundant, but Jenkins (1932, 59) was highly sceptical of whether it would provide sufficient protection for the humpback and more particularly, the right whale.

The Norwegian Government published a draft bill in late November 1928, broadly based, as the Norwegian Minister of Commerce acknowledged, on the licensing system long used by the Falkland Islands government. The Norwegian Storting unanimously approved the measure in June 1929, breaking all precedent by extending the kind of regulation previously confined to territorial waters to expeditions wherever they operated and therefore to the high seas. As Tønnessen and Johnson emphasise, its overriding purpose was to regulate, as opposed to curtail, further expansion of the industry. More particularly, the Act forbade taking right whales, females accompanied by a calf, and blue and fin whales of less than 60 and 50 feet, respectively. As well as requiring the extraction of oil from the entire blubber, head, flank, tongue and tail, the Act prohibited more whales from being killed than could be shortly processed. Bonuses were to be paid to gunners and crews according to how many whales were taken, their carcase size and the volume of oil extracted. There was provision to prohibit whaling within prescribed limits in both the tropical and subtropical waters and for increased 'telegraphic and telephone communication between the whalers and the floating factories or stations on land' when this made better use of the resource. There was provision for a 20 øre per barrel duty to be imposed on the oil producer as a contribution to both 'scientific researches of interest to Norwegian whaling' and to the costs of an inspector being appointed to every factory ship, the establishment of a Whaling Board to administer the Act, and for the appointment of an Institute of Whale Research and a Committee for Whaling Statistics. This last body, established in 1930 was to curate and publish annual summaries of the international whaling statistics, under the direction of Sigurd Risting (Tønnessen and Johnsen 1982, 364–346; Hart 2020, 367).[11]

The initiative might have passed to the Norwegians, but it was also highly significant, in regard of the British position, that Sir Cecil Hurst should have reacted so angrily in the Foreign Office to what he saw as the failure of the so-called experts to demand considerably more of their governments. Those on the ICES were so anxious to keep to 'the scientific aspects of whaling' that Hurst doubted whether they would support the initiatives required. Pelagic whaling was now so universal there must be international agreements. Time was of the essence in acting both before there was further expansion and before more draft conventions were drawn up of the French kind, which 'might not be at all to our liking'. Now the chairman of the League's Whaling Sub-Committee, Chapman, saw the Norwegian Act as the obvious model for an international convention. A drafting committee should be appointed straightway. Unless the United Kingdom Government came forward with proposals, there was serious risk of 'our being rushed later into' something far less suitable.(12)

An interdepartmental meeting, convened by the Colonial Office on May 1929 acknowledged that there was no practical alternative to some form of convention. An obvious difficulty for the Home Office was that any contravention on the high seas was obviously outside the ordinary jurisdiction of the courts; nevertheless, it would have to be tried within the United Kingdom. Sir Claud Schuster (the Permanent Secretary to the Lord Chancellor and Clerk to the Crown in Chancery) sympathised but contended such objection must be overidden as the only way of meeting the policy objective. There was precedent to be found in the North Pacific Fur Seals Convention of 1911.(13)

Chapman (now the President of the League's Economic Committee) chaired the Committee of Experts charged with drawing up the whaling convention. Other members included Maurice (as president of ICES), Darnley, and Borley representing the United Kingdom and Dominion governments, respectively. Acutely aware that much stronger measures were needed, those attending the Expert Committee's meeting in Berlin, in April 1930, also realised that only the most minimal of standards stood any chance of adoption. As Howard Smith minuted within the Foreign Office on 22 July 1931, however deficient the draft convention might be, priority had to be given to preventing companies from registering their vessels with non-subscribing governments. Lord Bledisloe wrote in a similar vein as the Governor-General of the New Zealand Government. The hope was, as the Dominions Office expressed it, that by laying down an elementary standard of conduct, a more effective system might emerge.(14)

Lord Cecil formally signed the Convention, as the head of the United Kingdom Delegation at Geneva on 24 September 1931, as the eighth of 26 delegate nations to do so. The Dominion governments, France and Norway did so five days later. The Convention followed the general lines of the Norwegian Act by banning the taking and killing of calves, suckling whales and females accompanied by calves or suckling whales. The High Contracting Parties (the respective governments) were charged with obtaining from those vessels flying their flag the most complete biological information practicable of the date, place of capture, species, sex, length and, where present and practicable, foetus. Every factory, whether afloat or on land, was required to provide returns of the number of whales of each species treated, the amounts of oil of each grade obtained, and the quantities of meal, guano and other products derived from them. Such statistical data were to be forwarded by the respective governments at least once a year to the Bureau for Whaling Statistics at Oslo, identifying the name and tonnage of each floating factory, the number and aggregate tonnage of its catchers, details of any land stations that had operated and the respective number of whales processed. Governments agreed to take appropriate measures to ensure the application of every provision and to punish infractions. The remuneration of gunners and crews would reflect for the most part the size and species of whales killed and the quality and quantity of the oil obtained. No vessel should engage in whaling without a licence granted by the appropriate government (Parliamentary Papers 1933–1934; Fitzmaurice 2015, 16–17).(15)

It had taken over a decade for the notion of an international convention to be realised, during which time there had been both striking advances in whaling technology and a failure to manage output according to demand. However obvious the shortcomings of the Geneva Convention, the principle of the international regulation of common property resources had been established for the high seas. A minimal standard had been set for any aspiring whaling nation to meet and for the established whaling nations optimally to excel (Brown 1963, 559–560; Gambell 1993, 98).

## 13.2 The Whaling Industry (Regulation) Act

The prospect for whale stocks had become bleaker than ever, as Sir Sidney Harmer had warned in his second presidential address to the Linnean Society, two years after his first in 1928. The global production of whale oil had risen from 780,000 barrels in 1925–26 to 1,600,000 barrels in 1928–29 and now exceeded 2,500,000 barrels, as a direct consequence of pelagic whaling (Harmer 1928, 1930). Stanley Kemp warned similarly that any postponement of effective controls for even a single season would enable vested interests to become even more entrenched.(16)

The International Whaling Convention was to become effective once eight of the signatory powers (including Norway and the United Kingdom) had ratified it. Norway had done so on 18 June 1932 and the United States on 7 July, but, as George E. Boyd Shannon (an Assistant Principal) wrote despairingly in the Dominions Office on 16 March 1933, the United Kingdom Government had still gone no further than to publish the draft Bill required for ratifying the Convention. Whale oil now constituted one-third of the fats used in the manufacture of margarine and cooking fats, but whaling counted for little in the economic and political turmoil of the early 1930s (Schwitzer 1956, 125). A further complication was the need for any United Kingdom legislation to be aligned with what was already in place for the dominions and colonies (Dorsey 2013, 54). The Polar Committee had appointed a sub-committee for the purpose in August 1931, but not until January 1934 did Darnley (as its chairman) sign its report, remarking in a covering letter that there had been numerous 'constitutional complexities' to overcome.(17)

No government department wanted to sponsor the ratifying bill. Other departments had assumed it would be the Colonial Office, but 'the Falklands interest in the whaling industry' was diminishing so rapidly only one whaling licence had been issued that season. Darnley's sub-committee proposed that the Board of Trade should act as the 'umbrella' department, with both the Ministry of Agriculture and Fisheries and the Scottish Office acting as 'the UK licensing authority'. Worries of whether there would be parliamentary opposition to the Bill were reinforced by a meeting of Sir Harry Batterbee (the Chairman of the Polar Committee) with Norman C. Watt (the managing director of the Southern Whaling Company). Watt, having just returned from Antarctic waters, ridiculed what he described as the scientists' concern for whale stocks. A good harpooner was still able to shoot as many whales in a week as 10 years previously. They all came from the same population moving around the Antarctic, seeking the best feeding grounds along the edge of the ice barrier. In Watt's view, the government's priority should be to use the money presently spent on the Discovery Investigations for helping to develop new markets for the sale of whale oil. Borley pointed to the futility of doing so without assurance of a continued supply of the oil.(18)

Other governments became impatient that the United Kingdom should be so slow in ratifying the Convention. The United States Ambassador enquired 'quite informally' when the

United Kingdom would ratify. The Norwegian Minister in London warned in March 1934 of 'the dissatisfaction amongst the Norwegian whaling people', stating that his Government might soon be forced to withdraw from the Convention, however regrettable the consequences. The Foreign Office, in circulating the letter among departments, emphasised the 'pressing importance' of being able to assure the Norwegian Minister that the necessary legislation would soon be in place. Henry Maurice was particularly embarrassed. As Travis Jenkins had written in the mid-1920s, Maurice's appointment as Fisheries Secretary in 1919 had brought immense improvement to a department that had previously been understaffed and out-of-touch with the fishing trade (Jenkins 1927). But as Maurice explained in a volume on the Ministry of Agriculture and Fisheries, the Minister's jurisdiction was confined to territorial waters even though deep-sea fishing now represented more than 90 per cent of the total British catch. Deep-sea fishing could only be regulated by Acts of Parliament intended to ratify and embody international conventions into law (Maurice 1927, 273–280).[19]

Maurice acknowledged, 'We are certainly in an awkward position', and despite his best efforts, he had failed 'to foist the legislation upon the Dominions Office'. The Minister, Major Walter E. Elliott, had accordingly agreed with the Scottish Secretary of State, they would jointly sponsor the necessary bill and take responsibility for its administration. The Norwegian Minister in London was assured that every effort was being made to expedite its enactment, but space had to be found in the legislative programme. More particularly, Maurice had come to recognise that the most obvious way of conserving stocks was to follow the Norwegian example of shortening the whaling season. The draft Bill was accordingly extended to enable the United Kingdom Government to adjust the length of the whaling season.[20]

Maurice and Borley drafted the accompanying Cabinet memorandum, explaining that the Mercantile Marine Department of the Board of Trade should be made nominally responsible for the measure, but the Fisheries Department and the Fisheries Board for Scotland should grant the necessary licences, appoint a whaling inspectorate and prosecute offences. Both the Colonial and Dominions Offices considerably extended the draft Bill to cover the respective colonies and dominions, causing J. J. Paskin (a Principal in the Colonial Office) to warn that as many as six ministers might have to sign the eventual Cabinet memorandum. The number was eventually whittled down to Walter Elliott and the Secretary of State for Scotland, Godfrey Collins. They signed on 21 May 1934, with the concurrence of the other interested departments. Maurice had meanwhile succeeded in having the Bill first introduced to the House of Lords, where at least it could be 'aired' by the end of the parliamentary session. This might help to allay 'the anxieties of the Norwegians and keep them quiet'. Johan Mowinckel (the Norwegian Prime Minister) had warned of 'real indignation' in his country. Hjort feared his reputation would be 'smashed' and the Convention abandoned altogether.[21]

The Cabinet Memorandum stated that Norway and 15 other governments had already ratified the Convention. The Foreign Office had been compelled to say semi-officially the Bill would be enacted that summer. The Cabinet finally considered it on 6 June 1934, with the Prime Minister Ramsay MacDonald in the Chair. Elliot emphasised its urgency in the face of mounting Norwegian anger and the threat that 'they might not continue to observe the Convention unless we acted'. Elliot thought the case for passing the Bill was 'unanswerable'. Sir John Simon (the Foreign Secretary) supported him. The reason for the Bill's length and complexity was the need for the British courts to have the powers to impose the necessary penalties on actions taken beyond British territorial waters. The Cabinet approved the Bill in principle and followed the normal course of referring it to the Home Affairs Committee for closer scrutiny on the presumption that parliamentary time would be found for its passage.(22)

On 13 June 1934 the Cabinet approved the recommendation of the Home Affairs Committee that the Whaling Industry (Regulation) Bill should be introduced forthwith to the House of Lords. MacDonald spoke of how

> *the passing of this Bill was necessary as a matter of international honour, and he would feel bound to make a serious effort to secure it, notwithstanding the difficulties of the Parliamentary position.*(23)

Lord de la Warr (the Parliamentary Secretary to the Ministry of Agriculture and Fisheries) moved its Second Reading on 21 June 1934, briefly recounting that the Convention could not be implemented until the United Kingdom had ratified it. The Norwegian Government was particularly critical, pointing out that the United Kingdom Government had been 'largely responsible for the initiation of the Convention'. The Bill was passed by both Houses without debate and received the Royal Assent on 31 July 1934 (PD, Lords, 93, 117–123; Whaling Industry (Regulation) Act, 24 & 25 Geo. 5, c.49).(24)

The Bill followed the Norwegian legislative example by including the additional power of deciding the length of the whaling season. The more Maurice had come to study the Norwegian precedent of a statutory whaling season, the more he saw it to be the key to tackling 'the whaling problem'. While recognising the risk of companies transferring to another flag, he doubted whether this would happen on any large scale. The Office of the Parliamentary Council had forwarded a fresh draft of the Bill on 4 June 1934, making it possible for the licensing authority to impose a closed season.(25)

Eighteen governments had ratified the Convention by the time the Secretary-General to the League of Nations was able to announce the deposit of the United Kingdom's instrument of ratification on 31 October 1934. The Convention accordingly came into force on 16 January 1935. Twenty-six governments had eventually signed it, despite some nine-tenths of the industry continuing to be in Norwegian and British hands (Antonello 2018, 305). Despite fears of the practice becoming commonplace, only Salvesen transferred the registry of their factory ship *Salvestria* to another flag: the Eire government in Dublin, and it was for the 1935–36 season only.(26)

CHAPTER 14

# *The London Conferences*

The long-term significance of the International Whaling Convention, the later London International Whaling Agreement of 1937 and Protocol of 1938 should be measured not by their effectiveness, which was minimal, but for the legal framework they established for regulating the industry (Fitzmaurice 2015, 27). The Convention brought no sense of finality. The ICES went no further than to acknowledge that 'some excellent work' had been done to limit catches, but further statutory regulation and 'biological and statistical investigations' were required. Malcolm MacDonald appealed, as the Secretary of State for the Colonies, in May 1936, for further measures to prevent further 'undue depletion' of stocks (Mackintosh 1965, 151–152; Went 1972, 87).[(1)]

The total catch in the Antarctic had risen from 406,056 tons in the 1935–36 season to 443,018 tons in 1936–37 and 556,722 tons in 1937–38, but fell to 470,129 tons in 1938–1939. The number of catchers had meanwhile risen from 118 in 1932–33 to 153 in 1934–35, to 256 in 1937–38 and 281 in 1938–39. The result, as Tønnessen and Johnsen (1982, 457) write, was 'a fearful slaughter' of fin whales. Twice as many (28,000) were killed compared with the previous season. Fears that they too were becoming over-exploited seemed confirmed in 1938–39, when only 7,200 were caught even though there were three more factory ships and 25 more catchers. The even larger fleet of 34 factory ships and 270 catchers had caught an aggregate 38,356 whales. Whereas the annual catches between 1932–33 and 1936–37 had been held to an average of about 31,000 whales, as many as 146,039 were recorded for 1937–38 (Hart 2020, 192–193).

In the meantime, the market for whale oil had collapsed. The price paid for whale oil at the start of 1937 had been £22 10 shillings per ton; it had fallen to £17 per ton by October, reflecting both the unprecedented volume of whale oil and oleaginous seed on the market. There was also 'a very substantial shift of emphasis away from Norway', which had been so badly affected by the consequences of overproduction and collapse in oil prices in the earlier 1930s. Salvesens had survived by drawing on their capital reserves and diverse business interests. *Saragossa* (one of the last floating factories and lost to fire in 1932) was quickly replaced by *New Sevilla*. Not only was the new vessel pelagic and twice the size of *Saragossa*, but the effect of its purchase was to eliminate the previous owner and rival – the Norwegian Sevilla Company. Another rival company, Strombus, was similarly eliminated by the purchase of its sole factory ship *Strombus*. For Harold Salvesen, it was 'a policy not of greed but of economic common sense' (Jackson 2005, 204–07).

The United Kingdom had become the leading whaling nation in oil production by the 1937–38 season, but the proportionate size of the British/Norwegian catch had fallen by about

40 per cent compared to other whaling fleets (Fitzmaurice 2015, 26). The composition of the British fleet had become strikingly different. Whereas Salvesen and the Southern Whaling Company had previously owned every factory ship, they now held only one-third of the total. The Hector Whaling Company's *Hektoria*, which first operated in 1929–30, achieved a catch of 13,917 tons, rising to 20,350 tons the next season, when it was joined by the Kerguelen Whaling Company's *Tafelberg*, obtaining 18,667 tons and 20,833 tons in 1932–33. They were joined by the Star Whaling Company's *Svend Foyn* in 1934–35, which achieved 30,725 tons but dropped to 21,717 tons the next season and 15,783 tons in 1936–37, in which season *Hektoria* had obtained 11,767 tons and *Tafelberg* 13,730 tons. Two further British-registered vessels had appeared that season: United Whalers' *Terje Viken* and the Union Whaling Company's *Uniwelco*, securing 17,000 tons and 12,358 tons, respectively. The same Bremen shipyard had built both to the same plans. Where Unitas Deutsche Walfang Gesellschaft operated *Unitas* and was therefore registered in Germany, *Terje Viken* became the largest factory ship ever to be flagged and operated from Britain (Jackson 2005, 206 and 248–249).

The rebalancing of British and Norwegian interests had become an integral part of the strategy required by individual companies to survive the consequences of over-capacity in market demand. The 1937–38 season had closed with the largest surplus of oil since 1930–31. Rupert Trouton lost £53,000 on the *Terje Viken*, and a further £28,000 on *Hektoria*, and had begun to negotiate their sale to Germany or Japan when war broke out. The Anders Jahr group had moved entirely into the tanker trade – both *Kosmos I and II* were disposed of as opportunity afforded (Jackson 2005, 203–04) The South African Trade Commissioner in London, Francois J. Du Toit, complained that companies operating under the flag of 'non-Agreement countries' – most obviously Japan – were still permitted to sell their oil to signatories of the International Whaling Convention. Henry Maurice raised the possibility with Clement Davies (the legal director of Unilever and Liberal Member of Parliament) of using Unilever's market dominance to threaten a boycott of Japanese whale oil. Davies dismissed the idea for its likely ineffectiveness. Sir Robert Craigie (the new Ambassador in Tokyo) reported, in late January 1938, the endeavours of both the German and Norwegian ambassadors to obtain some degree of constraint. So too did Mitsubishi Shoji Kaisha, the handling agent for Japanese oil sales in Europe. As the sales agent, David Geddes, reported, sales had been so disappointing there was a possibility that the factory ships under construction in Japan would be diverted to the mineral oil trade. The hope was that the Japanese had become so chastened as to be more amenable to restriction of whaling.(2)

'The only certainly effective method' of regulation was for all countries to limit the number of their expeditions. However, the German and Japanese Governments were certain to want what they saw as a fair share of the catch, which could come, if stocks were to be preserved, only from a reduction in British and Norwegian catches. There would be endless arguments with little prospect of a solution. Further inter-company agreements might conceivably work, but only if companies were convinced that they were their only chance of survival. It would entail some 'very hard bargaining'. Henry Maurice warned that governments could go no further than to provide a general policy within which the companies must themselves determine their future. As Kirkpatrick Dorsey writes, it was within this market uncertainty that modern whaling diplomacy was born as both companies and governments strove to hold firm on some issues and compromise on others. Not only did they differ on where 'the balancing point' between hunting and conserving might be, but the Japanese largely ignored what little was negotiated (Dorsey 2013, 50–51).(3)

## 14.1 The International Whaling Agreement 1937

Given Norway's long-standing dominance of the industry and its prominence generally in marine fisheries, Oslo was the obvious venue for an international conference, but, as Vera Schwach writes, a new Labour Government had been elected. The Minister for Foreign Affairs, Halvdan Koht, was keen to impose a more assertive national fisheries policy against a background of increasing militancy by the trade unions (Schwach 2002, 43–44). The idea of an Oslo conference quickly unravelled. The United Kingdom assumed that London would be more convenient for those from other continents. The German Government rejected Berlin, anxious that the first German expeditions should have returned before there was any further discussion of regulation. Laurence Collier suggested that the Norwegian and United Kingdom delegates might nevertheless first meet in Oslo to agree on general principles and an agenda for the London Conference. Both Maurice and Colban warned that the other governments would assume the purpose of the London Conference was simply to ratify what 'Norway and the constituent members of the British Empire' had already decided. Invitees must have the chance of attending both meetings (Tønnessen and Johnsen 1982, 448–449).[4]

Sir Ronald Lindsay (the British Ambassador) informed Cordell Hull (the United States Secretary of State), in March 1937, that the meeting's purpose was

> *to bring about an agreement between all the countries interested in whaling on the measures to be taken beyond the restrictions included in the Whaling Convention of 1931, in order to prevent excessive and wasteful exploitation of whale fisheries in the Antarctic in the season 1937–1938.* (FRUS *1937, 1/22 (920–1))*

As there was 'at least one factory ship and a number of whale catchers' operating under the United States' flag, Lindsay hoped the United States Government would participate. Herschel V. Johnson (the First Secretary at the United States Embassy in London), Dr A. Remington Kellogg (of the Smithsonian Institution) and a member of the United States Coast Guard were nominated to attend. Graham Burnett has drawn a valuable comparison between how whale conservation developed in both Britain and Norway, principally through the impetus of fisheries biology and the ocean sciences, and the position in the United States. Arthur Remington Kellogg had, in Burnett's words, 'found his way to whales' through 'a general interest in zoology and vertebrate palaeontology'. He had played a major part in founding the Council for the Conservation of Whales – the first association of its kind anywhere in the world devoted entirely to 'saving the whale'. For both Kellogg and Alfred Brazier Howell (the Secretary of the American Society of Mammalogists), science provided the rationale rather than simply the means whereby individual cetacean species might be saved. The Council, in raising general awareness of the impact of modern whaling on whale stocks, had already succeeded in having Kellogg appointed as the United States' representative in the various discussions leading up to the International Whaling Convention (Whitmore 1975; Burnett 2012).

William S. Morrison (the Minister of Agriculture and Fisheries) gave the opening speech at the London Conference, eventually held in May 1937. In it, he interpreted the large number of delegates present as signifying wide recognition that 'the future of the great whaling industry is in jeopardy'. In recalling how quickly the northern whaling industry had been reduced 'to a few small stations', the Falkland Islands Government and 'we in this country' had endeavoured to prevent such an occurrence in Antarctic waters. The number of shore stations and catchers had been restricted and encouragement given to 'research into the life history of whales'. Before any of this could get under way, pelagic whaling had enabled the industry to expand far beyond the Dependencies – indeed, to wherever whales might be found. Far greater stocks had been discovered than could ever have been imagined. Factory ships were now being built in ever-increasing numbers. The whale catchers had become faster and more powerful than ever. 'A vast destruction of whales' was taking place. The Discovery Investigations and the Norwegian statistical data 'collected from the factory ships' warned that at the present rates of destruction, there would soon be no 'profitable commercial venture'. The 'magnificent Blue Whale' might well be exterminated. The International Whaling Convention had brought 'a very modest degree of protection'. Although the whaling companies had entered voluntary agreements, the rise in oil prices and access to 'increasingly powerful and efficient factories and catchers' meant a critical juncture had been reached. The purpose of the conference was to agree to the further measures of protection required to prevent the endeavours of some countries from being defeated by the enterprise of others. Every government had to work toward a common purpose if the industry was to have a future. Morrison concluded his address by recognising that 'the path to conservation is beset by many difficulties', but an agreement founded on reasonableness and practicality might encourage those 'who are not with us to-day' joining 'us in the near future'.(5)

Herschel Johnson found British officials 'sceptical of achieving entirely satisfactory results'. Their anxieties had increased in the absence of any indication that the German and Japanese governments would attend. The Foreign Office telegraphed Sir Robert Clive (the British Ambassador in Tokyo) to stress the urgency of a Japanese response. Clive's report, some ten days later, that a further factory ship was joining the two already operating, implied that the Japanese were intending to pursue southern whaling on a large scale. A semi-official statement, of mid-April 1937, implied that the Japanese government would neither join the International Whaling Convention nor attend any conference at which the Norwegian and United Kingdom Governments would 'secure their own dominant position'. Japan must first attain its own competitive position. Clive cautioned against applying further pressure. The German Ambassador had also been instructed to intervene – he was likely to make a far greater impression (*FRUS* 1937, 1/22 (922–923)).(6)

Maurice's other priority had been to secure German attendance – even to the point of implying that the purpose was to secure 'a gentleman's agreement' rather than extend what the League of Nations had begun; the German Government was never likely to agree to anything that politically 'smacks so much of Geneva'. In confirming Germany's attendance, a spokesman noted that the initial expeditions would hardly affect stocks. There would be a biologically trained observer aboard each of its factory ships. Maurice later described the atmosphere at the conference as excellent. The German delegate, Helmuth C. H. Wohlthat, through his outstanding ability, practical common sense and humour, had been particularly helpful.(7)

The conference itself ran from 24 May to 8 June 1937, with delegates from Argentina, Australia, Eire, Germany, New Zealand, Norway, South Africa, United Kingdom and United States. Canada and Portugal had sent observers. Although the conference was scheduled to last a week, the complexities of the subject matter and the need for delegations to consult their governments, meant it lasted more than a fortnight. The United States delegation was under instruction to demand an increase in the minimal length for taking blue whales to 75 feet, a restriction in the number of killer boats and what could be processed into commercial products within 24 hours. In anticipating these and other changes to the Geneva Convention, the Acting Secretary of State, Sumner Welles, indicated that every amendment would require Congressional approval for its ratification by the President (*FRUS* 1937, 1/22, 923–924).(8)

The International Agreement for the Regulation of Whaling was signed in London on 8 June 1937. It again forbade the taking or killing of calves or suckling whales, immature whales and females accompanied by calves or suckling whales, but it also prohibited taking or killing any blue whale of less than 70 feet; fin whales of less than 55 feet, and humpbacks or sperm whales of under 35 feet. Using a factory ship or catcher for taking or treating whales in any waters to the south of 40° South latitude was forbidden, except between 8 December and 7 March, or from any land station for more than six months in any continuous 12-month period, thereby making universal the restrictions that had previously applied only to the South Georgia stations. The fullest use should be made of every carcase. Unless the carcase was intended for human or animal consumption, its oil must be extracted from every part of the carcase except for the internal organs, whale bone and flippers.(9)

Every factory ship or land station was required to submit to the government, under whose flag it was operating, details of the whales of each species caught, the amounts of oil produced of each grade and the quantities of meal, guano and other products obtained. Particulars were to be provided of the date and place at which each whale was taken as well as its species, sex, length and the presence of a foetus. Every government was required to take sufficient measures to apply the Agreement and punish infractions. To that end, at least one inspector was to be appointed to each factory ship. The remuneration of gunners and crews would be largely determined by such factors as the species, size and yield of whales taken rather than simply the number of whales caught. Pelagic whaling would be banned throughout the year to the north of 40° south latitude in the Atlantic and in the Davis Strait, Baffin Bay and the Greenland Sea; in the Indian Ocean north of 40° south latitude, and in the Pacific east of 150° West longitude between 40° degrees south latitude and 35° north latitude, and additionally west of 150° West longitude between 40° south latitude and 20° north latitude (Fitzmaurice 2015, 20–23).(10)

The Final Act of the Conference (the Appendix to the Agreement) recommended that further studies be made of the optimal number of catchers assigned to factory ships and land stations and of whether whaling should be entirely banned from specific whaling grounds. Governments were called on to prevent vessels being transferred to other flags by prohibiting their nationals from working on them. Measures should also be taken to reduce wastage. Often whales were fatally injured, but their carcases were lost in the sea. There should be further experiments with explosive charges and more particularly electrically charged harpoons, which might incidentally also reduce the undoubted cruelty involved with present methods. The German delegation reported success with an electric harpoon designed to kill the whale within seconds (Borley 1938, 63).(11)

A further conference was expected in a year's time. Laurence Collier thought the Norwegians had done 'really quite well' – Erik Colban had acknowledged they could hardly have done better. Reaction in Norway was more mixed; Norwegian interests had benefitted only where it had suited the British. The Oslo newspapers *Tidens Tegn* and *Morgenbladet* thought the Conference was an important beginning, particularly as Germany had participated. Both the Fisheries Department and British companies remained concerned lest there be further militancy from the Norwegian trade unions. Despite protest from the Norwegian delegation, Maurice obtained a clause relieving governments of any obligations where they were threatened by industrial action. Halvdan Koht expressed surprise and pain when both the British and German Ministers in Oslo had handed him identical notes threatening not to ratify unless the Norwegian delegation agreed to the clause.[(12)]

Maurice readily acknowledged that 'the weak spot' had been the absence of Japan, but as Aubrey S. Halford (a Third Secretary in the Foreign Office) minuted, its delegation would have faced 'a great deal of friction', particularly at reports of its expeditions continuing to kill undersized whales, suckling calves and even the right whale, while other national fleets abstained. Maurice was sure the Japanese Government would have heeded the 'veiled threat' he had deliberately inserted into the Final Act, warning of the possibility of the entire industry collapsing. The whole of Japan's 'big programme of building factory ships and catchers' would be wasted if the other whaling nations did 'their worst without restriction'. Maurice made no apology for including the threat. A more particular embarrassment for the Dominions Office had been the reluctance of the Union Government of South Africa to sign the conference documentation. Its shore stations were likely to be hard hit by the imposition of size limits on the whales taken (*FRUS* 1937, 925).[(13)]

The ICES saw the London Agreement as 'a great step forward', but whaling historians have generally dismissed it as 'a stop-gap measure' (Dorsey 2013, 68). No one doubted that it was only a first attempt to secure more effective protection under the League of Nations' International Whaling Convention. Encouragement was drawn from another set of London Conferences, those held in November 1936 and March 1937 to agree to a fisheries convention. As President of ICES, Maurice proudly declared that the Convention had 'set the seal of administrative achievement' on all the scientific work directed by the International Council (Went 1972, 90) The same accolade could not possibly have been extended to the London Whaling Agreement of 1937.

## 14.2 The International Whaling Protocol 1938

In May 1938, Ministry officials rehearsed before the whaling companies many of the issues a further London Conference might have to address. Maurice cited the early findings from the Discovery Investigations in pressing for a ban on catching humpbacks. Mackintosh warned of the increasing evidence of a decline in every species. There was unanimous support for the proposal, made by the representative of the Southern Whaling Company, for a minimal length of 65 feet for catching blue whales. Maurice acknowledged that any further shortening of the whaling season would render whaling uneconomic. Rupert Trouton thought there was now so great a range in vessel sizes and horsepower that it would be iniquitous for any limit to be set on the ratio of catchers to a factory ship or indeed tankers to an expedition. There was similarly no chance of the Germans and Japanese agreeing to any reduction in the number of expeditions per country. The meeting acknowledged the cruelty of existing catching methods. Harold Salvesen described trials with an 'electric harpoon' which, despite costing £3,000, had proved disappointing. Maurice acknowledged that it was far too soon to consider the compulsory use of electric harpoons. Turning to the manning of the individual expeditions, Maurice assured the companies, that although it could not be stated officially, they had the Ministry's sympathy for

wanted both governments to lodge a formal protest. Collier thought (correctly as it turned out) that the anticipated German protest would carry much greater weight. As Kodaki confirmed with Thomson, in February 1939, the German minister's intervention had meant no further humpbacks would be caught. There were expected to be six Japanese expeditions in the coming season. Whereas output had risen from 26,000 tons in 1936–37 to 65,000 tons the next, the target, according to newspaper accounts, was now 120,000 tons. The greater military activity in China had increased demand for whale meat and indeed for all whale products generally in Japan's drive for economic self-sufficiency.(27)

There was large-scale disruption in the European markets. Dormer reported from Oslo in May 1939 that negotiations between the Norwegian companies and German purchasing department had broken down. Norway had intended to sell almost the entire Norwegian catch (some 78,000 tons of whale oil) to the German market, but the German negotiators had offered the Norwegian sales pool £14 15 shillings per ton. Tønnessen and Johnsen (1982, 467–468) recounted that the companies wanted to sell the oil to the United Kingdom Government for cash rather than Reichsmarks but were afraid of German reprisals. With the Norwegian Government still hesitant to intervene, British buyers succeeded in buying the entire Norwegian catch for £15 10 shillings per ton shortly before a German offer was received of £16. The Norwegian Ministry for Foreign Affairs was reportedly highly embarrassed – the Germans had clearly wanted the Norwegian oil.(28)

Not only had the London Conferences failed to achieve any measurable improvement in 'stock conservation', but Dobson was afraid that any further meetings might jeopardise what little had been accomplished. Tønnessen and Johnsen (1982, 469) claim that it was German pressure that caused it to be held. The smaller German catch that season was attributed to 'declining stocks' caused by 'the ruthless catching undertaken by the Japanese, starting as they did four or five weeks before the others and ignoring the ban on catching humpbacks or minimum size limits'. It remained far from clear whether Japan would attend a further conference. Collier found Kodaki ambivalent, speaking of 'people in Tokyo who would be only too glad of an excuse to back out if they could'. Dobson was even more alarmed to learn from 'a most secret but entirely reliable source' that the Japanese authorities had come to realise their 'whaling technique' was so primitive that any move to adhere to the regulations of 'the well-established industries' would be 'the death-blow' to their industry.(29)

Alongside Kellogg and Wohlthat, Birger Bergersen had emerged as a key figure in the conference discussions. Bergersen held a doctorate in seal morphology and currently the post of professor of anatomy at the Oslo School of Dentistry. He had succeeded Johan Hjort as the Chairman of the Norwegian Whaling Council in 1936 (Walløe 2009). Robert S. Hudson (the Secretary of State for Overseas Trade) had met him during a visit to Norway and had found him so very pro-German as to suggest that a small demonstration should be laid on for him to show the strength of the British armed forces. The Secretary of State for Air went a step further by indicating that an air display should be laid on for all the delegations attending what was, by now, scheduled to be a further London Conference. Frank K. Roberts (a Second Secretary in the Central Department of the Foreign Office) acknowledged that 'the Germans and Japs . . . might with advantage be impressed', but they could also complain of its irrelevance to the purpose of the intended conference. Hudson persisted, and Ivone A. Kirkpatrick (the First Secretary of the Foreign Office's Central Department) temporised by writing that although 'whaling experts' were 'perhaps not ideal subjects for air propaganda', some good might come of it. Sir Orme Sargent (an Assistant Under-Secretary of State) thought it 'rather a waste of time and energy'. It was William Morrison (the Minister of Agriculture and Fisheries) who prevented it happening lest it should disturb 'the harmony of the conference'. The Parliamentary Under-Secretary for Foreign Affairs, Richard A. (RAB) Butler ruled that Morrison's view must prevail.(30)

There were far more significant machinations in the use made of Helmuth Wohlthat's presence in London as the leader of the German delegation. The Prime Minister, Neville Chamberlain, used it as another opportunity to probe the German Government's mind on the future of Poland, and more specifically Hitler's demand for the restoration of the Danzig Corridor. Wohlthat's meetings with Chamberlain's emissary, Sir Horace Wilson, took place (with both Hudson and the German Ambassador present) between the more formal sessions of the further whaling conference (Watt 1989, 399–401; Haslam 2012, 315).

Tønnessen and Johnsen perceive the International Whaling Agreement of 1937 and Protocol of 1938 as the work of diplomats and marine biologists. Although the Norwegian Storting had enacted a further measure in June 1939 to bring previous legislation into line with both the International Whaling Agreement of 1937 and the Protocol of 1938, it remained far from clear which countries 'had ratified what, and what at the moment was a binding agreement'. The first item on the agenda of what had become a third London Conference of July 1939 was one of clarifying more exactly what had been accomplished, as a first step, in determining what further action was needed to save both the industry and the stocks on which it depended (Tønnessen and Johnsen 1982, 462 and 469). President Franklin D. Roosevelt had formally initialled the instructions given to the United States Delegation to avoid any further promulgations which required his formal ratification and therefore reference to the Senate. The focus must be on making more effective what had already been agreed. Far from it being an academic exercise, the 'commercial types of whales' would, at present catch rates, soon become extinct. Unless more effective ways were devised for enforcing what had already been agreed to, delegates should be warned that the United States Government might have to reconsider the value of its further participation. The priority must be 'the more effective conservation of whales'. Attention should again be drawn to the International Fisheries Commission, which both the Canadian and United States Governments had agreed in January 1937 for the purpose of preserving the halibut fishery of the North Pacific and Bering Sea. It was a model of how sanctuaries could be established where whales were known to breed, and a mechanism by which the maximum catch per year was calculated. The present International Bureau of Whaling Statistics might similarly act as the body responsible for keeping the world's whaling fleets under surveillance and for closing the grounds once the predetermined limits to the season's catch had been reached (*FRUS* 1939, 2, 38–42).

The Whaling Conference, or more correctly the informal meeting, was held between 17 and 20 July 1939 ostensibly to make closer provision for the 1938–40 whaling season. It was obvious from the data brought by the Norwegian delegation that fewer whales had been caught (except for sperm whale) and less whale oil produced, despite the greater number and tonnage of the factory ships and catchers employed. Although the weather had shortened the season by eight days, the conference acknowledged the smaller catch must be the consequence of 'further depredations on whale stocks, which rendered it essential that the position should be watched from year to year'. There were delegations from Germany, Japan, Norway, the United Kingdom and the United States, with observers from Canada, Eire, New Zealand and the Union of South Africa. Tønnessen and Johnsen (1982, 469–470) note that Wohlthat dominated the proceedings. He had even threatened that unless the Japanese came 'into line with the other governments by next season', there must be a further conference to discuss 'what steps would necessarily have to be taken vis-à-vis Japan'.[(31)]

Dobson later minuted the Permanent Under-Secretary of the Ministry of Agriculture and Fisheries that the immediate purpose of what he described as 'the informal conference' had been attained; this was a Japanese undertaking whereby it would adhere to the 1937 Agreement and 1938 Protocol. Herschel Johnsen similarly reported to the State Department that 'within the limits of the power of the Conference these objectives were obtained, and in addition thereto other ends were accomplished which it is believed should further the conservation of the whale stocks' (*FRUS* 1939, 2, 41 and 49–50). While the Japanese delegation had assured the Conference that their Government would have acceded by the 1939–40 season, it was clear from 'a more meticulous statement', issued later in Tokyo, that the enabling legislation could not possibly be in place before the end of the 1940–41 season. As Dobson wrote impatiently, the Japanese delegation 'knew perfectly well that if they did not come into line now', there might be a German and Norwegian boycott of their oil. They were also fully aware of the very tentative suggestion that all purchases from 'non-adhering producers' would be banned. A particular difficulty for the United Kingdom Government was that whereas the German and Norwegian Governments had the legislative powers to do so, the Board of Trade did not possess them – nor would it be keen to acquire them for the possible repercussions in other sectors of trade.[(32)]

The informal London Conference, as adjudged by Tønnessen and Johnsen (1982, 471), was both disappointing and a step in the wrong direction in its concessions to win Japanese adherence. The most important resolution was to extend the Whaling Protocol of 1938 by a further year and to agree to the concessions sought by Japan. These were the exclusion of the Pacific grounds between 40° north latitude and 52° 30' north latitude, and between 159° east longitude and 140° west longitude; secondly, the waters between 52° 30' north latitude and 72° north latitude, and from 150° east longitude eastwards to 140° west longitude, and thirdly the Okhotsk Sea northward of 52° 30' north latitude. War shortly followed. As Dobson complained to Collier, some two months later, the Japanese continued to give the impression of wanting 'to wriggle out of their undertaking to accede'. More fundamentally, there continued to be no physical assurance of the longer-term viability of the whaling industry. The London Agreement remained extremely tenuous. There was no permanent body to enact and supervise what delegates had agreed to (Dorsey 2013, 70). There was nothing to prevent the United States Government, citing the example of its own fishing grounds, from assuming the mantle of custodian of the world's oceans. Its increasingly self-conscious policy of conserving the natural resources of the American continent – and now the continental shelf – would encompass the entirety of the oceans, in terms of whale stocks (*FRUS* 1939, 2).[(33)]

CHAPTER 15

# The further disruption of war

War had come as no surprise. Henry Maurice had resigned as President of the International Council for the Exploration of the Sea in May 1938, ostensibly on his retirement as the Fisheries Secretary for England and Wales. The Council's historian writes that with war so imminent, Maurice had thought it better that the president should come from a neutral country. Johan Hjort succeeded him (Went 1972, 91).

While there was no longer demand for glycerine from the armaments industry, the United Kingdom Government had taken full advantage of the low market prices to build up large stocks of whale oil – a move that won the fulsome tribute of Lord Woolton, the wartime Minister of Food (Woolton 1959, 237; Jackson 2005, 211). As well as purchasing a further 90,000 tons of whale oil from Norway and 60,000 tons from Japan, the Ministry contracted to buy what proved to be the entire British catch of 130,000 tons – not because the whaling companies were too frightened to put to sea but because they might decide not to do so because of the low market price. Alban Dobson found himself having to apologise to Laurence Collier for bothering him with whaling matters only three days after the declaration of war on 3 September 1939. With every whaling vessel still awaiting Admiralty instructions as to which convoy it should join, the expeditions would be late arriving on the whaling grounds. The companies had accordingly asked for the season to be extended to compensate for their late arrival.(1)

The Foreign Office intimated that even in wartime, at least six months' notice must be given from 1 January of any intended deviation from the International Whaling Agreement. The United Kingdom Government must particularly show 'regard to the decencies by consulting all the other contracting Governments', except of course Germany. Worse still, there appeared to be no 'orthodox method' by which any part of the International Whaling Agreement could be suspended without denouncing the whole Convention – something 'the Minister would certainly not wish to contemplate'. An obvious ploy was to argue that any catch made after the season had ended would be no more than what the German whaling fleet would have caught had the Royal Navy not prevented its putting to sea. However, there was no way of knowing what that catch would have been. It was not until 12 October 1939 that the ten British expeditions finally left, eventually joining ten Norwegian expeditions; six Japanese, and one each from Argentina, Panama and the United States (Tønnessen and Johnsen 1982, 476; Hart 2001, 292–293).(2)

Erik Colban had meanwhile confirmed that the Norwegian Government was opposed to any extension of the season. The United States Government was no more helpful, drawing particular attention to the most striking aspect of the previous season's catch: the marked reduction in catch size per catching unit. Even some of the more experienced gunners were now warning that the industry might soon be 'practically extinct'. With the prospect of higher wartime prices, the State Department saw no need for any extension. Dobson remained under considerable pressure from both the Ministry of Food and whaling companies to allow whaling beyond 7 March 1940. There was dissension within the Foreign Office. While the Northern Department continued to raise 'political' objections, warning of how 'our moral reputation would be damaged', Laurence Collier thought an extension entirely reasonable. The Ministry of Food should first establish 'the exact statistical position'. The letter seeking that information warned that any deviation from the International Whaling Agreement would have a detrimental effect on public opinion in the neutral countries and probably beyond.(3)

Sir Henry French (the Permanent Under-Secretary of the Ministry of Food) acknowledged the Foreign Office's reluctance to jeopardise the whaling agreement but was sceptical of any effect on public opinion in neutral countries. As he explained, whale oil had become 'our great stand-by against shortage of fats'. The present stock of 240,000 tons, which had been built up before the war, would have fallen to 160,000 tons by May 1940, when the forthcoming expeditions should have returned with a further 140,000 tons – or rather 170,000 tons if the season was extended. It was hoped an additional 80,000 tons could be purchased from Japan, and 60,000 tons from the Norwegians, thereby creating a total stock of about 470,000 tons. Although this additional 30,000 tons, which could be expected from an extension of the season, might appear small, butter imports from Holland, Denmark and the Baltic countries were already severely reduced and might cease altogether. Vegetable-oil supplies were also at risk from enemy disruption. Should all the projected whale oil be obtained, fewer expeditions might be required in the next season, with all the substantial savings this would entail.(4)

Collier and Dobson found Sir Henry French's 'arguments and facts' sufficiently convincing to agree that the Whaling Advisory Council, appointed by the Ministry of Food, should inform the five British companies of the extension to the season. The Foreign Office also wrote to the Japanese and United States Ambassadors, the Norwegian Minister in London and the Union Government in South Africa. Dobson thought the blow might be softened by retaining the ban on catching humpbacks 'if the other Governments will do the same'. Herschel Johnson 'returned to the charge', protesting at how the whole conservation effort was threatened. An additional catch of 1,000 to 1,500 whales was bound to have serious effect, even without the German expeditions. The delayed departure of the British expeditions had been 'one of the exigencies of war from which all commerce is suffering'. The price of whale oil had already risen from £15 to £40 per ton and was likely to go higher.(5)

Although Dobson thought the United States was 'making rather a mountain out of a mole-hill', he nevertheless minuted his unhappiness at the thought of presiding at future conferences. He came to regret the concession even more on learning that the weather had been so bad that the eight British expeditions had succeeded in whaling for only six-and-a-half days of the 24-day extension permitted. Far from making up the shortfall of 30,000 tons, caused by their late arrival, only 7,000 tons of additional oil were obtained. Whaling had otherwise proceeded in the manner of previous seasons. Francis (Dick) Ommanney, having written up his Discovery Report at the British Museum (Natural History), had taken the post of whaling inspector and biologist aboard Salvesen's factory ship *Salvestria*. He found it very much business as usual, spending most of his time recording as closely as possible the species, length, sex and sexual condition of the whales caught, and more particularly evidence of a foetus. He had also been required to make an immediate and formal protest whenever any undersized whale was caught or a 'nursing cow . . . when obviously escorting a calf' was shot. He suspected many more breaches occurred in those hours when he was asleep (Tatham 2008, 404–406).

It was the receipt of a Note from the Japanese Embassy, on 11 December 1939, that particularly alarmed Dobson for its baldly stating that 'every government must hesitate to bind itself to an Agreement . . . manifestly based on normal conditions'. The Japanese Government was accordingly postponing 'its adherence to the Agreement and Protocol until more normal conditions'. Kodaki particularly impressed upon Dobson his personal reluctance to deliver the Note, as he had hoped to the last that his Government would accede. The six Japanese expeditions accounted for half the aggregate catch of 1.1 million barrels obtained that season (Tønnessen and Johnsen 1982, 476).(6)

There had been increasing pressure from the Ministry of War Transport as it tried to take over as many factory ships as possible as 'much-needed tankers', with the obvious implication that their catchers might be assigned for use by the Admiralty. Even worse, Dobson was appalled to learn that the Ministry of Food intended to disband almost the entire whaling fleet once it had returned from the 1939–40 season. The Royal Navy had been so successful in intercepting the large volumes of vegetable oil assigned to the German ports that the Ministry of Food had simply run out of storage space. There was no longer need for a 1940–41 whaling season, particularly if the imports from Norway and Japan could be maintained. Harold Salvesen protested straightway, from his position on the Ministry of Transport's Whaling Advisory Committee, drawing particular attention to the loss of the skilled manpower that would result from this action. Reginald H. Dorman-Smith (the Minister of Agriculture and Fisheries) wrote straightway to William Morrison (the Minister of Food), on 28 February 1940, of how difficult it would be to reassemble a force of some '2,000 highly trained seamen and others'. Their loss to the British skilled labour force would be little short of disastrous. Although recognising the importance of buying all the Norwegian and Japanese whale oil available lest it fall into German hands, as many as five or six expeditions should be sent. Both Collier and Sir Orme Sargent (the Deputy Under-Secretary) thought the proposed transfer was short-sighted, but, as Sir Alexander Cadogan (the Permanent Under Secretary) minuted, 'Our eyes have to be focused on immediate needs nowadays', to which Lord Halifax (the Foreign Secretary) added, 'I rather fear so'. Two expeditions were eventually sent – the *Southern Empress* and the Star Company's *Svend Foyn* – both managed by Salvesen (Tønnessen and Johnsen 1982, 479–481; Somner 1984, 53–54; Jackson 2005, 212–213; Hart 2020, 193–195).(7)

The six Japanese expeditions only obtained 622,413 barrels from the 9,992 whales caught, compared with a British catch of 3,116 whales, from which 229,780 barrels were produced and a Norwegian total of 203,317 barrels from 2,387 whales killed. Grytviken produced 44,498 barrels from 868 whales, making for an aggregate total of 1.1 million barrels from 16,363 whales, of which 56.6 per cent were Japanese (Tønnessen and Johnsen 1984, 483). Norway, despite its declaration of neutrality, was invaded by German forces on 9 April 1940. A Norwegian pilot had boarded 'the big German whaling factory *Jan Willem* (flying the United States flag)' as it approached Narvik. But instead of unloading barrels of oil, Carl J. Hambro wrote later that troops and munitions had emerged under the orders of the newly appointed German commander-in-chief for Narvik (Hambro 1940, 178). The entire Norwegian whaling fleet of 13 factory ships and 105 catchers was diverted on its return to British waters, where it eventually joined the thousand or so vessels under the Norwegian Government-in-Exile, eventually being transferred to *Nortraship* (the Norwegian Shipping and Trade Mission in London). Most were assigned to the Atlantic convoy system, operating under charter to the United Kingdom or United States Governments (Basberg 1991).

Whatever the merits of sending further whaling expeditions, they proved highly vulnerable to enemy surface raiders. Every published account of wartime whaling tells that the German raider *Pingvin* surprised and captured the two Norwegian factory ships *Ole Wegger* and *Pelagos*, as well as their 11 catchers and the transport *Solglimt* to the east of South Georgia in January 1941. The *Solglimt* had just taken aboard 20,400 barrels of oil from the factory ship *Thorshammer*. Only the *Thorshammer* and three catchers managed to escape. The armed merchant cruiser *Queen of Bermuda* was despatched to escort the British expeditions back to Europe. The fortnightly summary, which the Secretary for Air regularly submitted to the War Cabinet, summarised, on 25 September, that Boston aircraft from Bomber Command had ten days earlier sunk the prize ship *Solglimt* at Cherbourg; air photographs showed the vessel 'lying on her port side with water covering her decks'. The *Queen of Bermuda* was thereafter sent to search for evidence of the enemy's presence in the Southern Ocean, with a specific instruction to destroy facilities which might be of use to the enemy at Deception Island. The largest vessel (22,575 tons) to have navigated its very narrow entrance, the ship's logbook recorded how 'we had no chart of the entrance except for a two-inch square let-in on a large chart' provided by the captain of one of the factory ships – it was not funny' particularly as a small iceberg had initially blocked the passage. The large coal dump was set ablaze and the piping associated with the oil tanks punctured. The shore station itself was left intact (Roskill 1954, 286; Jones 1974; Tønnessen and Johnsen 1982, 481–482; Hart 2001, 293; Headland 2009, 304–305; Johnson 2015; Dudeney and Sheail 2019, 248–249).[8]

The Ministry of War Transport eventually refitted each of the requisitioned factory ships as tankers, their strengthened decks proving ideal for conveying tanks, locomotives and aircraft across the Atlantic. Almost every one of their catchers was similarly assigned to the various convoys. Only Grytviken of the shore stations continued to operate. As the Magistrate for South Georgia, William Barlas, reported that there had been only seven blue whales among the 883 whales processed (compared with 65 the previous summer). Erik Hamilton warned of both an increasing scarcity of fin whales and a very substantial reduction in the proportion of pregnant whales of every species. The whale herds were becoming so thinned out there was little chance of their replenishment. The 1941–42 catch was no more encouraging. Of the 12,189 fin whales caught from Grytviken (585 male and 604 female), 60 per cent were juvenile. With only a quarter of the females pregnant, Hamilton wrote, 'One cannot take an optimistic view of the future of this species'. The whales were so scarce and the weather so bad that Arthur I. Fleuret (the Magistrate) was able to report a much greater use of the carcases in the 1942–43 summer. The 1944–45 catch proved one of the most productive in the 40-year history of Pesca, comprising 128 blue, 987 fin, 60 humpback, 76 sei and 45 sperm whales. With all German surface raiders sunk, *Thorshammer* had operated in the 1943–44 season, catching 837 whales and producing 82,000 barrels of oil. A shortage of catchers meant that only *Sir James Clark Ross* went south in 1944–45, producing 148,000 barrels and 12,000 sacks of meat meal. As remarked in the Fisheries Department, there was little sign of any wartime replenishment of whale stocks (Tønnessen and Johnsen 1982, 484; Hart 2001, 294).[9]

## 15.1 Readiness for peace

A post-war whaling policy began to emerge as part of a wider fisheries policy. An interdepartmental committee of scientists was appointed in 1940, charged with avoiding the scale of overfishing that had occurred in the seas around the British Isles after the First World War. The Committee, with Edward Russell (the Director of the Lowestoft Fishery Laboratory) as its chairman, recommended that the combined fisheries fleet for the first three post-war years should not exceed 70 per cent of what had been the effective tonnage in 1938. A permanent commission should be appointed to determine the longer-term size of the fleet. A third recommendation was to limit the time each fishing vessel spent at sea, and a fourth to adopt a minimum net mesh (Lee 1992, 136).

It was while looking through these papers in January 1942 that Robert Hudson (now the Minister of Agriculture and Fisheries) remarked that 'we ought to be prepared for action in regulating whaling in the Antarctic'. Dobson assured him whaling had been by no means forgotten, but 'we have not the slightest idea whether there will be any factory ships left'. As he further remarked, 'We may have to pinch some of the Japanese factory ships.' The shipping losses became even greater. Harold Salvesen reported to Dobson in August 1942 that the *New Sevilla*, *Salvestria*, *Terje Viken* and *Univaleco* had been sunk and *Svend Foyn* severely damaged. *Tafelberg* and *Anglo-Norse*, and possibly *Hektoria*, *Sourabaya*, *Southern Empress* and *Southern Princess*, had been stripped of their whaling gear to make more cargo space. Every modern catcher had been requisitioned. Eight had been sunk and others reportedly handed over to the Russians. Harold Salvesen warned that once the war was over, the priority must be to retrieve and refit every surviving vessel. The Smith's Dock Company of Middlesburgh had already enquired of Salvesen how this might be organised.(10)

Dobson welcomed the preparatory thinking but was alarmed by Harold Salvesen's insistence that every restriction on whaling should be jettisoned for the period taken to replenish global oil stocks. There was a pressing urgency for this to be addressed. Dobson responded by writing to Neil Mackintosh in early September 1942 that he would find it easier to secure the interdepartmental opposition required to any weakening of the whaling regulations if the Discovery Committee (which had continued to meet sporadically) submitted a formal resolution to that effect. Mackintosh soon submitted the desired resolution which, modified by Dobson, urged the Government, with 'other interested foreign governments as may be available', to take control of the whaling industry once the war ended. The modified resolution formed an essential part of a submission to the Secretary of State for the Colonies on 30 September 1942, emphasising that the pre-war measures had been perfectly sound, but they had not gone far enough in checking 'the progressive depletion of stock'. Concerted action was now required to devise a plan which would further reduce the annual catch ready for when the companies began to reassemble 'their equipment and personnel, now scattered on account of the war'. Everything must be done to prevent any resumption of whaling on its pre-war scale.(11)

A procedure was agreed whereby discussions would first be held inter-departmentally and with the whaling companies, and then with the Norwegian and United States Governments (with 'perhaps South Africa and Australia thrown in'). A Fisheries Department official, A. M. Lowe, thought 'a distinction should be drawn between measures required to protect stocks in the longer term and what he called a 'special more or less cut-and-dried' plan 'to cope with

## 15.2 Recovery amidst crisis

The Norwegian and United Kingdom Governments have been strongly criticised in the whaling literature for their failure to use the severe disruption caused by the war to re-balance 'the economic needs of whalers with the biological and reproductive capabilities of whales'. They allowed 'the same short-term economic priorities that had so far driven the capital-intensive era of modern whaling' quickly to re-emerge, 'signalling an ominous and unmistakable return to business as usual for the whaling industry' (Heazle 2006, 38). The commentaries show little appreciation of the crisis faced by governments. Whale oil prices had more than tripled from £40 to £100 per ton, caused by the obvious disruption to global fat and oil supplies and threat of starvation and worse in the newly liberated countries of Europe.

The First World War provided little guidance to the Second. The effects on the whaling grounds were very different. Whaling had continued, albeit at a lower intensity, throughout the First World War whereas there was comparatively little whaling once the 1939–40 summer was over. The devastation inflicted on fish stocks, once the North Sea was reopened in 1919, was a stark warning to those charged with managing the fishing and whaling industries after the Second World War. A demobilised workforce had found after the First World War both employment and an insatiable market for all the fish that could be caught – a volume far outstripping any wartime recovery of North Sea stocks (Heidbrink 2018). An analogous situation might easily develop in the fishing and whaling industries in the aftermath of the Second World War.

The priority for the companies was to recoup the costs of reconstituting their whaling expeditions, a prospect made all the more alluring by what was expected to be an insatiable demand for fats and oils. Harold Salvesen wrote a 'very strong letter' to the Ministry of Food, shortly before the D-Day Landings, demanding the removal of every constraint; but as Lowe in the Fisheries Department minuted on 1 June 1944, 'We have had to take into account the necessity of continuing the policy of trying to conserve the stocks of whales'. It was impossible to ignore the warnings 'by scientists who had made a careful study of the position and were well acquainted with Antarctic conditions'. The pre-war conferences had endeavoured 'to steer some kind of middle course', and 'with their eyes open' had come up with 'the best practicable compromise between the conflicting objects'. Although 'a pity that their decision differs so widely from the views held by Captain Salvesen', it did 'not necessarily follow . . . that the Conference was in error!' They had sought 'the best practicable compromise'. Although whales would have to be sacrificed if it could be shown that 'the supply of whale oil was the only thing which stood between the peoples of Europe and starvation', the Fisheries Department was far from convinced things would ever become 'as bad as that'.(19)

Tønnessen and Johnsen (1982, 486) wrote of the general presumption that the Norwegian and British companies would retrieve their near monopoly, as the German and Japanese industries would have been eliminated. One of the earliest moves made in the reconstituted Norwegian Storting was a provisional order of December 1945 prohibiting Norwegian citizens from joining expeditions mounted from a country with no pre-war involvement in whaling. British companies that had previously employed Norwegian nationals would therefore be unaffected. As the Norwegian Minister for Foreign Affairs assured Laurence Collier, it was the Norwegian intention to maintain 'their predominant position and therefore preserve whale stocks'. There were nevertheless rumours of both Russian and Swedish expeditions being prepared and, as

Collier wrote, of an increasing Dutch interest. Birger Bergersen warned that the industry would be ruined if every country suffering from the fats shortage took up whaling.(20)

Of more immediate concern to the British companies was the endeavour of the Norwegian unions to reassert their previous dominance of the labour force. After the amicable discussions affecting the first post-war season, the companies were shocked by the refusal of the four Norwegian unions even to negotiate with Charles Jarman (the General Secretary of the National Union of Seamen and the Navigators and Engineers Officers Union), who had flown to Norway for the purpose. Collier cabled from Oslo on 30 July 1946 that he had protested to the Norwegian Prime Minister at 'this blackmailing attempt to take advantage of British need of whale oil'. The Oslo newspaper *Arbeiderbladet* published a statement, following the return of the Chairman of the Norwegian Seamen's Union from London, saying that his delegation had been given a special guarantee whereby Norwegians would receive the same treatment on British vessels as a Norwegian one. The branch offices of the various unions in Vestfold were instructed to allow their members to join the British vessels. The British whaling companies agreed for their part that preference would be given to those previously employed.(21)

So desperate was the shortage of basic fats and oils that the Minister of Food, Colonel Jay J. Llewellin, forewarned the Minister of Agriculture and Fisheries, Robert Hudson, on 12 March 1945 that global supplies of oils and fats were 'likely to fall short of world requirements for a considerable time to come'. An obvious way of minimising those shortages was to let the weather determine when the whaling season should end rather than to set 'any arbitrary date embodied in an international agreement'. There would be so few expeditions that even if the grounds were worked for an entire 12 months, whale stocks would hardly be affected. It was absurd, Llewellin wrote, that at a time when the world was so desperately short of fats and oil 'we should hamper ourselves by the letter of an Agreement when the spirit of it is in no danger of being imperilled'. Hudson acknowledged, in his reply of 14 March 1945, that 'we have all along fully realised the urgent need for whale oil'.(22)

The other relevant department was the Ministry of War Transport, which, in June 1945, notified the whaling companies that as many as six to 10 British and Norwegian factory ships might be released and perhaps two German vessels. The Whalers' Section of the Chamber of Shipping responded by demanding their unfettered use. While acknowledging that the whaling agreement was a matter for the Minister of Agriculture and Fisheries, Llewellin noted it was also his responsibility, as the Minister of Food, to organise the forthcoming expeditions and unless an extension was agreed to, he would have to 'put the matter before the Cabinet'. Lowe forwarded Llewellin's letter to Dobson, remarking, 'I have always felt that it would be rather difficult to resist a proposal of this kind if it were pressed strongly'. Hudson agreed, on 23 June 1945, to seek the agreement of the other parties to the Protocol being extended to the first post-war whaling season due to the 'exceptional circumstances'; he remained highly sceptical of whether any substantial addition of oil would be obtained, but 'we need not worry colleagues' in Cabinet.(23)

## 15.3 The wartime Protocols

Alban Dobson, Fisheries Secretary, pressed in the spring of 1943 (at what proved to be the mid-point of the war) the value of an informal international meeting to discuss 'certain amendments in agreeing with Salvesens and United Whalers on 'their intended layout and equipment'. There was not the slightest chance of securing the labour to fit the liver extraction and meat meal plant which the companies demanded. The two vessels had already been given a greater priority 'than any other Merchant Vessels under construction on the North East Coast'. The instruction by the Ministry of War Transport to install only what was essential for 'the production of Whale Oil' caused Harold Salvesen to protest vigorously at what he called this 'mischievous' rebuff of 'our well-meaning activities'. Some 6,000 tons of dehydrated whale meal (as well as liver oil containing vitamin A) might be obtained. Did the Ministry really want all this uncooked meat thrown overboard? As an official of the Fisheries Department phrased it, 'We know perfectly well that the plant enabling them to make whale meal would be extremely valuable'. Whale oil was the overriding priority; the installation of the liver and meat-meal plant should be postponed to the 1946–47 season.[36]

The hopes of oil tankers being refitted as factory ships were soon dashed. The burgeoning demand for mineral oil, as military forces pressed ever deeper into Europe, meant that an even greater priority was given to its supply. As officials of the Ministry of Food were quick to point out, the provision of fats and oils was an equal priority for their Minister, one of them confessing that 'the oils and fats position frightens me more and more every day'. The position with regard to catchers was no better. An Admiralty phone call warned of 'a distinct bottleneck in the reconditioning yards' and of the possibility of a six-month delay. Undated Ministry of Food papers include a draft Cabinet memorandum that anticipated a global deficit of 1.5 million tons of oils and fats. Little could be done to bring about a rapid expansion, say, of dairy farming, but the large-scale resumption of whaling could provide over 500,000 tons of whale oil in a good season. Even a single factory ship with sufficient catchers might produce 25,000 tons of whale oil (the equivalent in output of 300,000 cows).[37]

The factory ships *C. A. Larsen*, *Jan Willen*, *Unitas*, *Walter Rau* and *Wikinger* were thought to have survived the war. Another, *Ole Wegger*, had been used as a block-ship near Rouen. The Norwegian Government and the Ministry of War Transport sought its removal to an English port. The Supreme Headquarters Allied Expeditionary Forces (SHAEF) refused lest 'she might founder in the Seine channels' (Tønnessen and Johnsen 1982, 486). Harold Broadley in the Ministry of Food was so furious that the Inter-Allied Reparations Agency had allocated every 'ex-enemy banana boat to the Americans' that he pressed for ministerial intervention at the highest level to achieve a more equitable distribution of the enemy whaling fleet. Every German whaling vessel should be earmarked for 'the only two United Nations countries' previously engaged in whaling. Sir Cyril Hurcomb refused to go further than to seek those ships that were in better condition, adding that 'we have too much at stake in other classes [of shipping] to do other than stick to the fair principle of partition by reference to losses'. It was not until mid-November 1945 that the Ministry of Shipping was able to confirm that the Norwegian vessels *C. A. Larsen*, *Norvhal*, *Pelagos*, *Sir James Clark Ross*, *Suderoy* and *Thorshammer* would be available for Norway, and, for the British part, the two factory ships under construction and 22 catchers, along with the transport ship *Saluta*. Salvesen was also allocated *Empire Victory* (previously the German vessel *Unitas*) for the forthcoming season; there was still no decision on the ship's eventual ownership (Somner 1984, 54).[38]

The war was hardly over, but the industry's longer-term management had already begun to dominate the interdepartmental exchanges. The Whalers' Section of the British Chamber of Shipping stated, in a letter of 6 March 1946, that no more than 16 factory ships would be able to operate economically within the 16,000-BWU aggregate set by the 1944 Protocol. The Norwegian expeditions had already reached their full share. The British fleet was nowhere near even a half share. The Norwegian intention to employ a further vessel meant either the industry would soon be 'overloaded' or, through British concession, there would be an 'undue Norwegian predominance'.[39]

CHAPTER 16

# *The United States – the superpower*

It was fitting for the United States' newly won status as a global superpower that it should host an international meeting to determine the future of whaling, despite having no substantial industry of its own. Kirkpatrick Dorsey attributes the initiative to Remington Kellogg, who had dismissed the pre-war conferences as 'junkets'. Although achieving his goal and chairing the proceedings, Kellogg came to play a diminishing role as the other goals set by the State Department took precedence, and none more so, in the whaling context, than the resuscitation of a post-war Japanese whaling fleet. But even if the Washington Conference is adjudged to have extended the United States' 'progressive tradition' to the global whaling grounds and the scientific conservation of stocks, the International Whaling Commission was fatally flawed. The 'objection clause', designed to protect national sovereignties, became the most damaging flaw in the whole post-war edifice (Dorsey 2013, 114–115).

It was a time of unprecedented interest in the economic potential of the Antarctic and its waters. Brian B. Roberts of the Scott Polar Research Institute at Cambridge (and a part-time officer in the Foreign Office's Research Department) published a paper in *Nature*, in which he identified a dozen or more proposed expeditions, some explicitly in search of the 'rich deposits of uranium and other valuable minerals' thought to be present on the Antarctic continent. At sea as many as 15 factory ships operated in the austral summer of 1946–47, of which seven were Norwegian; three British, two Japanese, and one each from Argentina, the Netherlands, South Africa and the USSR. Roberts (1947) emphasised the inherent difficulties of allocating the Continent's resources in any equitable manner.

There was a wider fisheries context for what the Washington Conference set out to achieve. An International Overfishing Conference was held in London in April 1946, in which both Alban Dobson (the Fisheries Secretary) and Edward Russell (the Director of the Lowestoft Fisheries Laboratory) had a leading role. The Fisheries Department sought what had become known as 'a Total Allowable Catch', the imposition of close seasons and appointment of fishing trusts vested with monopoly rights to the various specified grounds. As with the Washington Conference, the outcome was considerably more modest. The British delegation pressed for no more than a limit to the total fleet tonnage, and even this the conference rejected. Even the compromise resolution that there should be no increase in fleet size was rejected for taking too little account of the particular circumstances of each national fleet. Nothing more was achieved than a Convention for the Regulation of the Meshes of Fishing Nets and the Size Limits of Fish – something that had been agreed on in 1937 but had never been fully ratified because of the war (Lee 1992, 137).

Within an equally urgent but discouraging context, a longer-term policy for whaling was considered. The 1945–46 catch proved to be one of the poorest on record. There had been 'the worst-possible weather'; the blubber of even the whales caught was often 'quite thin and dry'. The catch of 1,913 whales comprised 1,456 fin, 238 humpback, 82 sei, 80 blue and 57 sperm whales; they were caught generally some 20 to 90 miles offshore. Grytviken processed 1,913 and Leith Harbour 350 whales. Tønsberg reopened Husvik Harbour, which had been closed since 1931, and processed 481. The 1947–48 season proved considerably more productive. The three shore stations processed an aggregate 2,949 whales, comprising 2,142 fin, 125 sperm, 609 sei, 46 blue and 24 humpback whales, making for a total of 163,651 barrels of oil, 137,173 bags (100 kg) of guano and 20 tons of baleen, as well as six tons of meat extract. The whales were reportedly in very good condition, caught mostly 70 to 130 miles offshore. The longer intervals between the carcases' delivery meant there was a better chance of their being fully processed. An even higher output might have been obtained had the weather been better.[1]

Erik Hamilton (who had continued as Government Naturalist) published a brief review in *Nature* in June 1948, which he began by citing an article published by *Norsk Hvalfangst Tidend.* Not only was a greater proportion of mature females caught globally in 1945–46 than immediately before the war, but there was a marked fall in the proportion found to be pregnant. It was for Hamilton irrefutable evidence that the habitual killing of immature whales was proving to be 'a particularly and obviously efficient way of exterminating a species'. The increasing use of the Blue Whale Unit (BWU) was encouraging a complacency both among companies and governments. As there would always be a proportion of the smaller whale species in any catch, the use of the Unit gave an entirely false impression of how stocks were being affected. In Hamilton's words, 'An industry of this sort cannot be regulated by reference to abstractions, when it depends entirely upon a stock of wild animals over the lives and breeding habits of which there is no control'. However easy it might be to assume the impossibility of injuring, still less destroying, large animal populations, Hamilton warned that

> *the stocks of blue whales are showing clear signs of reduction: 1) in the rarity or absence of the huge animals of earlier seasons; 2) in the reduced proportion in the total catch of this species; and 3) in the reduced percentage of mature females which were pregnant.* (Hamilton 1948, 913–914)

The fin whale was also starting to show a parallel reduction in number. It was salutary to recall how the North American bison had been brought to the verge of extinction, and how the passenger pigeon was exterminated despite their having once thrived in almost incredible numbers.

Meanwhile, in the straitened circumstances of post-war recovery, an immediate priority for the Ministry of Agriculture and Fisheries was to maximise the use of each whale caught – and therefore whale meat. The prospects had never been good. The Department of Scientific and Industrial Research had commissioned trials in the use of whale flesh as food following the First World War. Salvesen had provided the Household and Social Science Department at King's College for Women in London with quantities of whale flesh from the shore station in the Shetland Islands. The meat when fresh was hardly distinguishable from veal except for its greater tenderness. It would be highly suitable for canning and as 'galantine or camp pie'. Cooking trials, however, confirmed 'a grave defect'. Although the meat's high oil content

who make up the crews'. It must be a source of great satisfaction to delegates that, 'in taking measures to conserve whale resources you are protecting the interests of these brave men'. The conservation of wildlife resources provided an excellent opportunity for successful concerted international action. Davidson particularly cited the recent conservation of halibut stocks on the continental Pacific shelf of North America. The conservation story was essentially the same all over the world, but the United States had experienced particular difficulty in both learning and applying it. As a young country, stretching across an entire continent, 'we possessed resources which our ancestors felt were so ample that they could never run out'. That same abundance had led to waste and over-commercialisation. Resources continued to be over-exploited, even when their decline had become obvious. There had been a general reluctance to do anything effective by way of conservation until the resources were 'so endangered as to make conservation practices either useless, or long and arduous'.(15)

More prosaically in his report to the Minister, Dobson paid particular tribute to Mackintosh, who, with Thomson, had been 'as usual a tower of strength on the scientific side'. Salvesen had been 'helpful and cooperative' to the extent of leaving a congratulatory note with the delegation, signed on behalf of the Whalers' Section. The Ministry in turn praised Dobson for having 'once again steered the U.K. ship successfully through mine-sown international waters and brought it safely to harbour'. His experience and skilled negotiation had largely contributed to the success of the conference. A formal report to the Cabinet's Overseas Economic Policy Committee recounted that the delegates had unanimously signed the final act codifying all previous agreements and appointing an International Whaling Commission, comprising all the contracting governments, charged with both examining and amending the agreement as 'the condition of the industry and need for conservation of stocks may require'. The conference did not accept a British proposal for its headquarters to be in London lest the commission be later brought within the United Nations Organisation. The United Kingdom Government would, however, call the first meeting, at which the Commission would decide where and when further meetings would be convened. (16)

The Washington Embassy, in forwarding the conference documentation, saw the purpose of the new Convention as one of 'conserving and developing the whale resources, and thus make possible the orderly development of the whaling industry'. As well as setting out the regulations governing whaling operations, it would provide for the establishment of an International Whaling Commission, comprising one representative from each contracting government, which might

> *independently or in collaboration with other agencies encourage, recommend, or organise studies relative to whales and whaling; collect and analyse information concerning whale stocks, etc; study, analyse and disseminate information concerning methods of maintaining and increasing whale stocks.*(17)

The conference was adjudged within the Fisheries Department to have been notable for three things. It had reached 'agreement on a complicated matter within two weeks'. The Soviet Delegate had made an effective and practical contribution. And the proceedings had been 'conducted not merely without heat but in almost complete harmony'. The exception had been the Dutch Delegation's strong objection to the Norwegian decree of December 1945, prohibiting Norwegian nationals from taking part in expeditions from those countries that had not participated in whaling before the war. It had required the most careful negotiation by the chairmen of the special committee and of the conference, to avoid a situation where 'the Norwegian delegation could well have withdrawn from the Conference'.

Whereas the focus of the Washington Conference was on the future governance of the whaling industry, such bodies as the International Emergency Food Organisation (IEFO) continued to highlight 'the world's very limited supply of fats and oils'. The 'dietary quantities' in many of the war-devastated countries remained too low for them to achieve 'the proper productivity of the labour force'. Although not wanting to 'liberalize' catch limits too greatly, the IEFO was confident that whale stocks had obtained sufficient respite during the war years for the aggregate catch to be increased by 'as much as one-third above the present limit in the forthcoming season'. There would be need for little, if any, additional manpower, equipment and capital to secure that potential. A brief that Dunn prepared for delegates to an FAO council acknowledged that it was still impossible 'to estimate accurately the number of whales in the sea'. Scientists warned, however, that any increase in the limit of 16,000 BWU would incur 'undue risk to the long-term preservation of the stocks'. As Dunn acknowledged, it was common practice for the breeding stock of farm animals to be reduced to meet temporary crises for more meat. It did not follow that whale stocks could similarly be reduced to meet obvious shortages in fats and oils. As whales were wild species, the industry did not have the same degree of control over their breeding pattern. There was no assurance that stocks could be restored even by such measures as the imposition of closed seasons.

The 15 expeditions taking part in the 1946–47 season had caught 15,307 BWU. It was highly likely the anticipated 17 factory ships would attain the 16,000 BWU limit in the 1947–48 season. There was no certainty that even if the limit was raised, more whales would be caught. The average daily production per catcher had been falling for over a decade. Whereas it had been 163 barrels in the 1933–34 summer, it had fallen to 141 barrels in the 1936–37 season, and only 119 barrels in 1946–47, despite the average engine power of the catchers having been increased from 883 to 1,263 horsepower over that period. As Dunn warned, not only might any substantial increase in the season's catch be at the expense of later years, but the re-imposition of any previous limit would have become even harder, more particularly for the extent to which these further companies, and indeed countries, had invested in whaling. That had long been the view of His Majesty's Government, and 'nothing had occurred since to alter it'.[(18)]

## 16.2 The International Whaling Commission

There is no explicit recognition, within the Departmental files, of the significance of the Washington Conference in the sense of the United Kingdom Government ceding its leadership role for the management of global whale stocks to an International Whaling Commission. Although still vitally concerned with the attainment of that end, British responsibilities for the Falkland Islands Dependencies no longer required it to play so prominent a role. The documentation available at The National Archives at Kew ceases to have the depth and comprehensiveness that had previously made it so remarkable as a data source for the whaling industry.

Another two years elapsed between the Washington Conference of November 1946 and the first meeting of the International Whaling Commission in London in the spring of 1949.

A Note drafted for the Minister of Agriculture and Fisheries, on that occasion, recalled that the Convention had come into force on 10 November 1948, when it had been ratified by Australia, France, the Netherlands, Norway, South Africa, USSR, the United Kingdom and the United States. The Governments of Iceland, Panama and Sweden had indicated their adherence. The convention was to provide for the proper conservation of whale stocks to prevent further depletion of certain species of whales which have for some time been subject to overfishing, and to thus make possible the orderly development of the whaling industry.

The Commission would include one member from each of the contracting governments, with one or more advisers, charged with the study and investigation of present and future problems. It would have the power to amend the Schedule annexed to the Convention, setting out the specific regulations concerning protected species; open and closed seasons and waters (as included the designation of sanctuary areas); the size limits set for each species; the timing, methods and intensity of whaling (including the maximum catch for any one season); types and specification of gear and apparatus, and appliances which might be used; the methods of measurement and curation of catch returns and other statistical and biological records. Finally, the Commission was empowered to take action, whether independently or with governments and public or private agencies, to encourage, recommend, or, if necessary, organize studies and investigations; collect and analyse statistical data, and study, appraise and disseminate information on methods of maintaining and increasing stocks. The United Kingdom Government, and therefore the Foreign Office, was charged with calling the first meeting and consulting with other governments on where the Commission might be optimally located. The Commission would, in the course of doing so, decide whether it should become a United Nations' agency. Both the Foreign Office and the Ministry of Agriculture and Fisheries pressed for its headquarters to be in London. Whilst the Norwegians lobbied for Oslo, several governments opposed the move, recalling the 'one or two unsuccessful attempts' by the Norwegians to prevent further countries from entering the industry and more particularly the hostility shown by the Norwegian unions toward foreign vessels.[(19)]

The first meeting of the Commission was held in London between 30 May and 7 June 1949. Dobson's appointment as its Secretary meant, as was minuted in the Fisheries Department, 'we can afford to let the Norwegians have the Presidency', for which Birger Bergersen (now Norway's Ambassador to Sweden) was 'highly qualified'. Remington Kellogg was the obvious choice for the Vice-Presidency. Among the actions taken by the commission at that inaugural meeting, the dates of the whaling season were changed from 15 December – 1 April to 22 December –7 April. Another relaxed the total ban on hunting humpbacks, enabling as many as 1,250 to be caught south of latitude 40° south in each of the two seasons 1949 to 1951. Neil Mackintosh was appointed chairman of a standing Scientific and Technical Committee, with representatives from Australia, Canada, France, Norway (Bergersen), South Africa, United States (Remington Kellogg) and USSR (Professor A. S. Bogdanov, Director of Moscow's Fisheries Institute).

Every government (except Panama) that had ratified or adhered to the convention, attended the inaugural meeting, together with observers from Argentina, Brazil, Chile, Denmark and New Zealand, as well as the FAO and General MacArthur (the Supreme Commander, Allied Powers in Japan). Dunn was formally appointed the United Kingdom Delegate and a member of the commission. His expert advisers were identified as the Fisheries Secretary, the Chief Inspector and two whaling inspectors (drawn respectively from the Fisheries Department and the Scottish Home Department). They were joined by both Mackintosh (representing the Discovery Committee) and Harold Salvesen (the British whaling industry). The Commission was to be located in London, with its financial and accounting aspects handled by the Ministry of Agriculture and Fisheries on a costs recovery basis. The Commission's total running costs (estimated at less than £1,200 per annum) would be apportioned equally among governments.

Delegates agreed to a further Protocol extending what had been agreed for the previous season to the 1947–48 one and another affirming the annual limit of 16,000 BWU set by the 1944 Protocol for Antarctic waters, emphasising that holding to that figure would considerably help scientists in calculating annual whale stocks. The Norwegian and United Kingdom delegates pressed for the decisions of an International Whaling Commission to be made binding, but the Dutch, French, Russian and United States delegates insisted on an opting-out provision as a condition of their adhering to the Convention. While the concession came later to be seen as fatally undermining the Convention, the Norwegian and United Kingdom delegates had also miscalculated by assuming their expeditions' fleets would continue to dominate the industry (Fitzmaurice 2015, 30–31).

The shortcomings of the International Whaling Commission have attracted an immense literature. Gordon Jackson perceives two major flaws, the first being to set an annual quota limit that whaling scientists thought was already too high. Of the 1,241,970 whales 'officially' killed in the 60 years between 1905 and 1965, more than half were killed in the 20 years following the Second World War. Oil production in those same 20 years exceeded the inter-war total by 14 per cent. The second flaw proved to be the failure of the commission to set any limit to 'the number of wastefully-competing factory ships'. As Jackson (2005, 216–218) writes, 'The post-war trade was not the carefully-planned search for raw materials that might have been expected perhaps at the end of the war in which governments had organised so much'. It was left to capitalist principles and 'the freedom of the seas' to decide how the catch should be apportioned. The stage was set for what Tønnessen and Johnsen (1982, 487) characterise as the industry's 'last and decisive era *the struggle for quotas*'. The last, and what proved to be the most damaging, flaw of the Convention, was the freedom accorded governments to ignore amendments made by the commission to the Convention.

Of the many critiques published of both the Washington Conference and the Whaling Commission, all have focused on the continued use of the Blue Whale Unit (BWU) as the basis for managing the industry. J. L. McHugh (a former United States commissioner and later commission chairman) wrote much later that there should be no need 'to explain why the BWU is illogical as a management unit' with its 'totally-unbalanced' emphasis upon a single species. But even if this could have been resolved, the maximum permissible catch of 16,000 BWU would still have been too high. The IWC's Scientific Committee came to regard a 9,000 BWU as a more reasonable aggregate limit, but there were those at the time who calculated even 3,000 BWU as too high. As McHugh (1974, 309) writes, 'These things were not as clear in 1946 . . . when economic considerations were given much prominence'.

Michael Heazle writes that the Convention caused every season to become a free-for-all, as expeditions strove to secure the largest weekly catch possible before the Bureau of International Whaling Statistics at Sandefjord closed the season. The whalers themselves described it as 'the whaling olympics'. Whereas previously the goal had been to catch more than other companies, now it was to maximise the company's share of the prescribed quota as ruthlessly as possible – a contest that pressed ever more heavily on what was already a dwindling stock of whales.

catch of 16,000 Blue Whale Units a year must in all probability be maintained for some years'. Although the international regulations introduced in 1937 and 1938 had accorded some protection to species and classes of whale, they had been no more than 'a palliative and insufficient to check a further depletion of stock'. The near cessation of wartime whaling and the sinking of so many factory ships meant 'a new start was possible'. A figure of 16,000 units (about two-thirds of the average annual catch of the six seasons prior to the war) was considered to be 'the only hope of preserving the stock of Whales'. In as much as the whaling companies were required to invest heavily in 'materiel' to operate within that limit, it was obvious how extremely difficult it would be to restore that figure, if once raised.

In January 1949, the Foreign Office's German Section reported signs of a shift in the stance taken by General Clay. He was reported to have privately remarked that it was 'competitive rather than security motives [that] were behind the opposition of other whaling countries to German participation'. The United States could not afford to carry on providing foreign exchange for what Germany was increasingly capable of producing. The head of the Norwegian Military Mission had meanwhile warned that any move to meet Germany's request could only be 'at the expense of other people'. The Norwegians had retrieved their position as the provider of more than half the world's catch (Jackson 2005, 215–216). A striking feature in all the debate, whether by correspondence or other means, was the lack of reference to the resumption of Japanese whaling. As a Foreign Office official confessed in April 1947, it seemed somewhat unreasonable that 'the Japanese should in this respect be treated more leniently than the Germans'. To the port of Hamburg the disparity appeared even more indefensible, particularly for the remark made by Dean Acheson (as the United States Secretary of State), during the International Whaling Conference: 'The whale stock of the world is an international fat reserve, which belongs to no single nation and to no group of nations'.(23)

## 16.4 Allied dissension

The Washington Conference and inauguration of the International Whaling Commission are commonly seen as the post-war United States legacy. Much less attention has been given to the part the United States Government played in re-establishing the Japanese pelagic whaling industry. Despite the strenuous efforts of other governments to prevent its happening, the United States used the influence of its newly found superpower status in this instance to considerable effect. The United States physical occupation of Japan lasted until April 1952, with General Douglas MacArthur as the Supreme Commander for the Allied Powers (SCAP) in the Far East. Particularly as the State Department came to focus on the Soviet takeover of Eastern Europe and the plight of the West European economies, so General MacArthur 'reigned as a minor potentate in his Far Eastern domain'. He had 'decisive command over all aspects of the Occupation' (Dower 2000, 73–75 and 79–81). Far from pressing for a peace treaty and the end to physical occupation, his priority had become by late 1947 one of securing Japan as a politically stable bulwark against the further spread of communism (Schonberger 1989, 47 and 72–74).

The trappings of Allied government were there. The Moscow Conference of Foreign Ministers had agreed, in December 1945, to form a Far Eastern Commission of the Allied Powers, which would meet in Washington and be dissolved on the signing of the eventual

Peace Treaty with Japan. An 'effective three-step formula' was adopted whereby, as one American official described it, the first step was to grant the 11 Allies a partnership in policymaking, the second was to reserve all operational matters to the Supreme Commander and the third was to treat everything that was important as operational. Any initial power the other Allied Governments might have exercised soon waned as Cold War tensions grew. However much the Commission was intended to operate by consensus, this was increasingly circumvented by MacArthur's use of 'interim directives' (Cohen 1987, 56; Schonberger 1989, 47; Eiji 2003, 98). Just as important, General MacArthur, as the Supreme Commander, was severely restricted by the almost complete lack of linguistic and technocratic skills required for achieving the ambition of establishing a post-war Japanese democratic government. Rather than adopt the post-war German model, MacArthur ruled through what continued to be an essentially Japanese Government, and, by doing so, made it considerably easier for those officials to preserve what they saw as important. The post-war whaling story is an outstanding example of how the 'SCAPanese model' of post-war development proved so effective (Dower 2000, 212–213 and 558; Tsutsui 2013).

Something more than brute force was required to secure a post-war Japanese democracy. As George Atcheson Jr (MacArthur's political adviser in Tokyo) telegraphed in a report to President Harry Truman of 4 January 1946,

> *One fundamental factor that we must not lose sight of in dealing with Japan is the inescapable relationship of politics and economics. In the dire straits of the Japanese people today, political development depends upon solution of the problem of food shelter and clothing. Today political education and experience in the practice of democracy are impeded by the preoccupation of the people with their economic distress. Left unsolved, these problems may encourage extremism of types we did not desire. (*FRUS *1945, 6, 87–89)*

There were nevertheless differences in how far the various Allied Governments felt able and willing to provide for the civilian population of a defeated Axis power. Those differences were perhaps most strikingly revealed by MacArthur's ultimately successful endeavour to restore the Japanese whaling industry, both for the physical sustenance and the self-esteem of the Japanese people in that critical post-war period.

There were many reports of people in Japan dying in the streets from starvation. As many as 1,300 deaths were reported over a six-month period in the Tokyo/Yokohama area alone. Increasing numbers of workers in mines, textile factories and steel mills were too weak to work. It was within that context that Atcheson stressed that the United States Government had taken full responsibility for feeding the Japanese population, and it was not for other Allied Powers represented on the Far Eastern Commission to interfere. As he wrote, in April 1946,

> *The food is being provided by the United States alone, not through an international organization concerned with the distribution of contributions by various countries. The burden falls on the United States. (*FRUS *1946, 211–217)*

While those other governments acknowledged this in principle, the Foreign Office was concerned when it received telegrams from the Washington Embassy, in early October 1946, reporting that SCAP had definitely authorised a pelagic whaling expedition as an emergency measure for one season only. In doing so, the United States Government was 'in complete accord with the view that any long-range consideration of the rights of Japan to conduct [a] whaling operation' was a matter for Allied discussion and decision. The one-off expedition would be responsible directly to the Supreme Commander; a catch of 1,500 BWU was anticipated. The Foreign Office telegraphed its preparedness to waive all objection on the assurance that 'the future of Japanese whaling is for Allied discussion and *decision* – we attach great importance to the latter word'. The Australian Government was less easily mollified, being particularly worried that the expedition would become a precedent for 'the rehabilitation of the Japanese whaling industry'. It must be made clear to 'the Japanese people and to the world that it is not a Japanese expedition'.(24)

There was no disguising the embarrassment felt within the United Kingdom Government. The Whalers' Section of the Chamber of Shipping had continued to lobby for any surviving Japanese factory ships and catchers to be transferred to British ownership. As the Washington Embassy impressed upon the State Department, 'the United Kingdom government are much perturbed by the information which reaches them as to the present policy of food collection and distribution in Japan'. The per capita supply to the Japanese people was already 'much greater than that available to the people of territories in South East Asia for which the U.K. is responsible'. It seemed furthermore anomalous that the Japanese population should be better fed than during the war, whereas 'the people of the U.K. are in a worse position and even faced with the prospect of further reductions' (*FRUS* 1946, 379).(25)

A SCAP press release of January 1947 recounted that whaling was to be resumed not only in the traditional shore stations but also in the Bonin Ogasawara and Kazan Islands, some 520 miles southwest of Japan. The refitting of two 1,200-ton oceangoing transport ships, *Nisshin Maru* and *Hashidate Maru*, had been approved, which together achieved an aggregate catch of 1,165 whales (690 blue, 474 fin and one sperm whale) in their first season. Their primary objective was to obtain the largest quantity of whale meat possible. As the *Fisheries Newsletter* later reported, under the headline, 'Whale Meat before Oil: Japs Break Agreement,' the *Hashidate Maru* and its six catchers had obtained as little as 9.57 tons of whale oil per BWU, compared with the 19 tons achieved by the British and Norwegian fleets that season. Altogether, a further 3,000 tons of oil could have been obtained. An Australian observer reported whole backbones and whale sides being thrown overboard. The Australian delegate at a specially convened meeting of the Far Eastern Commission, in January 1947, demanded that all future Japanese whaling would be confined to the shore stations. Both the New Zealand and United Kingdom representatives supported him. As Neil Mackintosh acknowledged shortly afterwards, the shore stations might process a far higher proportion of humpbacks and immature whales than the factory ships, but their aggregate catch was generally much smaller. Atcheson countered from the chair that the SCAP measures were purely temporary and set no precedent for the future. Any intervention made on the part of the Commission would 'unduly hamper and complicate fishery administration' (*FRUS* 1947, 161–164).(26)

There was Japanese pressure, even before the first expeditions had returned, for MacArthur's consent to 'the usual end-of-season repairs'. Dunn, at the Fisheries Department, drew the attention of the Foreign Office's Control Office for Germany and Austria, in April 1947, to a report of MacArthur taking 'strong exception to proposals for limiting Japanese whaling', claiming that 'the Japanese expeditions had complied fully with International Whaling Agreements'. They had provided 'gratifyingly large quantities of whale products' as 'a useful addition to Japanese food supply and a substantial sum in foreign exchange'. It was clear from the tone of the message that MacArthur intended to allow further expeditions. If so, Dunn wrote, they could only do so (within the 16,000-BWU limit) at 'the expense of Allied factory ships'. German commentators were already citing the renewal of Japanese whaling as 'a precedent for the renewal of German whaling'.[(27)]

Lord Inverchapel (the United Kingdom's Ambassador in Washington) was instructed in late April 1947 to remind the State Department of its definite assurance that the previous expeditions were 'an emergency measure and for one season only'. Any longer-term consideration of 'the right of Japan to conduct whaling operations was a matter for Allied decision and discussion'. This had become all the more important because any Japanese catch in the 1947–48 season would require 'a diminished catch for Allied vessels', if the overall 16,000-BWU limit was to be adhered to. However compliant the present Japanese expeditions might be when they were closely supervised, Inverchapel was instructed to warn that 'their present good behaviour cannot be regarded as evidence that they will continue to comply with International Agreements'. Not only should the Japanese vessels be allocated as war reparations, but the longer this was postponed, the stronger would become 'the vested interest of Japan in whaling'. The Foreign Office telegram concluded, 'Further expeditions should not (repeat not) be allowed'.[(28)]

The State Department essentially ignored the *aide memoire* by authorising a further Japanese-manned expedition. Atcheson insisted, at the Far Eastern Commission, that his Government had 'never given a clear-cut commitment which would prohibit another operation'. Whaling afforded the 'most efficient employment of indigenous Japanese resources by use of available ships, trained crews, supplies and readily obtainable petroleum products'. As Atcheson telegraphed,

> *Not to take advantage of these resources would be tantamount to deliberately wasting opportunity of profitable employment and result in void which American taxpayer would be called upon to fill by supplementing critical shortage of protein foods in Japan. (*FRUS *1947, 212)*

The Japanese would interpret the denial of another expedition as evidence that

> *SCAP and US are unable to withstand selfish and venal pressure of governments which, in fact, are making little if any direct contribution toward attainment of allied objectives in Japan and whose main objection is based upon a desire to maintain monopoly of whaling industry and profit therefrom at our expense.*

The expeditions had met a third of the Japanese meat requirement and gone a long way to compensating for the loss of the pre-war Manchurian soya-bean supplies, which had formed so large a part of the pre-war protein diet. It was in any case United States policy to oppose the kind of industrial monopoly that the British and Norwegian Governments were endeavouring to impose. It would be for the intended peace treaty to decide the longer-term size and disposition of the Japanese whaling fleet (*FRUS* 1947, 213–214).

The State Department's Legal Adviser confirmed that the Supreme Commander had 'complete authority', under the Potsdam Proclamation (as invoked through the Japanese Instrument of Surrender), 'to issue all orders for occupation and control of Japan'. Although he was required to consult Allied Governments, the Supreme Commander was the 'sole executive authority for the Allied Powers in Japan' and had full authority to order and administer a further expedition.

So informed, Major-General John H. Hilldring (an Assistant Under-Secretary and chief of the Civil Affairs Division of SCAP) called the representatives of Australia, New Zealand, Norway and the United Kingdom to his office on 27 May 1947 to discuss a second expedition. He began by explaining that the War Department was required to divert $325 million from its annual budget for supporting Japan. He then acknowledged the United Kingdom's *aide memoire* of 30 April, which had shown 'a fundamental difference of philosophy'. The United States Government did not believe that Japan, as 'a defeated enemy', should be prevented from whaling because 'she is a defeated enemy'. Whaling agreements were 'not articles of association of a club which could blackball new members'. There should be a second season, Hubert A. Graves (a counsellor at the Washington Embassy and the United Kingdom representative on the Far Eastern Committee) responded. While the United Kingdom Government would 'do almost anything to give a viable economy to Japan so as to reduce the burden on the United States budget', it was impossible to accept 'the view of the United States Government'. As Graves recalled later, only 'a wall of opposition' had stopped Hilldring from formally announcing a second expedition. He had bluntly warned that 'Uncle Sam could not be asked to dip into his pocket for oil and meat for Japan when there was this ready means of getting the proteins for yen'. Without a further expedition, dollars would have to be spent 'to provide the oil to prevent disease and unrest'. Graves accordingly applied for instruction as to 'whether we are to force the Americans to the point of veto' (*FRUS* 1947, 214–219).[29]

Others similarly lobbied the State Department. The Australian Ambassador, Norman O. Makin, protested, in the course of a meeting with Dean Acheson (the Under-Secretary of State) on 29 May 1947, that MacArthur's Natural Resources Section was attempting to revive both Japanese pelagic whaling and a phosphate-rock industry on the former Japanese possession of the Palau Islands (Eiji 2003, 188–189). Acheson rejected all notion of there being any threat to international security. The United States taxpayer could not be expected to pay for the import of proteins, which the Japanese were fully capable of providing for themselves. As to the Australian offer of its pelagic whaling fleet, the American taxpayer would still have to meet the cost. The Norwegian representative on the Far Eastern Commission, Knut Lykke, similarly warned Hilldring, in early June 1947, of the damaging effect that any further Japanese whaling would have on global supplies of oils and fats. As the 16,000 BWU limit would soon be met and the Japanese fleet had obtained a 'substantially lower output of oil per unit', there would in effect be a significantly lower aggregate output than if the British and Norwegian fleets alone had operated. The final figures from the Bureau for International Whaling Statistics showed the SCAP expedition to have achieved 78.5 barrels of whale oil per BWU. The Norwegians had obtained 116.5 barrels, which was nearly half as much again and comparable to what their average had been for the last four pre-war years (*FRUS* 1947, 220–221).[30]

## 16.5 The resumption of Japanese whaling

The State Department endeavoured to reassure SCAP, on 10 June 1947, that there would be a second expedition, but strenuous protests had delayed any public announcement until the Allied governments had first been notified. The intended *aide memoire* would warn that any further penalisation would only 'retard the reorientation of Japan as a peace-loving and law-abiding member of the family of nations'. The United Kingdom Embassy in Washington was instructed, on the receipt of the *aide memoire*, to protest strongly at the absence of any meaningful consultation. The second expedition should be either cancelled, or consent for it should be withheld (*FRUS* 1947, 6/8, 222). [(31)]

An eventual press release of 22 June 1947 announced 'a second SCAP-controlled and Japanese-manned Antarctic whaling expedition'. The first had obtained 'some 21,000 metric tons of much-needed protein food' as well as 12,000 tons of whale oil and 11 tons of vitamins A and D oil – a saving of more than $10 million in supporting the Japanese economy. The sale of the whale oil was expected to bring a further $6 million in foreign exchange. There had been 'no undue diminution of the whaling potential' as there had been so little whaling during the war. The Chief of SCAP's Natural Resources Section would again supervise the expeditions, with full daily radio reports and close monitoring by Allied observers (*FRUS* 1947, 6/8, 235–236).[(32)]

A paper drafted by the Foreign Office's Economic and Industrial Production Section on 'the Future of Japanese Pelagic Whaling' warned that it promised to be 'an extremely contentious issue in the Peace Treaty Negotiations'. The British and Norwegian companies wanted a permanent ban. The Australian and New Zealand governments wanted assurances on security grounds. But, as with Germany, there had to be a viable Japanese economy. The Japanese loss of whaling grounds, as a consequence of the war, had made those in the Antarctic all the more valuable. Although the Japanese pre-war record was indisputably bad, the author of the Foreign Office paper thought this would hardly warrant their permanent exclusion.[(33)]

Hubert Graves relayed, on 26 June 1947, that the strength of opposition on the Far Eastern Commission and 'the weight of our several diplomatic protests' had clearly made a considerable impression on the Secretary of State. However, Graves remained sceptical of whether the second expeditions would be withdrawn. The Australian and New Zealand delegates intended to press the issue for a vote on the Far Eastern Commission, making it very likely that the United States would use its right of veto for the first time. The Australian Ambassador, Norman J. O. Makin, had read an excellent statement at the ad hoc meeting of the commission, preparatory to moving a motion to ban all Japanese deep-sea whaling before the intended Peace Conference. Sir Carl Berendsen (the New Zealand Ambassador) had seconded the motion. In doing so, both drew particular attention to their wartime exposure and the proximity of the Antarctic whaling grounds to their countries. The minutes of the ad hoc meeting of the Far Eastern Commission record that Graves intervened at that point both to congratulate Makin on his cogent arguments and to warn that any vote taken by the commission at that point might disrupt the diplomatic negotiations that were taking place. Graves accordingly recommended, as he had been instructed to do, that any formal resolution should be postponed. The Canadian and Indian representatives, with the French (whom Groves described as always helpful), regretted the unilateral action that had been taken by the Supreme Commander but agreed that there should be a postponement of any vote.[(34)]

The Foreign Office's Economic Supply and Relief Distribution Section had taken charge of the question, telegraphing the Washington Embassy on 30 June 1947,

> *You should now see Mr Marshall and urge him to reconsider the decision to authorise the second expedition, at least until there has been an opportunity for further discussion between the interested Governments, which include Australia, New Zealand and the United Kingdom. The desirability or otherwise of Japanese whaling*

Another American *aide memoire* followed on 13 April 1948, announcing a third and even larger expedition to meet 'the acute food shortages and maximise foreign exchange resources for Japan'. It would now comprise three factory ships, seven carrier vessels and 15 catchers. Adequate steps would be taken to ensure adherence to international whaling regulations – Allied observers would once again be welcomed. The Glasgow-based brokers, David Geddes and Company, warned the Ministry of Food of the encouragement this would give the Germans in their endeavour to rejoin the industry. Whatever the nationality of the particular fleet, it 'could only result in increased competition for the same number of whales and a less economic production of whale oil'. The Whalers' Section of the Chamber of Shipping had meanwhile begun to assail the Ministry of Food, protesting that the Japanese catch of 1,321 whales was almost exactly the equivalent to what the other expeditions would have caught in the eight days denied them in the 1947–48 season by being required to keep below the 16,000-BWU limit. Although ostensibly it was still for the peace treaty to decide the issue, the impression had already been implanted in 'Japanese minds' that their former prominence in Antarctic whaling had been restored. Far from raising the BWU limit to accommodate this, the Norwegian representative on the Far Eastern Committee warned that the limit might have to be reduced to accommodate a further decline in whale stocks.[42]

In warning that further expressions of opposition would 'only harden the Americans' in their attitude, Graves saw the key question to be the number of Japanese expeditions. MacArthur was keen to authorise three factory ships, as opposed to the previous two permitted. The State Department went no further in discussions than to agree that the third post-war Japanese expedition should comprise two factory ships and their attendant vessels. Remarking on the press reports of the departure of the *Hashidate Maru* for the Antarctic, an official in the Ministry of Transport noted that the presence of thousands of cheering people (including organised student rallies) seemed all too reminiscent of the pre-war years. There was no doubting that any exclusion of Japan 'from Antarctic whaling would be bitterly received and cause considerable ill-feeling'. Further exchanges of *aides memoire* followed, including one of mid-November 1950 protesting MacArthur's approval of a new factory ship. The order went ahead, despite the 'quite firm language' of others on the Far Eastern Commission. It was through piecemeal concessions that the much-postponed peace treaty had become an irrelevance. Japan had already joined the IWC, pledging its adherence to the International Whaling Convention. As the Depositing Government, the United States had formally notified the IWC of Japan's faithful observance of 'all obligations and responsibilities set forth in the provisions'. There followed, in mid-May 1951, a *note verbale* in which the IWC was informed that Akiro Kodaki (an official of the Political Affairs Bureau of Foreign Affairs in Tokyo) had been appointed the Japanese Commissioner.[43]

**Figure 22** A selection of images of whaling activities at South Georgia. © British Antarctic Survey, UKRI.
(a) The Grytviken whaling station in 1929 from the foot of Mount Hodges, in the *lower right-hand corner* is the Norwegian Whalers Chapel. The snowy peak just showing in *the upper right-hand corner* is Mount Paget.
(b) Two whale carcases alongside a whale catcher.
(c) A rorqual on the plan at Grytviken showing the baleen plates in its mouth. Note the flensing knife in action to the *right* of the carcase.

**Figure 22** (continued)
(d) A large rorqual being hauled up the slipway to the flensing plan at Grytviken in 1929.
(e) A whale floating off the slipway waiting to be flensed in 1958. Note the large gathering of Cape Pigeons around the carcase. The whaling industry provided a rich source of food for a variety of scavenging sea birds.
(f) Men loading blubber onto the conveyer belt, which transported it to the cookers to be rendered down to oil.
(g) The manager's villa at Grytviken in 1958. This building became the whaling museum established by Nigel Bonner and was later renamed the South Georgia Museum.
(h) Falkland Islands Government Settlement at King Edward Point on South Georgia with Mount Hodges (605 m) and the Grytviken whaling station in the background.

# PART V

# *A scientific overview*

CHAPTER 17

# *The Discovery Investigations – scientific opportunities, successes and failures*

The previous chapters have provided a diplomatic history of the part Britain and its Empire played in the first half century of whaling in the Southern Ocean. The 20th century opened with Britain taking a central role in controlling and regulating the development of the industry with protection of the whales a core part of its strategy. Fifty years later Britain's leadership had all but disappeared and this book rightly draws the analysis of the diplomatic history to a close at the point when a comprehensive international regime for the management of whaling had been agreed. This was the International Whaling Commission under the International Convention for the Regulation of Whaling that was signed in Washington DC on 2nd December 1946. There has been much written about how the IWC has performed over the subsequent time but that is very well documented elsewhere. Now, this book will turn to the subject of the development of whaling science and in particular the scientific programme of the Discovery Investigations. These began in 1925 and ran in parallel with political developments but its story is perhaps best described in the style of a scientific review. Fifty-five years passed between the start of the Investigations and publication of the last report, but despite the great political and commercial changes that the industry experienced over that time, including the Second World War, there is virtually no evidence of these in the scientific record. Furthermore, because the science programme continued well into the IWC era it links into late 20$^{th}$ century science and to later research with its technologies that could not have been imagined in 1925. This review therefore briefly looks at scientific developments beyond the Investigations and into the 21st century.

The Discovery Investigations were the first scientific programme to investigate whale biology in relation to whaling, and their primary purpose was to provide information for sustainable management of the whaling industry. Beyond that, the Investigations were to gain an understanding of the ecological system sustaining the large populations of whales in Southern Ocean waters. There has been criticism of the failure of the Investigations to provide the scientific evidence needed to regulate the whaling industry and the supposed culture that developed among the 'hip-booted' scientists and the unhelpful relationship between scientists and whalers as they worked together on the whaling stations and factory ships (Burnett 2012). More has been written highlighting the role of the Investigations in sustaining sovereignty claims in the Antarctic and sub-Antarctic, at a time when the sun began to set on the British Empire (Roberts 2011). These are compelling historical aspects, and there is no doubt that every attempt to regulate the industry had failed when stocks became so depleted by the 1960s that Antarctic whaling collapsed and never recovered.

Data were to be collected on how environmental variability causes changes in food resource and, in turn, whale numbers – particularly of vulnerable young – as well as changes in the movements and distribution of adults.

Kemp also identified the need for research on the natural history of the elephant seal (which was exploited commercially but not threatened by over-exploitation at the time) and seabirds at South Georgia. He also wrote that the commercial potential of finfish resources, particularly on the Patagonian Shelf, should be explored. Finally, hydrographic surveying was needed to improve charts of the waters of the Falkland Islands Dependencies, in particular for the benefit of whalers. These objectives did not include any assessment of the size of the exploited whale populations nor determination of reproductive rates. There was no method at that time for measuring either of these parameters that were needed for understanding of the dynamics of the exploited whale populations. Population dynamics theory and methodologies for practical data collection from fish stocks developed more or less independently of whale science during the course of the Investigations. Fisheries science and whale science would only come together late in the era of Antarctic whaling.

Scientific equipment carried on *Discovery* and *William Scoresby* included vertical and horizontal plankton nets equipped with depth gauges and closing mechanisms, young fish trawls, scientific trawls and dredges, hand nets and rakes of the type used to harvest mussels. They also carried harpoons and a small harpoon gun for collecting porpoises and dolphins. For marine chemistry and hydrological research they carried water samplers, Secchi discs (to measure transparency of seawater) and a variety of hydrological instruments and coastal survey equipment. Scientific accommodation in *Discovery* included a laboratory for biology and one for chemistry, a darkroom, two holds and a locker. In *William Scoresby* there was a single scientific laboratory below deck. Ashore on South Georgia, Discovery House had science laboratories as well as accommodations for the Discovery Investigations staff.

The scientific staff appointed at the start of the Investigations would each have had some influence over the research during the early years. Stanley Kemp was science director. Although he had no experience of whale biology he had started his scientific career as a fisheries researcher where his mentoring by Sidney Holt shaped his professional life (Wilkins 2020). His own research had been with exploratory fishing off the west of Ireland with a specialist interest in crustacean taxonomy. Fisheries science generally was still in its infancy at the start of the Investigations so there was much to be discovered, especially in relation to whales. Kemp, who was highly respected for his energy, leadership and intellectual capacity was the science leader until he left in 1936 to become director at the Marine Biological Association, Plymouth (Hardy 1946).

Kemp's initial team of scientists included 13 zoologists and three hydrologists. Most were recent graduates, but some had prior scientific experience. Only one had experience of whales or whaling. Several provided substantial intellectual contributions during and after their employment with the Investigations. These included Neil Mackintosh, J. F. G. Wheeler, Alec Laurie and Leonard Harrison Matthews who focused on whale biology. Others, including Alister Hardy and George Deacon, made notable contributions to the wider areas of marine science during and subsequent to the Investigations.

The only zoologist with any prior experience of whaling was James Hamilton, who had by now spent several seasons in the dependencies. However, during his secondment to the Investigations, he had argued with Kemp and failed to get on well with his colleagues.

Kemp opposed his appointment as senior scientist in 1928, recommending that he return to colonial employment. He had taken strong positions on both the conservation of fur seals and striated caracara in the Falkland Islands. One of his regular reports on whale catches at South Georgia in 1942 particularly alarmed officials in the Colonial Office by warning of a dramatic decline in the numbers of adult whales in the catch and the implication for survival of the stocks. The Colonial Office responded by commissioning a report by C. F. Hickling and Hamilton, which similarly warned that southern whale stocks were on course to suffer the same fate as those of the north (https://falklandsbiographies.org/biographies/hamilton_james).

Where available, summary biographic details of scientists mentioned here and later in this text are in the Appendix.

## 17.1 Whale science, 1925–42

### *Initial research at Discovery House*

Scientific research began on 5 February 1925 when the recently arrived team at Discovery House started work on their first whale carcase. It ended in 1980 with the publication of the last Discovery Report, on cephalopods in the diet of sperm whales by Malcolm Clarke, who was born five years after the Investigations had begun (Rodhouse et al. 2014). The scope of the Investigations over 55 years stretched from the biology of single species of whale to the biological, chemical and physical oceanography of the Southern Ocean. The primary objective was to gather information for rational regulation of the whaling industry.

At least 90 per cent of a whale's life is spent submerged in the vastness of the oceans, so its population ecology could not be studied directly (Laws 2017). Apart from inferring distribution and abundance from spouts, there were only two ways of gathering data: the first was firing tags into the blubber of whales at sea, which might later be recovered when the animal was killed by whalers. This might provide information about migrations and perhaps growth and abundance. The second was to work on carcases of whales to study feeding, age, growth and reproduction. This was the case throughout the Investigations, and today, assessing the status of whale populations is still problematic. Whale science continues to be limited by the difficulties of determining numbers in present populations, by the long periods needed to assess population trends and the challenge of determining the structure of populations, especially when breeding grounds are unknown.

Led by Neil Mackintosh, the first scientists to arrive at Discovery House – Wheeler, Clowes, Harrison Matthews and the laboratory assistant Saunders – spent three austral summers between 1925 and 1927 working on whale carcases and collecting samples at the Grytviken whaling station. During the intervening winters, they moved to the Saldanha Bay whaling station on the Cape Coast, South Africa, and continued their research, particularly to compare aspects of whale biology between the two areas and to investigate whether any relationship existed between them. Early in 1928 there was a change of staff at Discovery House and work was continued by F. C. Fraser and George Rayner, later joined by Dick Ommanney in 1929. In 1931, research at Grytviken and Discovery House ended.

## *Research results 1925–1942*

The years 1925–31 produced a large body of data on the external characteristics of the whales, their food, blubber, parasites, reproductive organs, breeding and growth. In the first two years alone some 1,683 carcases were examined, the majority of which, 1,577, were blue and fin whales in roughly equal numbers (Mackintosh and Wheeler 1929). The series of observations revealed mating time and foetal development. Timing of births and nursing enabled growth from conception to birth and weaning to be determined with some certainty. In fin whales, gestation was shown to take about one year, calves weaned at around six months and female sexual maturity was thought to be reached after a little over two years. Some of these data would be extended and the results revised, mostly from later work on factory ships, but this period at Discovery House marked the beginning of modern whale science.

Originally, no morphological differences could be found between blue and fin whales examined at South Georgia and South Africa, and there was circumstantial evidence of migration between the two areas because gaps in the size frequency of the South Georgia whale catch were matched by the presence of those sizes in the South African whales. The data revealed that the proportion of immature whales being killed was unduly high at South Georgia and especially off South Africa. Mackintosh and Wheeler warned that this was economically extravagant and, although on a relatively small scale, maximum damage to the stock was being inflicted relative to profit gained.

Blue and fin whales appeared to have a protracted breeding season. It was not known where they went to breed in the austral winter when the lack of exploitation coupled with a high rate of growth to maturity suggested the stocks of both species might be able to maintain themselves under exploitation. However, this observation was countered by the slow rate of propagation of both species, which would make them susceptible to experiencing diminishing numbers.

Mackintosh and Wheeler seem ambivalent at this point in the Investigations. At a time when it was already clear that the humpback population had been over-exploited to a considerable degree and when the blue and fin whales were already under pressure, they seemed to be warning that these species were at risk and then following up with caveats about uncertainty, which would have undermined any case for restraint by the industry. Having drawn the conclusion that for both species, the whales at South Georgia were of the same stock as those off South Africa, they then argued it was possible that, after reducing the stock in one area, it could be replenished from the other.

## *Determining age and growth*

Knowing the age structure of Antarctic whale populations and how it changes under exploitation would provide a reliable indication of the influence of whaling on the stock. Populations experiencing unsustainable rates of exploitation suffer a progressive loss of older animals until a point is reached when there are too few sexually mature animals remaining to produce enough offspring to balance those being killed.

Data on growth rate of animals are also used for estimating the rate at which populations can be exploited. If whales caught could be aged as well as measured, then plotting size against age would provide an average growth curve for each species. Something akin to the annual rings in the ear bones (otoliths) of fishes was needed, but at the time there were no known time markers to estimate the age of whales. Mackintosh and Wheeler had visited Cambridge to meet Francis Marshall, who was a leading authority on mammalian reproduction. From this collaboration a possible time marker was identified. A female mammal is born with a lifetime store of eggs in the ovary and when she reaches reproductive age these are released, one or more at a time. Each egg leaves a scar, or corpus, at the site of attachment (the corpus luteus which later shrink and are then termed corpora albicantia). In some mammals an egg is released only after mating but in others at more or less regular intervals. It became clear that baleen whales do not need the stimulus of mating to ovulate because two females, each with one corpus in the ovary, had immature mammary glands and unbroken vaginal bands. There was therefore a chance that the number of corpora could be used to estimate the number of years since the attainment of sexual maturity of females.

At Grytviken, routine data collection from whales on the flensing plan included counts of corpora in the ovary. These are about the size of a grapefruit at first and shrink to a diameter of about 20mm over time. Once a large data set had been gathered, it was possible to discern frequency patterns in the numbers of corpora present. Some numbers were more common, a pattern that enabled the average number of ovulations between breeding cycles to be counted, whether or not they had resulted in pregnancy. This in turn enabled the approximate age of female fin whales to be determined from the number of corpora. Wheeler (1931) reported that female fin whales with one to four corpora in the ovaries are in their first or second year after sexual maturity; those with five to nine are in their third or fourth year, and those with ten to fourteen are in their fifth or sixth year, and so on. Only a small number of whales showed signs of multiple ovulations, and very few pregnant females were carrying twins, so it seemed that ovulation generally took place one egg at a time. Additionally, there was no sign of menopause in whales up to age 20 years, so corpora could be expected to be added regularly at least to this age. The researchers concluded that the number of corpora in the ovaries of sexually mature females could be used to estimate age in the whale's early years, taking into account the observation that females breed every two years, and that ovulation would be halted during pregnancy and nursing.

The first estimates of growth of blue and fin whales were published by Mackintosh and Wheeler (1929) and Wheeler (1931). No growth estimates could be made for males, and the results for females were approximate because they depended on an assumption about the number of ovulations per year derived from the poorly defined plots of the frequency of numbers of corpora across the population. In fact, the conclusions were revised some 30 years later by Laws (1961), whose research resulted in the age of sexual maturity in female fin whales being revised upwards to an average of four to five years and the frequency of ovulation based on corpora counts to be revised downwards. Together these revisions meant that fin whale age estimates in the earlier study had been underestimated. Nevertheless, the early data from Discovery House and Saldanha Bay, collected at considerable effort, had provided a useful first estimate of growth rate of females following sexual maturation.

Corpora provided the only means for estimating whale age during the early years of the Discovery Investigations and, although less precise and accurate than later data from annual laminations in the whale's ear plugs, they provided useful insight into the adverse effect of exploitation on the age structure of populations. They are no longer used today as a sole means of estimating age. Growth layers in ear plugs have become the most widely used and reliable means of ageing baleen whales.

At the start of the Investigations, it was known that the degree of fusion of the spinal column increases with age up to the time the young whale becomes physically mature (Flower 1864). It was thought that this might provide a way to determine the age of younger whales. As it turned out, fusion was found to be poorly correlated with whale length. It was impractical in any case to determine the degree of fusion routinely in the time available during the flensing process. Complete fusion was found only in fin whales with more than 15 corpora present in the ovary and it was concluded that physical maturity, defined by spinal column fusion, was reached at some six to eight years of age (Wheeler 1931).

### *Use of age data to demonstrate the consquences of exploitation in fin whales*

The first published assessment of the influence of whaling on the stocks of fin whales was by Wheeler (1934), who drew on the age data based on corpora counts in the ovaries of females over the seven seasons of research based at Discovery House. Age composition of the catch indicated that the stock was being reduced by exploitation at a rate of 26 per cent over the period of the study. About 50 per cent of the expected numbers of older whales were missing from the population and sexually immature females comprised 32 per cent of the total catch. Wheeler considered this to be an alarmingly high proportion for the health of the stock. If the trend of an ongoing reduction in number of sexually mature females was continued, the rate of replenishment would slow and eventually cease altogether.

Wheeler was compelled to add some caveats. First, he questioned the extent to which the catch at South Georgia might represent the whole stock and second, whether the same stock was represented in the fishery in different years.

Another element of Wheeler's (1934) report was evidence of repeat migrations from warmer waters in the numbers of scars left on the skin by external parasites. These attach in warm water but do not remain once the whales migrate to cold seas leaving scars behind that increase in number with the age of the whale. Initially there had been some hope that counts of parasite scars on the skin might provide a simple and rapid means of estimating age, but this proved to be impractical.

### *Use of age data to demonstrate the effects of exploitation in blue whales*

Alec Laurie undertook and published research on respiration in blue and fin whales (Laurie 1937), but his main contribution was on the age of blue whales and the effect of exploitation on their numbers. He first joined the Investigations as zoologist in RRS *William Scoresby* during 1929–30, then in RRS *Discovery II* from October to November 1930. Between 1930 and 1931 he worked at the Marine Biological Station, South Georgia. By that time pelagic whaling had become established and he had the opportunity to work on *Southern Princess* for the season 1932–33. Laurie had collections of ovaries from the 1925–31 South Georgia and Saldanha Bay work, from his time on *Southern Princess* and from factory ships operating on the southernmost whaling grounds. He also obtained statistics from other factory ships that had provided ovaries and summary statistics from Norwegian authors.

Laurie concluded that blue whales examined on factory ships in the Weddell Sea and around Bouvet and Kerguelen were identical – including in their reproductive behaviour – with those taken around South Georgia and off the African coast. Sexual maturity was attained at the same age and length, and the mating season and gestation time were identical. Most female blue whales were found to breed every three years, but some bred once in two years.

Newly matured females had an average of 1.91 corpora in the ovaries and in subsequent years the increment decreased to 1.13 per year. Laurie tentatively concluded that, as for fin whales, there was a correlation between age and number of corpora, with whales between two and three years old having two corpora and then for the number of corpora to increase by eleven every ten years. The oldest females were around 30 years old.

Most important, the large data set on the age distribution of adult blue whales proved that recruitment had slowed progressively in preceding years, and recruitment rate under the rate of exploitation was insufficient to maintain the stock. The length of females killed over the previous five years revealed that, on average, they had been killed before they had any chance of reproducing. In the 1936–37 season over 41 per cent of all females had been killed before they reached sexual maturity.

Laurie's final words in his report read: 'The evidence above all points to the acute hardship which is being suffered by the Antarctic stock of blue whales. The stock is already seriously depleted and further hunting on the same scale bids fair to make blue whales so scarce that they will cease to be a source of profit to the industry and so diminished in numbers that the stock even if completely protected may take many years to recover.' This was the clearest statement in any of the *Discovery Reports* up to that time that virtually irreversible damage was being done to the whale stocks by the whaling industry.

## *Applied whale science of other species*

Monographs on the biology and impact of whaling on four species of whale were published by Matthews (1938a, b, c, d). The species concerned were the humpback, sperm, southern right and sei whales. The reports were largely descriptive accounts drawing on data from the Discovery Investigations and elsewhere plus summary accounts of the commercial status of each species.

None of these species were of much commercial interest to the Antarctic whalers at the time. The humpback whale stock had been the predominant species killed when the Antarctic fishery started but the stock was rapidly reduced, thus the species became insignificant to the industry once the larger rorquals were targeted. There had been excessive exploitation, both on the feeding and breeding grounds, and the stock would only recover if given total, global protection for an extended period of time. If there were to be a recovery, any future fishery would have to be restricted to the Antarctic feeding grounds and the catch rate carefully controlled.

By the early twentieth century, petroleum products had replaced sperm whale oil for most applications, and the American sperm whale fishery had closed (Hohman 1928). Demand for sperm whale oil was low and of minor value to modern whalers as it could not be hardened

Two relatively brief *Discovery Reports*, contemporary with the fieldwork, dealt with distribution and movements of whales (Kemp and Bennett 1932; Hart 1935). The results from the marking programme to 1939 were reviewed by Rayner (1940), who later published a brief report on the distribution of blue fin and humpbacks at the time they were marked (Rayner 1948). Most showed only local movements for the reasons mentioned above. Blue whales at South Georgia and elsewhere were found to move within the Antarctic whaling season. They were dispersed around the Antarctic and although they tended to return to the same area between years, there was no evidence of segregation between regions. Fin whales at South Georgia were moving too, but they tended to spend more time there than blues. They mostly returned to the same area between seasons but, as with blues, there was some dispersal between South Georgia and Shag Rocks, some 110 miles to the west of South Georgia, as well as, on a larger scale, between the Bellingshausen Sea and Queen Mary Land. Brown (1962) later presented evidence that some fin whales migrated into subtropical waters during the winter.

Of the three species, humpback whales provided the best information about migrations. They were found to migrate from Antarctic to tropical waters, including to northwest Australia from Queen Mary Land, and to Madagascar from Enderby Land and Dronning Maud Land. These migrations appeared to be annual. Unlike fin and blue whales, there was little evidence of dispersal, and it was concluded that there are more or less self-contained groups that migrate between distinct summer and winter locations. It was evident from the high proportion of marked humpbacks killed that exploitation of the species was heavier than for blue and fin whales because of the increased whaling in tropical waters. Of the other two species it was evident that blue whales were being depleted faster than fin whales.

Given the effort expended on whale-marking, the rates of return were disappointingly small. The results gave more insight into the small movements of whales within a season than about large migrations and provided no useful data for growth studies or for estimating population size. Nevertheless, once integrated with catch data they provided details of whale populations and movements.

## *Progress reviewed in 1942*

A report on the southern stocks of baleen whales by Neil Mackintosh was published in 1942, in the middle of the Second World War and was followed by a more widely available review published immediately after the war (Mackintosh 1946). *Discovery II and William Scoresby* had been pressed into war service with the Royal Navy, *Discovery II* as an armed boarding vessel, and several staff with seagoing experience joined the navy or otherwise found wartime roles that used their experience (Coleman-Cooke 1963). Only Mackintosh, now director, was retained so that the Investigations did not end completely. As well as publishing his own research, he ensured that the research carried out during the last years of peace was published. About half of the 1942 report was a summary of the biological research published since the start of the Investigations in 1925, the main elements of which have been covered in the previous paragraphs. The rest used the catch and related data provided by the whaling companies to the Norwegian Committee for Whaling Statistics. The Committee had published since 1930 an analysis of the statistics for the Antarctic in annual reports (Hjort et al. 1933a, 1933b, 1933c–1938). Mackintosh used these statistics, together with data from the Discovery

Investigations, to provide an extensive account of migrations, distribution and relative abundance of the whale species exploited.

Of special interest is the section in the report on the status of the stocks. It stresses in the first paragraph that no method existed to make reliable estimates of actual numbers of whales in Antarctic waters, so the effects of whaling could only be seen from changes in the size and composition of the catch.

The percentage of blue whales in catches and catch per effort had been declining, whereas the proportion of immatures in the catch had been increasing over time. The average size of the whales had been decreasing. This was clearly affecting the industry, which was taking a greater proportion of less frequently caught species to compensate. There was less evidence that the stock of fin whales was being depleted other than an increase in the number of immatures in the catch. However, there was little doubt that the stock could not withstand the rate of exploitation that had been occurring up to the start of the war in 1939. The marked decline in humpbacks was again emphasised. Because the stock is divided into geographical groupings, the effect of exploitation had not been even around Antarctica, the most affected area being South Georgia. Humpbacks had initially been abundant when the pelagic fishery moved to the Indian Ocean sector, but numbers there soon declined.

When the report was published there were no reliable means of assessing the size of whale stocks and Mackintosh went to some length to discount theories about how the problem might be solved. Direct counts at sea from ships was discounted, largely because of errors introduced by variable atmospheric conditions and visibility. The 'mark recapture' approach, based on knowing the number of whales marked and the ratio of marked to unmarked whales in the catch, was also discounted because there would be too many unknown variables. Mackintosh particularly considered a statistical method of assessment, developed by Hjort et al. (1933a), based on the assumption that the rate of reduction of the catch is proportional to the fraction of the total stock killed each year. As for the other proposed methods, this approach was also discounted because it would have to make untested assumptions about rates of whale growth and propagation.

The review ends with a prescient proposal. Because there was no method to assess the size of stocks, or to calculate a sustainable level of exploitation, Mackintosh emphasised that all that could be done at the time to conserve stocks was to reduce the rate of depletion rather than determine a rational rate of exploitation. The London Conference of 1938 had agreed to prohibit the killing of whales in the Antarctic between 40° and 160° west for a period of two years. Mackintosh's proposal was to establish whale sanctuaries, particularly remarking that 'there seems good reason to believe that reserves of whales can be protected by means of sanctuaries . . .'; ' . . . it may be suggested that the term "sanctuary" is best applied to an area in which hunting is totally prohibited'. Fifty-two years later, in 1994, the Southern Ocean Whale Sanctuary was established by the IWC with 23 countries supporting the agreement and only Japan opposing it. Another 20 years elapsed before Japan accepted the ban in 2014.

## *Other whale science 1929–1942*

During the period to 1942, aspects of whale biology that had no immediate relevance to the applied problems of the fishery were published in several *Discovery Reports*. The science was mostly descriptive and took advantage of the large quantity of material available during processing of whales at shore stations and on factory ships. One study of *retia mirabilia* (vascular networks) in whales (Ommanney 1932a) had been recommended as an important subject for research in the Report of the Interdepartmental Committee (Parliamentary Papers 1920), apparently for its academic interest. Others were anatomical works including the urogenital system of the fin whale (Ommanney 1932b) and histology of the cetacean lung (Haynes and Laurie 1937). Laurie's 1933 report on respiration in blue and fin whales described pioneering research on whale diving physiology. Two other reports were descriptive accounts of single whole specimens of species about which very little was known: a rare porpoise (Hamilton 1941) and the southern bottlenosed whale (Fraser 1945).

## 17.2 The Investigations post–Second World War

Neil Mackintosh's *Discovery Report* of 1942 marked the end of the initial Investigations. It would be another three years before the war ended, after which the IWC was established with the responsibility of regulating whaling worldwide. The Discovery Investigations were transferred in 1949 to the Admiralty and incorporated into the new 'National Institute of Oceanography' (NIO). George Deacon and Neil Mackintosh became director and deputy-director, respectively. *Discovery II* and *William Scoresby* would return to the Southern Ocean for one more season in 1950–51. Henry Herdman was chief scientist on *Discovery II*, accompanied by young new recruits Peter David, R. A. Cox, Peter Foxton and three assistants. On its return, *Discovery II* was used by the NIO for research in other oceanic regions while *William Scoresby* was taken out of service and broken up.

### *Applied whale science*

The establishment of the IWC and NIO brought major changes. The only field science involving British scientists after 1951 was done by biologically trained whaling inspectors aboard the pelagic factory ships, who included Richard M. Laws, Nigel Merritt and Malcolm Clarke. Offshore marine science in the Southern Ocean from the 1970s was dominated in the United Kingdom by the British Antarctic Survey, an exception being in 1978 when the 100th cruise of a later RRS *Discovery* occupied a series of stations between South Africa and Antarctica, during which an Antarctic colossal squid, *Mesonychoteuthis hamiltoni*, was caught. It was the first live adult to be seen, although it is commonly found in the diet of male sperm whales in Antarctic waters (Rodhouse and Clarke 1985). By this time, NIO had become the Institute of Oceanographic Sciences (IOS). It is now the National Oceanography Centre (NOC), having merged with the Proudman Oceanographic Laboratory in Liverpool. In his autobiography, Laws (2017) describes the complex internal politics and some eccentric characters at NIO in the 1950s but also provides an account of the key research then carried out. Laws and seven of his colleagues would later be appointed Fellows of the Royal Society for their work.

Laws published his work on foetal growth, and adult age, growth and reproduction of southern fin whales as *Discovery Reports* (Laws 1959, 1961). For this he drew on a much larger collection of samples and data than had been available to Mackintosh and Wheeler in the 1920s; this allowed him to update much of that earlier work with age data derived from growth laminae in ear plugs. He was then able to plot growth from conception to full adult size and fit a von Bertalanffy (1938) growth function to the plot.

The ear plug of baleen whales had been discovered by Denis Lillie (1910) while he was working at the South Innishkea whaling station off the west of Ireland, before he joined the *Terra Nova* Expedition with Scott. However, it was not until some 45 years later that the structure of plugs was found to include laminations that could be used for determining the age of whales. The plugs, later found in other baleen whales and sperm whales, are formed from wax and keratin which, when sectioned longitudinally, reveal alternating light and dark bands. These were first described by Purves (1955), who examined Lillie's samples that had been deposited at the British Museum (Natural History). Roe (1967) later confirmed that each light and dark pair represented one year of age by examining monthly series of samples throughout the year. Other authors continue to confirm the correlation between corpora counts and growth layers in the ear plugs of other baleen whales (Gabriele et al. 2009; Read et al. 2018).

In his 1961 report, Laws also confirmed Mackintosh's (1942) observation that the rate of pregnancy of blue and fin whales from the Antarctic increased until the Second World War and then declined. Pregnancy rates had increased again after the war, suggesting the increase in whale stocks had reduced the biomass of krill and hence pregnancy rates.

Mackintosh and Brown (1956) published preliminary estimates of the population size of the larger Antarctic baleen whales and Ray Gambell (1972) later published a study of data from the sperm whale fishery off Durban, as well as an analysis of the fishery which is discussed below (Gambell 1972). Later, Mackintosh (1966) published a review of the distribution of southern blue and fin whales.

Research done in the 1920s at the Discovery House laboratory had shown that baleen whales in the Antarctic fed almost exclusively on Antarctic krill, *Euphausia superba*. Other zooplankton, such as the amphipod *Themisto gaudichaudii*, were often found in small numbers in whale stomachs, as well as other occasional planktonic and nektonic forms. The latter were inconsequential and certainly taken incidentally. Parts of Marr's (1962) monograph on krill included observations on the quantity of krill eaten by baleen whales, the effect on foetal growth of krill consumption by the mother, feeding migrations of whales in relation to krill distribution, swarming behaviour of krill and the effect of whales feeding on swarms.

Attention in the 1960s was turning to the effect of the reduction of whale biomass on the Southern Ocean ecosystem and especially on krill biomass. Gambell (1968) found that sei whales were advancing the time of their migration into the Southen Ocean and were also going farther south than in earlier years. He concluded that this may have been because there was less competition for krill from blue and fin whales. At this time krill consumption by baleen whales was thought to be about one-third of that consumed prior to the whaling era, and a few years later, Laws (1977) reviewed increases in populations of several species of seals and seabirds that appeared to be due to an increase in krill biomass.

in food resource and, in turn, whale numbers, particularly of vulnerable young, as well as changes in the movements and distribution of adults? Having found earlier that baleen whales in the Southern Ocean fed almost exclusively on Antarctic krill, *Euphausia superba*, this species became the prime focus for studies of whale diet. Yet, despite the considerable amount of research undertaken, the question of the annual variability in krill biomass is not addressed in the Reports. The research voyages of the ships were not planned to enable repeat sampling at regular intervals over time, despite this being needed to answer the question. The spatial and temporal scales of the sampling programme were not compatible with the scale of the process under investigation. A study of annual variability would have required krill to be sampled quantitatively and annually at specific stations relevant to the whale's annual migration cycle. This would have proved impracticable on voyages that were designed to meet a wide range of scientific objectives.

Why were more of the Investigation's objectives related to applied whale science not achieved? Part of the answer may lie in the cultural norms of the scientific community in the first half of the twentieth century. Scientific research was largely an individual activity or done at most by one person with a small group of assistants. Rutherford first split the atom at the Cavendish laboratory, Cambridge, in 1932 with a team of only two graduate students. It is also notable that, although scientific teams are much more common today, especially for complex interdisciplinary projects, the highest awards for scientific achievement continue to go mainly to individuals. A quick scan down the contents of the *Discovery Reports* reveals that only 11 of the reports had two authors, of which only three dealt with applied whale science. With hindsight, it is apparent that progress in applied whale science required a combination of biological and mathematical learning and skills, and that a group of young biology graduates working in a remote location in the earlier twentieth century, however bright, was unlikely to make the kind of advances required. It took Beverton's biological and Holt's mathematical expertise to produce their seminal work, developing their own interdisciplinary thinking and the thinking of others who had gone before them. Progress in the application of population dynamics was slow. It had taken from 1918 to the 1950s before scientists with the necessary focus, working across disciplines, provided the practical methodologies required by fishery scientists.

Perhaps equally important, there was little or none of the culture of 'science into policy' that exists today – and even in the twenty-first century, translating fisheries science into effective policy remains problematic (Daw and Grey 2005). The individual scientists of the Discovery Investigations, with scientific careers before them, would have had little personal incentive to invest time and effort in using their research results to influence policy on the regulation of whaling. None of them, and least of all Neil Mackintosh, seemed inclined to become embroiled in a political way, as did Sidney Holt once he had become engaged with whale science within an IWC context. It would today seem obvious that if the will had been there, a cogent argument could have been made publicly for what became known as the 'precautionary approach', citing Sidney Harmer's review of the nineteenth-century whale population disasters, the collapse of humpback whale stocks in the Antarctic and the evidence of Wheeler (1934) and Laurie (1937) of fin and blue whale stocks being exploited unsustainably. All of this was brought together by Mackintosh in his review of 1942, with his earlier proposal for whale sanctuaries in the Antarctic. However, it would be another 20 years before the publication of Rachel Carson's *Silent Spring* in 1962 gave the conservation movement generally the impetus it has today. Any earlier public campaign on behalf of the whales would probably have attracted little support.

There were also conceptual and practical limitations – the science of ecology was still comparatively young. The concept of ecological processes operating at a range of spatial and temporal scales is well understood today and has been applied widely, including in the context of whale ecology (Anderwald et al. 2012). But for much of the history of biology, the focus of research has been on understanding processes at ever-smaller scales, starting with the invention of the microscope and leading to the study of genetics at the molecular level. In the first half of the twentieth century there were no 'macroscopes' such as we have today with satellite remote sensing technology or sensors attached to drifters and moorings or autonomous surface and underwater vehicles, which allow the collection of synoptic data over large oceanic areas.

Of the above, the most significant constraint on achieving the original objectives of the Discovery Investigations was perhaps the most important: assessing the optimal rate of exploitation of the whale stocks. There was no clear means of achieving this – no complete conceptual framework; the science team had the wrong expertise for tackling the question, and even though some theoretical thinking was developing elsewhere, there would not be a practical means of applying it for several decades after the Investigations were launched. The science of population dynamics and its practical application had begun roughly at the same time as the Discovery Investigations but had developed in parallel. It was more than four decades before they converged. We have to conclude that when the Investigations were launched there was neither the theoretical basis nor the methodologies needed to achieve all the practical goals set by the research programme.

## 17.4 The wider science of the Investigations

The research considered briefly here was immediately concerned with understanding the ecological processes influencing the biological productivity of whales in the Southern Ocean, which supported large populations of whales prior to their exploitation and the near extinction of some species. Alister Hardy provides a popular account of the work of the Discovery Investigations up to the early 1960s in a volume largely concerned with the wider ecology of the Southern Ocean (Hardy 1967). A large proportion of the biological material was identified by the Discovery scientists themselves or by collaborating experts, notably at the British Museum (Natural History) but also in Belgium, Canada, Ireland, the Netherlands, New Zealand, Norway, South Africa and the United States. These large collections were recorded and organized to provide a broad foundation of knowledge concerning the biodiversity of the Southern Ocean and some adjacent seas. Much of it was not directly relevant to whaling but was published in the *Discovery Reports*, covering birds, seals, fishes, crustaceans, molluscs, sponges, echinoderms, foraminifera and other groups including terrestrial spiders.

### *Physical oceanography*

Some previous research had been done on the physical oceanography of the Southern Ocean, largely conducted by German scientists on Filchner's South Polar Expedition 1911–13 and the subsequent cruise of the *Meteor*. George Deacon's Discovery Investigations work would extend right around the Southern Ocean and provide a holistic view of the structure and dynamics of water masses from the continent northward to the edge of the tropical seas. Deacon's (1937) report was based largely on the second commission of *Discovery II*, which centred on a circumnavigation starting at Cape Town. It included 40 north/south vertical sections across the Southern Ocean and two east/west sections, showing temperature, salinity and oxygen concentration. It identified and described the characteristics and movements of the five water masses: Antarctic surface water, sub-Antarctic water, subtropical water, warm deep layer and Antarctic bottom water, showing that there are essentially two components to the movement of water in the Southern Ocean. In the Antarctic surface waters, the first of these is horizontal, from west to east, but there is also a south to north component. Where Antarctic surface water meets sub-Antarctic water, it flows beneath it at what was called the Antarctic Convergence (now the Antarctic Polar Front). There is also a northwards flow of Antarctic bottom water. At intermediate depths, the warm deep layer moves north to south and replaces the north-going surface and deep water.

Deacon's elegant descriptions of the large-scale movements of water in the Southern Ocean provided the three-dimensional physical background to the chemical understanding of plant nutrient cycling and for the biological understanding of the life cycle dynamics of the phytoplankton, krill and other zooplankton. It supported an understanding of the dynamics of the whole Southern Ocean ecosystem with the food chain from primary producers to whales.

## *Chemical oceanography*

A. J. Clowes was appointed as a hydrologist to the Investigations with George Deacon – his focus was on the dynamics of plant macronutrients, nitrate and silicate (Clowes 1938). He showed that inorganic and organic phosphate is carried northwards away from the Antarctic in surface water and then, at the Antarctic Convergence, it is carried down through the water column where much of it is transferred by vertical mixing to the southward flowing warm deep water. Mixing with Antarctic bottom water also transfers phosphate to the warm deep layer where it is also carried south. By these processes phosphate is cycled back towards the Antarctic, thereby maintaining phytoplankton production.

There are differences between the way phosphates and silicates cycle between the water masses. When phytoplankton is consumed by the zooplankton, most of the silica is excreted and much of the phosphorus is retained until the animals die and decompose. The greatest concentrations of silica are therefore found in the Antarctic zone where high concentrations of phytoplankton are consumed by zooplankton. The highest concentrations of phosphate are found where most zooplankton decomposition takes place in warm deep water of the sub-Antarctic, where it is returned towards the south. Antarctic bottom water is also rich in silicate from the rain of dead diatoms and zooplankton faeces from the surface water. Clowes concluded that phosphate was never likely to be a limiting factor in the growth of phytoplankton in the Antarctic, whereas silicate may at times become sufficiently depleted to limit phytoplankton (notably diatom) production.

## *Phytoplankton*

Alister Hardy was one of the first scientists employed by the Investigations, joining the voyage of RRS *Discovery* from 1925 to 1927 to lead its research on plankton. Antarctic phytoplankton was already known from the net hauls made by the *Challenger*, *Valdivia*, *Belgic*, *Pourquoi Pas?*, and *Scotia* expeditions, and had been described in several monographs. With much systematics done, Hardy was able to focus on the ecological aspects in his report (1935) on material from South Georgia, identifying five different phytoplankton communities around the island and major differences in the community to the north in sub-Antarctic waters. Hardy postulated that high phytoplankton productivity at South Georgia was related to being in an area where water from the Weddell Sea and Bellingshausen Sea mix. Although phosphate concentrations in seawater were lower where phytoplankton density was higher, phosphate was probably never a limiting factor, causing him to propose a general hypothesis that phytoplankton production is enhanced in areas where different water masses mix. It was already known that coastal upwelling elsewhere brings plant nutrients to the surface thus enhancing productivity, but this was not the cause of high production levels at South Georgia. He concluded that the effect might be caused by the mixing of water with different, but complementary, concentrations of plant nutrients, especially iron derived from a terrestrial source, and that seeding of resting spores of diatoms from one water mass to another might enhance the effect of nutrients. By the time Hardy's report was published, iron had also been identified as a factor regulating phytoplankton production elsewhere (Gran 1931).

Phytoplankton of the Southwest Atlantic and the Bellingshausen Sea were detailed by Hart (1934) based on sampling during cruises of *William Scoresby* and *Discovery II.* He dealt with the floras of the different water masses, seasonal succession, areas of concentration of phytoplankton, comparison of fertility of different areas and possible limiting factors. This study measured both phosphate and nitrate, demonstrating that they rarely became sufficiently depleted to limit phytoplankton production and indicating that iron might be the limiting factor to phytoplankton growth when macronutrients, including silicate, are not depleted. Accordingly, the richer phytoplankton flora in the Weddell Sea than in the Bellingshausen Sea might be explained by a higher presence of morainic bergs in the Weddell, enriching the seawater with iron.

Hart (1942) also used distribution of phytoplankton to delineate biogeographic regions in the Antarctic zone of the Southern Ocean with reference to the physical features of surface waters as described by Deacon (1937). Other substantial reports on phytoplankton included a systematic account of Antarctic planktonic diatoms (Hendey 1937) and an analysis of the annual periodicity of phytoplankton based on measurements of plant pigment concentrations (Hart 1942).

## *Krill*

Early work on krill, *Euphausia superba*, focused on the seas around South Georgia (Hardy and Gunther 1935). Five reports dealt with aspects of krill biology and ecology: Fraser (1936) and Bargmann (1937) covered the development of young stages and anatomy of the reproductive system, while Mackintosh (1972, 1973 and 1974) reported on life cycle, distribution and sizes of krill consumed by whales. Krill biomass increased by a factor of five in plankton samples between October/December and January/March due to recruitment of larvae from deep water, where they hatch, and increased growth of all year classes during the mid-austral summer.

The most extensive report on krill was by James Marr (1962) on the natural history and geography of the species. It is a long and diffuse document written in an eccentric style, and the controversy it created prior to publication seems to have been the reason that it was the only report prefaced by a disclaimer that 'the treatment and interpretation [of the data] are those of the author'. Nevertheless, it was a significant publication detailing the findings of over four decades of research on krill biology and biogeography (Laws 2017). Key subjects include swarming, vertical migration, distribution around the whole of the Southern Ocean and ecological role, especially in relation to whales. A major element of the report was working out the life cycle, which starts with female krill releasing fertilized eggs near the surface after which they sink to a depth of some 2,000 m during which cell division commences. The eggs hatch at depth and the first developmental stage (the nauplii) start to migrate upwards, passing through two more phases before arriving at the surface some two to four weeks after hatching.

### *Other zooplankton*

The foundations of Antarctic zooplankton science were laid by Hardy and Gunther (1935) during their two years spent on *Discovery*. They sampled around South Georgia and across the Antarctic Convergence (now known as the Antarctic Polar Front) between Antarctic and sub-Antarctic waters. Cataloguing the species and their vertical and horizontal distribution, this was a major study of zooplankton behaviour in relation to physical, chemical and biological aspects of the environment. The data sparked ideas about the driving factors behind the observations. Hardy noted an inverse relationship between the density of phytoplankton compared with zooplankton and proposed his 'animal exclusion' hypothesis. Discussion with a friend and colleague, Alan Gardiner (Hardy 1943), about the observation that dissolved phosphate concentrations are generally inversely related to those of phytoplankton led to the idea that phosphates are higher where there are more animals, including whales and krill, which excrete phosphate.

Prior to the 1925–27 *Discovery* voyage, Hardy had invented and built a prototype continuous plankton recorder (CPR) to sample phyto- and zooplankton along transects. After the voyage, the design was improved and described in detail (Hardy 1936). Samples and data collected at South Georgia were used to quantify the patchy distribution of plankton, and subsequently, the CPR was used routinely on *Discovery II* and later for plankton surveys in the North Sea on vessels of convenience operating out of Hull.

Other plankton research included Mackintosh's (1934) report on the species of near-surface macroplankton – mostly copepods, amphipods and gelatinous forms – and their horizontal distribution. His 1937 report describes the seasonal cycle of the three most abundant elements of the macroplankton community: the copepods *Rhincalanus gigas*, *Calanus acutus* and the arrow worm *Eukrohnia hamata*. During summer months, these are mostly concentrated in Antarctic surface water; during the winter they descend into the relatively warm deep water.

The northerly component to the direction of movement of the surface water and a southerly component to movement of the warm deep water is relevant here. The large scale of the vertical migrations of 400–600 m, and hundreds of miles being carried horizontally northwards and southwards at different depths, have the effect of maintaining the animals within a broad latitudinal band at high latitudes south of the Antarctic Convergence.

An analysis of Antarctic zooplankton biomass in the water column, based on closing vertical net hauls, was published by Foxton (1956). The seasonal vertical migration pattern of the total biomass correlated with that of the three species of macroplankton described by Mackintosh (1937). There was little longitudinal variation in standing crop around the Antarctic – the highest biomass was found in the West Wind Drift (now termed the Antarctic circumpolar current). A broad comparison with data from elsewhere showed that the standing crop biomass of Antarctic zooplankton was at least four times greater than in tropical seas.

One of the most common gelatinous forms in net hauls, sometimes in very large numbers, are salps, a planktonic group related to benthic tunicates or sea squirts. Peter Foxton published two *Discovery Reports* (Foxton 1961 and 1966) on the subject, one on *Salpa fusiformis*, the other on the distribution and life history of *S. thompsoni*, the most common species in Antarctic waters, and *S. gerlachei*. Both *S. thompsoni* and *S. gerlachei* were first described and named by Foxton. The most compelling aspect of the life history of salps is the way generations alternate between sexual and asexual reproductive phases. The solitary over-wintering phase rapidly produces chains of buds asexually in the spring, which remain aggregated. The aggregates then each produce a single egg which is fertilized sexually and develops into the solitary, free-living form. Foxton estimated that each of these can produce up to 800 buds, so numbers can increase rapidly when conditions are favourable in the summer.

## 17.5 A twenty-first century perspective

The Discovery Investigations are largely of historical interest today. The science has moved on, and the whaling industry is now largely confined to relatively small catches in Icelandic, Japanese and Norwegian waters, plus some aboriginal subsistence whaling. In Iceland, whaling was temporarily stopped during the COVID-19 pandemic and it may cease altogether. The biggest threat to cetaceans in the world's oceans today is their accidental capture in the fishing gear targeting other species. More than 300,000 cetaceans annually are estimated to be taken as by-catch and die. Other threats are ship strikes, ocean noise, marine debris, chemical pollution and the existential threat of climate change. Today the IWC and numerous other organisations are concerned with all aspects of cetacean conservation and have at their disposal data collected with scientific procedures mostly developed since the publication of the last Discovery Report in 1980.

During the Antarctic whaling era, scientists had access to large numbers of whale carcases for study at the whaling stations and on factory ships but were limited in their ability to discriminate between putative stocks or effectively to track migrations over large geographical scales. Today the situation has reversed; it is rare to have access to whale carcases, especially of the great whales, but new technology enables an individual whale's movements to be followed and geographical variation among populations to be analysed at the molecular genetic level.

Satellite transmitters are attached to whales without causing harm, and these enable large-scale movements to be followed over extended periods. Movements can be mapped in relation to environmental variables including sea surface temperature, chlorophyll concentration, near-surface ocean currents, gravity and magnetic field from satellite remotely sensed imagery (see, for example, Horton et al. 2020).

Large spatial and time-scale data on whale distribution can now be obtained using passive acoustics technology that detects and records the vocalisations of whales underwater. The technology can be deployed either on research vessels (e.g., Širović and Hildebrand 2011) or on moored sonobuoys (e.g., Poupard et al. 2022) and species can be identified from the characteristics of their calls.

Although at an early stage of development, use of very high-resolution satellite imagery of whales at sea promises to provide an innovative survey method. At the time of writing, realising the potential of this technology requires development of automated detection systems. Data have been published for southern right, humpback, fin and grey whales from areas including Argentina, New Zealand, South Africa, the United States and Mexico. These are available as a nascent open source library for training and testing machine learning models (Cubaynes and Fretwell 2022).

Molecular genetics research using gene sequencing technology can be performed on small biopsy samples collected harmlessly from cetaceans or DNA shed naturally into the environment, known as eDNA (environmental DNA). Molecular genetics data are used to analyse population structure, which provides information for stock assessments, species identification and research on taxonomy and systematics. Omics research, which involves sequencing DNA, RNA and proteins followed by complex computational analysis of data, is being used to identify fine scale structure in populations and to unravel taxonomic relations at higher levels.

Returning to analysis of gross anatomical features, twenty-first-century microscribe technology is enabling precise linear and three-dimensional measurements of skeletal structures to be made. As in palaeontology and anthropological research, these data may be used accurately to determine individual variability in cetaceans.

## *Fishery science today*

Before the end of the twentieth century, failures of policy and the over-exploitation of many fisheries revealed the practical weakness of using only the MSY approach and led to the introduction of the radically different 'ecosystem-based fishery management' (EBFM) approach (Pitcher and Pauly 1998). Put simply, a fishery exploiting a wild population of one species cannot be managed in isolation, ignoring the interactions between that species and the ecosystem, and in particular its food resources and the predators that are dependent on it.

EBFM is now being widely adopted. It was introduced in the Antarctic with the establishment of the Commission for the Conservation of Antarctic Marine Living Resources (CCAMLR). With the collapse of most whale stocks in the 1960s, commercial attention was turning to the possibilities of exploiting Antarctic krill – the main food of the whales, and other marine living resources in the Southern Ocean (Everson 1977). An ecosystem-based approach was devised as a means of managing krill fisheries in the context of the conservation of dependent predators. Although this would logically have included whales, there was agreement between the IWC and CCAMLR that the latter would not formally include whales in its remit.

EBFM recognises that the goal of fisheries management should not be concerned solely with maximising yield for human consumption but with maintaining a healthy marine ecosystem. Nevertheless, the theoretical basis that culminated with Beverton and Holt's synthesis and the MSY concept remains as a method in the EBFM approach. It is still necessary to know the size of an exploited stock and understand the effects of fishing. The difference is the recognition under EBFM that sustainability relates to the consquences of fishing on the ecosystem as well as on the exploited species.

In 2001 a new applied fisheries research station was opened next to Discovery House at King Edward Point, South Georgia. Commissioned by the government of South Georgia and the South Sandwich Islands, the station was built and is operated by the British Antarctic Survey for research on the commercial fisheries around the islands. Although South Georgia and the South Sandwich Islands are British sovereign territory, they fall within the CCAMLR area, and the fisheries within the territory's 200 nautical mile Maritime Zone are managed within the context of CCAMLR, and implemented through a range of domestic fisheries management and licensing measures. An ongoing dispute between Russia and other CCAMLR member states caused a lack of agreement on catch limits for the South Georgia toothfish fishery, which has resulted in the government managing this fishery solely under domestic regulations, while remaining in line with CCAMLR's objectives.

**Figure 24** The modern fishery research station at King Edward Point operated by the British Antarctic Survey on behalf of the Government of South Georgia and the South Sandwich Islands. © Paul Rodhouse.

### *The Discovery Investigations and ocean-ecosystem science today*

The Discovery Investigations laid the foundations of several areas of science that continue to have high relevance today. The Southern Ocean plays a vital role in regulating the global climate; it provides a physical link between the world's other oceans and it is the most important oceanic region for the uptake and storage of heat (Sallée 2018). This fundamental understanding of the physical dynamics of the Southern Ocean dates back in particular to George Deacon's research over eight decades ago.

Alister Hardy and Thomas Hart first showed that the plant macronutrients in the Southern Ocean were often not limiting factors for phytoplankton growth, but that iron was likely to be limiting. The role of iron in limiting planktonic primary producers in Antarctic waters has been a subject of continuing research in more recent years (de Baar et al. 1990; Martin et al. 1990) and iron enrichment has been proposed as a means of artificially increasing carbon drawdown in the oceans (Boyd et al. 2007). Alister Hardy and Alan Gardiner first discussed the idea that animal excretions in the water column might enhance phytoplankton production in the Antarctic. There is currently a renewed interest in the idea that whales may act as ecological engineers, stimulating phytoplankton growth by their excretions in a process analogous to the "green wave" driven by migrating herds of American bison across the prairies (Geremia et al. 2019). Phytoplankton ecology and whale ecology in the twenty-first century are converging in ways first envisaged in the 1930s.

Within ten years of the collapse of Antarctic whaling, commercial interest in developing an Antarctic krill fishery started to emerge. Given the role of krill in the ecology of the Southern Ocean as food for whales and numerous species of seabirds, seals and fish, the prospect of a krill fishery was a cause of concern for fishery scientists and conservationists. Extensive, uncontrolled exploitation of sub-Antarctic fish had already begun in the late 1960s, so in 1975 the Antarctic Treaty consultative meeting recommended that krill and fish resources should be protected, studied scientifically and exploited rationally. A new, international research programme, initiated by the Scientific Committee for Antarctic Research, was established to start new krill research. This research continues today and supports the ecosystem-based management approach adopted by CCAMLR.

The ability of salps to increase rapidly by asexual reproduction under favourable conditions has come into focus as regional warming has affected the Antarctic Peninsula. Krill overwinter under annual sea ice whereas salps tolerate warmer conditions. As warming reduces the winter habitat of krill, their susceptibility to competition from salps increases with consequences for the vertebrate predators of krill (Loeb et al. 1997). Long-term records suggest that krill stocks have declined in the southwest Atlantic sector of the Southern Ocean since the 1970s, accompanied by an increase in salps (Atkinson et al. 2004). Salps are not consumed in any quantity by whales, seals, seabirds or fish in the Antarctic. They are effectively a trophic dead end. A shift towards an increasing dominance of salps in the Antarctic may be part of a larger scale trend of increasing dominance by gelatinous forms associated with overfishing and climate change (Richardson et al. 2009).

Whereas the whale research programme of the Discovery Investigations was done initially in virtual isolation, the wider Discovery sciences were developing in the context of the physical, chemical and biological ocean sciences being pursued elsewhere, notably in Germany, the

United Kingdom and the United States (Mills 1989). Plankton research was amenable to study with the technology that was available at the time. The small size of phyto- and zooplankton meant that sampling was possible with a relatively simple apparatus, which could be operated with relative ease from small- to medium-sized vessels. Both phytoplankton and zooplankton could be kept alive for observation and experiment. Meaningful research on plankton could be undertaken at a small scale, and Alister Hardy's invention of the CPR made it possible to gather synoptic data on these small organisms over large geographical scales. One hundred years on the CPR continues to be used for research, including in the Antarctic (Pinkerton et al. 2020). The data obtained have become increasingly valuable over time, as the length of time series increases, and as the data reveal changes in the plankton associated with global climate change.

Observations on the role of iron, a micronutrient identified as a limiting factor for phytoplankton growth elsewhere (Gran 1931), appeared around the time Hardy, Hart and Clowes were publishing their *Discovery Reports* on plankton and plant nutrients. Alfred Redfield (1934) at Harvard discovered about the same time that the carbon/nitrogen/phosphorus ratio in both phytoplankton and seawater was remarkably consistent throughout the world's oceans at about 106:16:1. This is now known as 'the Redfield ratio', and it has become a fundamental tenet of ocean biogeochemistry. Plankton research in the Antarctic during the Discovery Investigations was in accord with international development of the field and continues to be an active topic today, particularly in relation to the role of the oceans in global climate change research.

## *Conclusion*

A century has passed since the early development of ideas that launched the Discovery Investigations, and over half a century since the Antarctic whalers brought many species of whales to commercial extinction, thereby putting themselves out of business. We are starting to see evidence that even the hardest hit whale species are showing signs of recovery. What is more, whale science and some of the curiosity-driven research that was encouraged and had its early beginnings during the Investigations, has converged in a way that would not have been originally anticipated. Rowland Darnley and Stanley Kemp clearly hoped and expected useful scientific developments to emerge, but perhaps not over the long time it has taken. New knowledge of the Southern Ocean ecosystem was expected at the time to lead to an understanding of the food resource of the whales. After the destruction of the whale stocks, it became clear that loss of these top predators must have brought about a fundamental change in the structure of the ecosystem. Notably this led to the 'krill surplus' hypothesis (Surma et al. 2014). Only recently has a more holistic view of the effects of over-exploitation emerged with ideas that there was considerable feedback in the Southern Ocean production system prior to exploitation. The whales were not only removing biomass but redistributing nutrients, releasing iron, a vital micronutrient, and sequestering carbon. Removing the whales was not only a conservation failure but a cause of large-scale biogeochemical disruption, making its own contribution to the global climate and biodiversity crisis being faced today. Now, as the whale stocks are eventually showing signs of returning, we have predictions that global climate change threatens a full recovery.

Antarctic whaling in the twentieth century is a case study of probably the largest ever removal and long-term loss of animal biomass, leading to the commercial extinction and almost total extinction of several species of whale. The full consequences of this devastation continue to be felt today, six decades after the collapse of the industry. It was a failure to prevent a foreseeable ecological catastrophe. Not only was it a failure to conserve whales but, as we now know, it brought about fundamental degradation of the structure and function of the Southern Ocean ecosystem with large-scale consequences. We now face the existential threat of global climate change causing widespread extinction in the sea and on land, including possible eradication of the human species (Kemp et al. 2022). Again, politicians have dithered, industry has obfuscated while climate change scientists have spent years frustratingly trying to deliver their urgent message. As Olson (2009) has pointed out, the global scientific community still needs to consider how to promulgate their most important messages in a totally convincing and compelling manner.

CHAPTER 18

# *Epilogue*

The political and scientific history of the Southern Ocean whaling industry charts an enterprise that nearly destroyed several species of the great whales and only came to an end when it became un-economical to continue catching the few that remained. Thankfully some 60 years since the industry collapsed there have been recent signs that some species are now recovering and there is even hope that populations might return to levels that existed prior to the whaling era. The whalers always asserted that there was no possibility that the industry would cause mass extinction before it was forced to close down on economic grounds. However, we doubt they expected a wait over half a century for the signs of recovery to be obvious! Hindsight has demonstrated that they very nearly did cause a mass extinction and for blue whales it is still uncertain whether, and on what timescale, they will show a significant increase in numbers.

The industry was started by Norwegian whaling entrepreneurs, particularly Carl Larsen, on the island of South Georgia in 1904 exploiting the most up to date technology – fast steam powered catchers mounting powerful harpoon guns. These allowed them to successfully hunt the fast-moving humpback, blue and fin whales that were to be found in great abundance in the seas around the island. Great Britain was near its peak of international and Imperial power with territorial ambitions for the islands of the Southern Ocean, including South Georgia and for the Antarctic continent itself. They were seen as dependencies of the Falkland Islands, although that was not formalised as such by letters patent until 1907. So, Britain saw both as its duty and to its advantage to require the whaling companies to accept its authority and abide by the regulatory regime it progressively created from 1905 onwards. An added complication was that there was a developing territorial dispute with Argentina over the sub-Antarctic islands and the Antarctic which had to be managed in the context of the long running dispute over sovereignty of the Falkland Islands.

For the first two decades or so of the whaling era a symbiosis developed in which the whalers – predominantly Norwegian – were content to accept control which required them to pay for leases and licences to operate, and to abide by management regulations set by Britain via the Falkland Islands Government. This was because, although the whales were hunted on the high seas, the whalers needed access to land for processing the carcases. The regime also provided a defined and level regulatory playing field by which the size of the industry was constrained so that long term investment could be made with confidence. There was some heightening of tension during the First World War because of the British need to stop oil reaching Germany and to get the maximum amount for the manufacture of explosives, but, for the most part, the relationship served both nations well. However, by the mid 1920s technological developments enabled pelagic whaling which was entirely independent of land based structures. This became the norm and Britain quickly began to lose both control and revenue.

From the outset Britain shaped its legislative and management approach to try to ensure what we would now describe as a sustainable whaling industry. The government was well aware of the danger of overexploitation and sought to limit the scale of the industry by strict control of the number of leases and licences issued and by legislation requiring that the whole carcase be used. Even before First World War the Colonial Office was keen to investigate how the populations of whales could be protected and what new knowledge would be needed to underpin this goal. The First World War stymied this initiative, but after the war the initiative was revitalized in a broader framework of economic development of the Falkland Islands and its dependencies with funding provided by revenue in the form of licence fees and duty charged on the whale products. The latter provided ample funding to establish a very substantial programme of research – the Discovery Investigations – to provide the scientific knowledge needed to underpin legislation to manage the whaling industry. This book has investigated to what extent it was successful in achieving this. Although it was a very major, multifaceted and sustained programme of marine research, and can be seen as the beginning of whaling science, its impact on the pace of whaling was minimal. The research was launched with no methodology to measure the size of whale populations, or how these were being impacted by exploitation. Nor was there any theoretical framework for understanding the dynamics of exploited populations. By the time methodologies had been developed and a theoretical knowledge of population dynamics under exploitation had been developed in the wider field of fisheries science it was too late to save the whales. There had been too many unknowns and too much resistance to conservation measures from the whalers themselves. The management of the research programme was from within the Colonial Office with little or no connection with the wider academic community and was beset by controversy over whether the science was targeted enough to meet the aims. Nevertheless, the Discovery Investigations founded the science of whale biology beyond systematics and skeletal anatomy, and left a lasting legacy of knowledge on the physical, chemical and biological oceanography of the Southern Ocean which underpins research today.

As pelagic whaling allowed the industry to free itself from control by the British Government, they and the Norwegian government endeavoured to counter the explosive growth in whale catches which it made possible by attempting to establish internationally agreed whaling management regimes. This book has detailed the several attempts that were made in the decade leading up to the Second World War and has shown that these largely failed to stem the slaughter. The Second World War disrupted both the whaling industry and attempts at international regulation. The influence that Britain exercised on the industry had waned with the growth of pelagic whaling and with the end of the war its overall influence in world affairs had been surpassed by that of the United States – the new superpower.

Although the USA was not by then itself a whaling nation, it now assumed leadership of the development of whaling legislation. As the post-war power in Japan it also over-ruled the Norwegian and British objections to the establishment of a Japanese whaling industry to help stave off the famine raging there. The USA now took the lead in the post-war negotiations for an international regime to govern whaling with the UK still much engaged but no longer having the final say. However, British scientists and the output from the Discovery

Investigations played an important part in shaping the International Whaling Commission, especially the work of its scientific committee, and the IWC headquarters were set up in Cambridge. The combination of the Discovery results with the new mathematical techniques developed for fisheries management finally gave some hope of providing a regime to sustainably manage whaling, but by then it was too late to save the industry or more importantly the whales. A tragedy of the commons, from which only now are the whales finally recovering.

# *Biographical notes on key scientific staff of the Discovery Investigations*

These brief notes, with reference to online sources, provide background information about the key scientists cited in the text, many of whom continued with distinction, to pursue scientific careers in the same or related fields of research after leaving the Investigations. The list is incomplete, as published biographies are not readily available for all the scientific staff. Coleman-Cooke (1963) provides summaries of many of the scientists' service at sea and ashore during the Investigations, including some that are not covered here.

Key to abbreviations: BANZARE: British Australian and New Zealand Antarctic Research Expedition; BMNH: British Museum, Natural History; CBE: Commander of the British Empire; DI: Discovery Investigations; FRS: Fellow of the Royal Society of London; ICES: International Council for Exploration of the Sea; IWC: International Whaling Commission; MBA: Marine Biological Association; NIO: National Institute of Oceanography; RN: Royal Navy; SSO: Senior Scientific Officer.

**Bennett, Arthur George** 1888–1954. Not appointed but contributed to *Discovery Reports* by Kemp (as co-author) and Norman. His meticulously collected data used in Harmer's published addresses to the Linnean Society 1928 and 1930 on over-exploitation of whales. Accomplished naturalist and taxidermist. 1913–21: summer seasons at whaling station on Deception Island, South Shetlands, variously as deputy collector of customs, chief customs officer, receiver of wrecks, whaling officer and postmaster; 1921–22 took part in relief of British Imperial Antarctic Expedition. Author: *Whaling in the Antarctic*, William Blackwood and Son (1931). https://www.falklandsbiographies.org/biographies/bennett_arthur

**Brown, Sidney** 1924–2010. Appointed 1951. Left school at 16. Joined Army, Royal Ordnance Corps 1942. After WWII studied zoology with marine biology at University of Hull and graduated BSc. On joining DI served on whaling factory ship *Balaena*. Whaling inspector at Grytviken whaling station, South Georgia 1963 (then operated by Japanese). Inspector Japanese whaler in Southern Ocean 1966. Member scientific committee IWC from 1956; co-ordinator IWC international whale-marking scheme from 1955. https://www.theguardian.com/environment/2010/mar/31/Sidney-brown-obituary

**Clarke, Malcolm Roy** 1930–2013. Appointed NIO 1958. Graduated BSc Zoology University College Hull 1955. Whaling Inspector in Salversen's *Southern Harvester* 1955–56, graduated PhD University College Hull in whale parasitology 1959, NIO 1958–72, MBA 1972–87, elected FRS 1981. https://doi.org/10.1098/rsbm.2014.0011

**Clarke, Robert Henry** 1919–2011. Appointed NIO 1948. Educated at St. Olave and St. Saviour's Grammar School (1932–38). Awarded government scholarship to read Natural Science at New College, Oxford. Studies interrupted 1939 by start of WWII and volunteered as a lieutenant in the RN Unexploded Bomb Department and the Admiralty R&D (India), with special ops in India, Burma, and Ceylon (1940–46). After of WWII returned to conclude studies at Oxford and graduated with Masters Degree (Hons). Second Inspector on the British whale factory ship *Southern Harvester* for the Antarctic season 1947–48. Scientific Committee of the IWC (1955–70). Retired from NIO 1971. Continued whale research in Peru, Mexico, Ecuador and Chile.

**Deacon, George Edward Raven** 1906–84. Appointed 1929. Graduated from Kings College, London in 1926 with a first-class honours in chemistry and in 1927 a diploma of education. Appointed director of NIO 1949. DSc London 1937, elected FRS 1944, CBE 1964, knighted 1977. https://doi.org/10.1098/rsbm.1985.0005

**Fraser, Francis Charles** 1903–78. Appointed 1925. Graduated from Glasgow University in 1924 having studied botany, zoology and natural philosophy. From 1933 to 1969 Cetologist at the BMNH. Elected FRS 1966. https://doi.org/10.1098/rsbm.1979.0010

**Gambell, Ray** 1935–. Appointed 1963 NIO. Educated Kingsbury County School; Graduated BSc and PhD Reading University. Whale Research Unit NIO 1963–76. Scientific adviser to UK delegation of IWC 1965–76. Secretary IWC 1976–2000. https://www.sciencedirect.com/science/article/abs/pii/0308597X77900860

**Gunther, Eustace Rolfe** 1902–40. Appointed 1925 having recently graduated from Cambridge. Promising career cut short: died 31 May 1940 on duty as 2nd Lieutenant, 72nd Searchlight Regiment. https://www.nature.com/articles/146123a0.pdf

**Hamilton, James Eric** 1891–1957. Seconded to Discovery Investigations from post as government naturalist, Falkland Islands in 1924. Graduated Liverpool University in 1913, spent two summers working at the Belmullet whaling station, Co. Mayo, Ireland. Awarded M.Sc. for work in oceanography in 1914, took posts as an assistant demonstrator at Kings College, London, and then Queen's University, Belfast. In 1916 joined the Royal Army Medical Corps as bacteriologist; in 1919, after WWI moved to Falkland Islands as magistrate for West Falkland Island and South Shetlands, appointed government naturalist in January 1921. https://falklandsbiographies.org/biographies/hamilton_james

**Hardy, Alister Clavering** 1896–1985. Appointed 1924 as zoologist to the first *Discovery* voyage. Started to study forestry at Oxford; WWI intervened, and after one month left to spend war years in the army using his scientific and artistic talents as a camouflage officer. An inventor; notably designed the continuous plankton recorder. An accomplished artist well known for his watercolours of marine life. First Professor of Zoology University of Hull 1928–42,

Professor of Natural History at the University of Aberdeen in 1942–46, Linacre Professor of Zoology, University of Oxford and Fellow of Merton College 1946–63; elected FRS 1940, knighted 1957. https://royalsocietypublishing.org/doi/10.1098/rsbm.1986.0008

**Hart, Thomas John** 1907–70. Appointed 1929 with a first-class degree in zoology from University of Leeds. Trained in zoology but became a distinguished authority on marine diatoms. http://192.171.196.76/files/documents/facilities/library/HART%2C%20TJ%20%28John%29%20COLLECTION.pdf

**Herdman, H. F. P.** Worked on *William Scoresby*, chartered whale catchers and *Discovery II*. Published a report on pack ice. At NIO researched the deep scattering layer in the sea. https://doi.org/10.1038/172275a0

**John, D. Dilwyn** 1901–95. Appointed 1925 after graduating from the University of Wales. In 1935 appointed at the BMNH and put in charge of the echinoderm collections. Territorial army officer prior to WWII and put on war service until 1946, then deputy keeper of zoology, National Museum of Wales in February 1946. Appointed Director of the National Museum of Whales in 1948 succeeding Sir Cyril Fox. https://www.nature.com/articles/161840c0 and https://cardiffnaturalists.org.uk/htmfiles/150th-74.htm

**Kemp, Stanley Wells** 1882–1945. Appointed director 1924. Graduated Trinity College Dublin 1903. Inspectors of Irish Fisheries as assistant naturalist under Ernest Holt and W. S. Green. Worked at the floating laboratory Ballynakill Harbour, Co. Galway, later in research cruises to the Atlantic slope with special interest in crustaceans. In 1910 joined staff of zoological section of the Indian Museum, participated in collecting expeditions to remote areas of the subcontinent and Southeast Asia. Later assistant director Zoological Survey of India. Completed 17 works on crustaceans. While in India in 1919 received a doctorate in science from Trinity College. Elected FRS in 1931. Appointed director MBA 1936. https://royalsocietypublishing.org/doi/10.1098/rsbm.1947.0012 and https://falklandsbiographies.org/biographies/kemp_stanley

**Laurie, Alec Hibburd** 1907–87. Appointed 1929. Educated at Sedbergh, read science at King's College, Cambridge. Taught biology briefly at Westminster School before joining the Investigations as biologist and chemist. During WWII involved in weapons research, as a civilian and later commissioned officer in the Royal Navy. Post-war, research in harbour protection, fish farming and sewage treatment. https://archiveshub.jisc.ac.uk/search/archives/d80a755c-e9cd-3f7b-97d0–4ab4a4c17994 and https://archiveshub.jisc.ac.uk/files/scottpolar/0466_2002.xml

**Laws, Richard Maitland** 1926–2014. Appointed NIO 1953. Graduated with first-class honours in zoology, Cambridge 1947. Joined Falkland Islands Dependencies Survey 1947, served as biologist on whale factory ship *Balaena* 1953–54 Antarctic season, member of IWC Scientific Committee 1955–61. Senior researcher, then Director at Nuffield Unit of Tropical Animal Ecology 1961–68. Head of Life Sciences Division, then Director British Antarctic Survey 1969–87. Master St Edmund's College, Cambridge, 1985–96. Elected FRS, 1980. https://royalsocietypublishing.org/doi/pdf/10.1098/rsbm.2015.0006

**Mackintosh, Neil** 1901–74. Appointed 1924 as zoologist to lead the advance party to establish the laboratory at Discovery House on South Georgia. Educated at Westminster school and studied zoology at Imperial College, London. Senior biologist and then succeeded Kemp as Director of the Investigations, 1936–49 when they became part of the NIO and he transferred as Deputy Director. Awarded the Patron's Medal of the Royal Geographic Society in 1954. https://www.nature.com/articles/249785a0.pdf

**Marr, James William Slessor** 1902–65. Appointed 1927. Graduated in classics and zoology University of Aberdeen in 1925. Selected to join the Shackleton-Rowett Antarctic Expedition aboard *Quest*, but this was cut short by Shackleton's death at South Georgia in 1922. Returned to complete MA in Aberdeen. Joined the British Arctic expedition as marine biologist collecting samples using a tow net and dredge 1925. On return, after a year as Carnegie scholar, University of Aberdeen, joined staff of the Discovery Investigations. Seconded to BANZARE with Sir Douglas Mawson 1929–31. In 1940 joined RNVR, promoted Lieutenant Commander, led Operation Tabarin. Established two Antarctic stations, over-wintered at Port Lockroy. Joined NIO 1949, as SSO, where he remained until his death. Awarded W. S. Bruce Medal 1936. https://www.falklandsbiographies.org/biographies/marr_james

**Matthews, Leonard Harrison** 1901–86. Appointed in 1924. Graduated with a first-class degree in biological sciences from Cambridge two years earlier. Appointed Scientific Director to the Zoological Society of London in 1951. Elected FRS 1954. https://royalsocietypublishing.org/doi/10.1098/rsbm.1987.0016

**Ommanney, Francis Downes** 1903–80. Appointed 1929. Joined the Investigations with a PhD, Imperial College of Science and Technology, London University. Had been working as an assistant lecturer in zoology at East London College (later to become Queen Mary College). Joined the RNVR in 1940 as a meteorological officer, saw action in the Mediterranean and North Atlantic; joined the Mauritius-Seychelles Fisheries Survey 1947–49: worked in Zanzibar and Singapore during six years with the Colonial Research Services: Reader in marine biology at the University of Hong Kong 1957–60. https://www.falklandsbiographies.org/biographies/ommanney_francis

**Wheeler J. F. G.** Appointed 1924. Came with an MSc and was one of Mackintosh's co-workers at Grytviken for the first season. Lecturer in zoology, University of Bristol 1931. 1932 Director Bermuda Biological Station (Anonymous 1932). 1943 Marine biologist on Mauritius employed to establish fisheries organization to improve nutrition of the islanders (Anonymous 1943).

4.4 THE FURTHER ORDINANCES

47 CO 78/122 (8400); CO 78/126 (5686): 48 CO 78/122 (8400): 49 CO 78/126 (22020): 50 CO 78/130 (13384): 51 CO 78/130 (13384); CO 78/137 (10922): 52 CO 78/193/12 (19232/2): 53 CO 78/119 (25913): 54 CO 78/119 (25913 and 31842); CO 78/123 (33832); CO 78/127 (30139 and 37057): 55 CO 78/193/12 (19232/2): 56 CO 78/122 (5322, 8400 and 18264); CO 78/127 (30139): 57 FO 371/1563 (14164); CO 78/119 (25913); CO 78/122 (8359; 8400 and 18264); CO 78/123 (27712, 27584 and 33835); CO 78/124 (22758); CO 78/128 (2762); CO 78/193/12 (19232/2): 58 CO 78/123 (33832); CO 78/125 (8762, 15871–2, 19976, 26580 and 35878); CO 78/126 (12354): 59 CO 78/126 (22968); CO 78/129 (18860)

4.5 THE POLITICAL DIMENSION

60 CO 78/113 (17194): 61 CO 78/117 (15659): 62 CO 78/125 (20712): 63 CO 78/120 (40952): 64 CO 78/129 (23410): 65 CO 78/133 (1383 and 3378); CO 78/153 (26074, 67992 and 69687); CO 78/158 (4327 and 6095): 66 CO 78/125 (61): 67 CO 78/125 (40270): 68 CO 78/125 (40270): 69 CO 78/129 (19059): 70 CO 78/129 (14090): 71 CO 78/129 (1726): 72 CO 78/129 (1762): 73 CO 78/129 (4672): 74 CO 78/129 (4672)

**CHAPTER 5 THE SAFEGUARDING OF WHALE STOCKS**

1 CO 78/131 (31908): 2 CO 78/131 (31908 and 31008): 3 FO/371/598 (23619): 4 CO 78/121 (7341): 5 CO 78/120 (14488): 6 CO 78/120 (14488): 7 CO 78/120 (324921): 8 CO 78/125 (1821); CO 78/129 (2543, 5004, 1889 and 21011): 9 CO 78/129 (21097 and 28415); 10 CO 78/129 (1889): 11 CO 78/132 (6624; 12605 and 25781): CO 78/169 (53409)

5.1 FOREIGN OFFICE INQUIRY

12 CO 78/149 (11080): CO 78/160 (52526): 13 CO 78/156 (20037): 14 FO 371/1564 (20027); CO 78/120 (34921); CO 78/125 (40565): 15 CO 78/120 (34921): 16 CO 78/120 (34921): 17 FO 371/1564 (16958); CO 885/22/5: 18 FO 371/1564 (16958 and 29475; CO 78/124 (13497): 19 CO 78/129 (21097); 20 CO 78/129 (21097)

5.2 THE WHALING AND PROTECTION OF WHALES COMMITTEE

21 FO 371/1564 (16958 and 29475): FO 371/1972 (4144 and 9522): CO 78/129 (7399 and 9427): 22 CO 885/22/5: 23 CO 78/124 (3128); CO 885/22/5: 24 FO 371/1564 (29475, 40842 and 45204); CO 78/132 (27924 and 33244); CO 885/22/5 (21); 25 CO 78/129 (10493): 26 CO 885/22/5: 27 CO 885/22/5: 28 CO 323/642 (24184)

5.3 THE WIDER FISHERIES CONTEXT

29 CO 885/22/5: 30 CO 885/22/5; CO 885/23/11: 31 CO 885/22/5: 32 CO 323.642 (24184); CO 885/22/5: 33 CO 323/642 (25761): 34 CO 323/642 (25761): 35 CO 323/642 (24184): 36 CO 323/642 (23998 and 25761)

**CHAPTER 6 WEAPON OF WAR**

1 CO 78/149 (39535): 2 CO 78/141 (53874); CO 323/710 (14450 and 17928): 3 CO 78 139 (7241); CO 149 (39335): 4 CO 78/143 (6839): 5 CO 78/149 (39535); CO 323/675 (41136): 6 CO/78 141 (10130 and 42237); CO 78/323/679 (55690); CO 323/727 (5776)

6.1 THE DANNEVIG PLEDGE

7 CO 78/140 (17656): 8 CO 78/140 (17656); 9 CO 78/140 (17656): 10 CO 78/140 (14083 and 22815); CO 78/141 (16989): 11 CO 78/145 (9057); CO 78/146 (18466 and 24781): 12 CO 78/149 (39535)

6.2 SALVESENS – 'THE STURDY BEGGARS'

13 CO 78/141 (9587, 12157 and 12804); 14 CO 78/137 (33259, 38995 and 40104); CO 78/140 (5583): 15 CO 78/137 (16549 and 38386); CO 78/149 (39535); CO 323/727 (5776): 16 CO 78/137 (20058); CO 323/727): 17 CO 78/137 (25878 and 33555); CO 78/141 (10130 and 27086): 18 CO 78/141 (15934): 19 CO 78/146 (21137 and 22913): 20 CO 78/143 (48305); CO 78/149 (39335): 21 CO 78/143 (52195); CO 78/147 (32282); CO 78/149 (39535)

6.3 THE WARTIME CATCHES

22 CO 78/135 (51849): 23 CO 78/138 (18813); CO 78/131 (31908); 24 CO 78/135 (8183 and 56437): 25 CO 78/138 (26742); CO 78/139 (47624); CO 78/142 (19392 and 48771: 26 CO 78/142 (48771): 27 CO 78/142 (48771): 28 CO 78/142 (48771): 29 CO 78/148 (1527): 30 CO 78/147 (32281); CO 78/148 (1527)

**CHAPTER 7 RESEARCH AND DEVELOPMENT OF THE FALKLAND ISLANDS**

1 CO 78/136 (32929); CO 78/148 (63004): 2 CO 78/138 (26742); CO 78/143 (19862): 3 CO 78/141 (50912); CO 78/145 (4191): 4 CO 78/147 (32285): 5 CO 78/147 (32285)

7.1 REALISING THE COLONY'S POTENTIAL

6 CO 323/728 (14927A): 7 CO 78/725 (40053): 8 CO 323/726 (53536): 9 CO 323/726 (61941): 10 CO 78/145 (31445): 11 CO 78/145 (31445); CO 78/146 (56705); CO 323/763 (41713): 12 CO 78/145 (31445): 13 CO 78/149 (15654); CO 78/152 (62457)

7.2 THE COMMITTEE'S FINDINGS

14 CO 78/145 (4191): 15 CO 78/159 (61161): 16 CO 78/160 (27858 and 54260)

**CHAPTER 8 THE HEYDAY OF DEPENDENCIES WHALING**

1 CO 78/153 (16720); CO 78/158 (36683): 2 CO 78/163 (37081): 3 CO 78/137 (20811; CO 78/153 (26074, 67992 and 69687); CO 78/158 (4327): 4 CO 78/137 (20811); 5 CO 78/153 (65561); CO 78/165 (37465): 6 CO 78/173 (15872); CO 78/175/13 (35055)

8.1 LOCAL DIRECTION

7 CO 78/150 (26136 and 64537); CO 78/151 (61579); CO 78/157 (8146 and 18332): 8 CO 78/147 (29696 and 58909); CO 78/154 (20464); CO 78/162 (41445 and 56934): 9 CO 78/162 (41445): 10 CO 78/159 (27445): 11 CO 78/159 (27445): 12 CO 78/159 (21478 and 27445); CO 78/164 (42438); CO 78/170 (40899); 13 CO 78/173 (23419); 14 CO 78/167 (33882); 15 CO 78/173 (23419): 16 CO 78/168 (16874 and 24518); CO 78/169 (49306)

8.2 FRAUDULENCE

17 CO 78/170 (21000 and 21003): 18 CO 78/179/5 (56787): 19 CO 78/179 (56787): 20 CO 78/165 (43009, 45536, 46267 and 56757): 21 CO 78/165 (45536 and 46267); CO 78/166 (58967); CO 78/169 (39701); CO 78/170 (40978): 22 CO 78/170 (21000): 23 CO 78/164 (42438)

8.3 THE ROSS SEA

30 ADM 116/2386: 31 ADM 116/2470 (M572): 32 FO 371/12644 (W1985); ADM 116/2470 (M2217): 33 ADM 116/2470 (M572)

## PART III PELAGIC WHALING 1925–1950

### CHAPTER 9 FREEDOM OF THE SEAS

1 CO 78/130 (20754); CO 173 (4032): 2 CO 78/126 (16046): 3 CO 78/170 (9471): 4 CO 78/170 (9741); CO 78/172 (46873): 5 CO 78/172 (43737): 6 CO 78/175/8 (35029); CO 78/179/9 (567793): 7 FO 371/13358 (7554); CO 78/179/9 (567793): 8 CO 78/179/9 (56793)

9.1 ICE-FRONT WHALING

9 DO 35/9 (9557): 10 FO 371/11860 (A3519); DO 35/9 (9557): 11 FO 371/13358 (W7694)

9.2 THE DIPLOMATIC RESPONSE

12 CO 78/173 (4046); CO 78/181/5 (56862): 13 CO 78/173 (15172): 14 CO 78/173 (23314): 15 FO 371/11860 (A3520 and W 7745); CO 78/173 (15172): 16 CO 78/173 (23314): 17 FO 371/12644 (W1410 and W 2197); CO 78/175/9 (35038): 18 CO 78/175/9 (35038); CO 78/181/5 (56862): 19 ADM 116/2470 (M386); CO 78/181/5 (56862); 20 FO 371/13358 (W1223 and W6429): 21 FO 371/12645 (10619); FO 371/13358 (W1223, W6429 and W 3879); CO 78/181/5 (56862): 22 MAF 41/76 (7568A): 23 DO 35/9 (D9557): 24 DO 35/9 (D9557)

9.3 THE EROSION OF AUTHORITY

25 ADM 116/2470): 26 CO 183/3 (66755): 27 CO 78/183/3 (66755): 28 FO 371/14088 (W10718); CO 78/184/1 (66811)

### CHAPTER 10 MARKET COLLAPSE

1 CO 78/185/9 (66880): 2 CO 78/188/9 (86767); MAF 41/79: 3 CO 78/188/9 (86767): MAF 41/80: 4 MAF 41/79–80; CO 78/188/9 (86767)

10.1 RE-APPRAISAL

5 CO 78/188/9 (86767): MAF 41/79: 6 FO 371/15657 (W2303): CO 78/188/9 (86767): 7 MAF 41/79: 8 MAF 41/79: 9 CO 78/188/9 (86767); CO 78/191/2; MAF 41/80: 10 CO 78/192/1 (19202/1); CO 78/196/5 (39227/1): 11 CO 78/192/2 (19202/2); CO 78/192/8 (19203); CO 78/192/12 (19208)

## CHAPTER 11 MARKET RECOVERY

1 FO 371 16403 (W2500); CO 323/1230/1: 2 FO 371/14921 (W10033; MAF 41/81 (FG 11040A): 3 FO 371/16404 (W2500); DO 35/149/1 (6079/138); MAF 41/81 (FG 11040A): 4 FO 371/16404 (W8104); DO 35/149/1 (6079/138)

### 11.1 QUOTA RESTRICTIONS

5 FO 371/17320 (W9643): 6 FO 371/18494 (W5185); DO 35/149/1 (6079/197); MAF 41/714: 7 FO 371/19625; MAF 41/714–5: 8 FO 371/19625 (W9733); MAF 41/729: 9 FO 371/20451 (W6290); CO 78/202/6 (88035/1); DO 35/149/3 (6070/316): 10 DO 35/149/3 (6079/310): 11 FO 371/20451 (W5900, W7249, W7533 and W8277); CO 78/202/6 (88035/1); DO 35/149/3 (6079/310–316): 12 CO 78/202/6 (88035/1)

### 11.2 THE NORWEGIAN BOYCOTT

13 FO 371/20452 (W13058); FO 371/20453 (W17329): 14 FO 371/20450 (W5358): 15 FO 371/20451 (W6016, 6045, 9056 and 9188): FO 371/20452 (W10659): 16 FO 371/20451 (W9660); FO 371/20452 (W10680): 17 FO 371/20451 (W9751): 18 FO 371/20541; FO 371/20451–2: 19 FO 371/20452

### 11.3 THE JAPANESE AND GERMAN WHALING FLEETS

20 FO 371/18495: FO 371/19624: DO 35/149/1–2 (6079): 21 FO 371/19625 (W6593 and W10730); FO 371/20450) (W3174): 22 FO 371/20450 (W3195); FO 371/20452 (W10481); FO 371/20453 (W16966); CO 78/201/1 (88007): 23 FO 371/21078

# PART IV THE DISCOVERY INVESTIGATIONS

## CHAPTER 12 ESTABLISHING THE DISCOVERY INVESTIGATIONS

1 CO 78/156 (4442 and 45964); CO 78/157 (10528): 2 CO 78/160 (56044); CO 78/163 (9905): 3 CO 78/156 (45964 and 57244); CO 78/2366); CO 78/163 (37967): 4 CO 78/157 (21800); CO 78/161 (2422); CO 78/193/11 (19232); CO 78/193/12) (19232); 5 CO 78/157 (23824 and 57477); CO 78/158 (23591 and 59545); CO 78/160 (17063); CO 78/161 (6654); CO 78/162 (10340): 6 CO 78/164 (28205 and 33026); CO 78/165 (47773): 7 CO 78/159 (2878): 8 CO 78/159 (2878): 9 CO 78/159 (2873): 10 CO 78/162 (31257)

### 12.1 THE DISCOVERY COMMITTEE

11 CO 78/167 (39667): 12 CO 78/165 (13864); CO 78/167 (39667); CO 78/172 (41150 and 41237): 13 CO 78/165 (22528): 14 CO 78/165 (55169): 15 CO 78/168 (10645): 16 CO 78/165 (55169): 17 CO 78/165 (55169); CO 78/167 (12457 and 32938); CO 78/168 (11198): 18 CO 78/167 (41202); CO 78/168 (10645, 16783, 25959 and 29040): 19 CO 78/169 (59260): 20 CO 78/169 (59260); CO 78/171 (7950)

### 12.2 THE DESPATCH OF THE *DISCOVERY*

21 CO 78/172 (35624); 22 CO 168 (17166 and 25959): 23 CO 78/169 (58010): 24 CO 78/167 (58145); CO 78/172 (19724): 25 CO 78/169 (59260); CO 78/172 (19724); CO

78/175/7 (35027): 26 CO 78/171 (13981): CO 78/172 (18553 and 21859): 27 CO 78/172 (19724): 28 CO 78/169 (59260); CO 78/172 (43117)

12.3 THE INITIAL INVESTIGATIONS

29 CO 78/169 (39701): 30 CO 78/169 (57695): 31 CO 78/160 (5275 and 16357): 32 CO 78/172 (31412): 33 CO 78/193/11 (19232)

12.4 CHALLENGED BY THE GOVERNOR

34 CO 78/170 (21000): 35 CO 78/181/4 (56858); CO 78/193/15 (19246): 36 CO 78/175/14 (35057): 37 CO 78/175/14 (35057): 38 CO 78/180/8 (56816): 39 CO 78/182/10 (56930): 40 CO 78/180/8 (56816): 41 CO 78/180/8 (56816): 42 CO 78/180/8 (56816): 43 CO 78/180/8 (56816): 44 CO 78/182/10 (56930): 45 CO 78/182/10 (56930): 46 CO 78/182/10 (56930)

12.5 CHALLENGED BY THE SECRETARY OF STATE

47 CO 78/194/5 (19267); T 164/128/1: 48 T 164/128/1: 49 CO 78/193/11 (19232): 50 CO 78/193/11 (19232): 51 CO 78/191/15 (96184); CO 78/193/11 (19232): 52 CO 78/193/11: 53 CO 78/193/11 (19232); CO 78/193/12 (19232/2): 54 CO 78/193/11 (19232); CO 78/193/12 (19232/2): 55 CAB 24/245 (CP295); T 161/666: 56 CAB 23/79 (CM 17(34): 57 CAB 23/77 (CM79); T 161/666: 58 CO 78/194/5 (19267): 59 CO 78/193/10 (19232)

12.6 THE NEAR-DEMISE OF THE INVESTIGATIONS

60 FO 371/12645 (W2001; CO 78/181/16 (56901); T 161/901: 61 CO 78/208/5 (88011): 62 CO 78/208/5 (88011)

**CHAPTER 13 INTERNATIONAL REGULATION**

1 FO 371/12645 (10904): 2 CO 78/165 (36410); CO 78/167 (39667): 3 FO 371/12644 (W1555); FO 371/14087 (W4482); MAF 41/76: 4 FO 371/12644 (4155); FO 371/14987 (4481); MAF 41/76: 5 ADM 116/2470; FO 371/12644; DO 35/0 (D9557; MAF 41/76 (7568A): 6 FO 371/13358 (W7819); CO 78/173 (23314); MAF 41/353

13.1 THE INTERNATIONAL WHALING CONVENTION

7 FO 371/15659l; FO 371/12645 (W8969): 8 FO 371/14086 (W393); CO 78/175/9 (35038): 9 ADM 116/2470; CO 78/78/175/9 (35038); MAF 41/353: 10 MAF 41/353: 11 FO 371/13358; MAF 41/353: 12 FO 371/14086 (W393; CO 78/184/3 (66814); MAF 41/75: 13 CO 78/184/3 (66814); MAF 41/77: 14 FO 371/15658 and 15659; DO 35/148/1; MAF 41/76: 15 FO 371/15659 (W8989 and W9945); FO 371/16403; CO 78/193/10 (19232); DO 35/148/2

13.2 THE WHALING INDUSTRY (REGULATION) ACT

16 FO 371/14921: 17 CO 323/1260/10 (30273/1); DO 35/149/1; DO 35/151/1 (6079C); DO 35/151/2: 18 DO 35/149/1: 19 CO 323/1260/10 (30273/1): 20 FO 371/18494; MAF 41/714: 21 CO 323/1260/10; MAF 41/714: 22 CAB 23/79; CAB 26/17; CAB 24/249): 23 CAB 23/79: 24 DO 35/151/3: 25 MAF 41/714: 26 FO 371/18495; DO 35/888/8

## CHAPTER 14 THE LONDON CONFERENCES

1 CO 323/1362/5 (1973/1): 2 FO 371/22271; CO 852/138/10 (15917); MAF 41/360: 3 MAF 41/364

### 14.1 THE INTERNATIONAL WHALING AGREEMENT 1937

4 FO 371/21078; DO 35/888/9: 5 FO 371/21078 (N2861): 6 FO 371/21078) (N1364): 7 FO 371/20450–2; MAF 41/357: 8 MAF 41/356: 9 FO 94/1326: 10 MAF 41/357: 11 FO 371/21080 (N3041): 12 FO 371/21079; MAF 41/357: 13 FO 94/1326; FO 371/21079 (N4792); FO 371/21080 (N4855); DO 35/888/7; MAF 41/357; MAF 41/358; MAF 41/359

### 14.2 THE INTERNATIONAL WHALING PROTOCOL 1938

14 FO 371/22272 (N2688): 15 MAF 41/364: 16 FO 371/22272 (N2649); MAF 41/360: 17 FO 371/22273; MAF 41/363: 18 MAF 41/360: 19 CO 852/138/10; MAF 41/363; MAF 41/365: 20 CO 852/138/10 (15917); MAF 41/363; MAF 41/365: 21 CO 852/138/10 (15917): MAF 41/364–5: 22 MAF 41/364–5: 23 FO 371/22272 (N2671 and N2698); FO 371/22273 (N2941 and N3758); MAF 41/364–5: 24 FO 371/22274; MAF 41/364–365: 26 MAF 41/363; MAF 41/365

### 14.3 THE INFORMAL LONDON MEETING 1939

27 FO 371/22274; FO 371/23660: 28 FO 371/23672 (N2370); 29 FO 371/23661: 30 FO 371/23661 (N2889): 31 FO 371/23661: 32 MAF 41/369: 33 FO 371/23661; FO 371/3352; MAF 41/369

## CHAPTER 15 THE FURTHER DISRUPTION OF WAR

1 DO 35/889/2; MAF 129/62: 2 FO 371/23661 (N4088 and N4224): 3 FO 371/23662: 4 FO 371/23662: 5 FO 371/23662; FO 371/24816: 6 FO 371/23662: 7 FO 371/24816 (N2674): 8 ADM 1/16103: ADM 116/4662; CAB 66/14/38: 9 CO 852/637/11: MAF 97/110; MAF 209/19

### 15.1 READINESS FOR PEACE

10 FO 371/32810 (N5657); MAF 41/1332; MAF 85/550: 11 FO 371/32810 (N5156); MAF 41/1332: 12 FO 371/32810 (N5156 and N5657); MAF 41/1332: 13 CO 852/455/7: 14 MAF 85/550: 15 FO 371/32810 (N6102); CO 852/455/7; MAF 41/1332; MAF 85/550: 16 MAF 41/1332: 17 FO 371/32810 (N6276); CO 852/455/7; MAF 41/1332: 18 MAF 41/1332–3

### 15.2 RECOVERY AMIDST CRISIS

19 MAF 41/1333: 20 MAF 83/593; MAF 83/1667: 21 FO 337/111; MAF 83/1667: 22 MAF 41/1333; MAF 83/592: 23 MAF 41/1333; MAF 83/592

### 15.3 THE WARTIME PROTOCOLS

24 MAF 41/1332–3: 25 CO 852 638/1; MAF 41/1333: 26 MAF 41/1333: 27 CO 852/638/1; MAF 41/1333: 28 FO 372/4380); MAF 41/1333: 29 FO 371/4061; FO 372/4379 and 4380; MAF 41/1333–4: 30 MAF 83/591–2: 31 FO 372/4379–80; MAF 41/1333–4;

& Taylor, B. L. (2019). Revision of fin whale *Balaenoptera physalus* (Linnaeus, 1758) subspecies using genetics. *Journal of Mammalogy, 100*(5), 1653–1670. https://doi.org/10.1093/jmammal/gyz121

Ardley, R. A. B., & Mackintosh, N. A. (1936). The royal research Ship "discovery II". *Discovery Reports, 13*, 79–105 Plates 3–8.

Arlov, T. B. (1993). Whaling and sovereignty. The role of whaling in the struggle for supremacy over Svalbard. In B. L. Basberg, J. E. Ringstad & E. Wexelsen (Eds.), *Whaling and history* (pp. 81–90). Commander Chr. Christensen Hvalfangstad Museum.

Atkinson, A., Siegel, V., Pakhomov, E., & Rothery, P. (2004). Long-term decline in Krill stock and increase in salps within the Southern Ocean. *Nature, 432*(7013), 100–103. https://doi.org/10.1038/nature02996

Baranov, F. I. (1918). On the question of the biological basis of fisheries. (in Russian). *Byulleten Rybnogo Khozyaistva* W.E. Ricker 1945. Unpublished mimeo, *1* (pp. 81–128). Transl. Bloomington, IN.

Baranov, F. I. (1926). On the question of the dynamics of the fishing industry. (in Russian). *Byulleten Rybnogo Khozyaistva* W.E. Ricker 1945. Unpublished mimeo, *5*(8) (pp. 7–11). Transl. Bloomington, IN.

Bargmann, H. E. (1937). The reproductive system of *Euphausia superba. Discovery Reports, 14*, 325–350 Plates I–V.

Barr, W., & Watt, J. P. C. (2005). Pioneer whalers in the Ross Sea, 1923–33. *Polar Record, 41*(4), 281–304. https://doi.org/10.1017/S0032247405004638

Bartholomew, G. A. (1974). The relation of the natural history of whales to their management. In W. E. Schevill (Ed.), *The whale problem. A status report* (pp. 294–302). Harvard University Press.

Basberg, B. J. (1985). Technological transformation in the Norwegian whaling industry in the interwar period. *Scandinavian Economic History Review, 33*(2), 83–107. https://doi.org/10.1080/03585522.1985.10408044

Basberg, B. L. (1991). Whaling or shipping? Conflicts over the Use of the Norwegian Whaling Fleet during World War II. *International Journal of Maritime History, 3*(1), 165–176. https://doi.org/10.1177/084387149100300109

Basberg, B. L. (1993). Survival against all odds? Shore station whaling in South Georgia in the pelagic era, 1925–1960. In B. L. Basberg, J. E. Ringstad & E. Wexelsen (Eds.), *Whaling and History: Perspectives on the evolution of the industry* (pp. 157–167). Sandefjordmuseene.

Basberg, B. L. (1998). The floating factory: Dominant designs and technological development of twentieth-century whaling factory ships. *Northern Mariner / le Marin du Nord, 8*(1), 21–37. https://doi.org/10.25071/2561-5467.648

Basberg, B. L. (2004). *The shore whaling stations at South Georgia.* Novus Förlag.

Basberg, B. L. (2006). Two hegemonies – Two technological regimes. American and Norwegian Whaling in the 19th and 20th Century. *Institutt for Samfunneøkonomi*, Discussion Paper.

Basberg, B. L. (2008). Whaling – A controversial industrial adventure. In J. Gunnar-Winther et al. (Eds.), *Norway in the Antarctic. From conquest to modern science* (pp. 26–43). Schibsted Förlag.

Basberg, B. L. (2016). Keynes, Trouton and the hector whaling company. *Business History, 59*, 471–496.

Basberg, B. L. (2019). Chr. Christensen and C. A. Larsen: A comparative analysis of two whaling entrepreneurs. *International Journal of Maritime History, 31*, 1–2.

Basberg, B. L., & Hacquebord, L. (2023). Industrial whaling in the Arctic and Antarctic. In A. Howkins & P. Roberts (Eds.), *The polar regions* (pp. 378–406). Cambridge University Press.

Basberg, B. L., Ringstad, J. E., & Wexelsen, E. (Eds.). (1993). *Whaling and history. Perspectives on the evolution of the industry* Kommander Crh. Christensen Hvalfangstmuseum.

Beddington, J. R., Rosenberg, A. A., Crombie, J. A., & Kirkwood, G. P. (1990). Stock assessment and the provision of management advice for the short fin squid fishery in Falkland Islands waters. *Fisheries Research, 8*(4), 351–365. https://doi.org/10.1016/0165-7836(90)90004-F

Bejder, M., Johnston, D. W., Smith, J., Friedlaender, A., & Bejder, L. (2016). Embracing conservation success of recovering humpback whale populations: Evaluating the case for downlisting their conservation status in Australia. *Marine Policy, 66*, 137–141. https://doi.org/10.1016/j.marpol.2015.05.007

Bennett, A. G. (1920). On the occurrence of diatoms on the skin of whales. *Proceedings of the Royal Society of London Series A, 91*, 352–357.

Bennett, A. G. (1931). *Whaling in the Antarctic.* Blackwood.

von Bertalanffy, L. (1938). A quantitative theory of organic growth. *Human Biology, 10*, 181–213.

Bertram, A. (1930). *The colonial service.* Cambridge University Press.

Beverton, R. J. H., & Holt, S. J. (1957). On the dynamics of exploited fish populations. Fishery Investigations Series II. HMSO, *19*.

Board of Agriculture and Fisheries. (1919). *British fisheries: Reconstruction.* HMSO.

Boardman, R. (1981). *International organization and the conservation of nature.* Macmillan.

Boldingh, J. (1969). Research. In J. H. van Stuyvenberg (Ed.), *Margarine* (pp. 167–225). Liverpool University Press.

Bonner, W. N. (1980). *Whales.* Blandford Press.

Boothe, J. N. (2011). *The storied ice. Exploration, discovery, and adventure in Antarctica's peninsula region.* Regent Press.

Borley, J. O. (1938). Antarctic whaling. *Polar Record,* 2(15), 62–64. https://doi.org/10.1017/S0032247400036603

Boyd, P. W., Jickells, T., Law, C. S., Blain, S., Boyle, E. A., Buesseler, K. O., Coale, K. H., Cullen, J. J., de Baar, H. J. W., Follows, M., Harvey, M., Lancelot, C., Levasseur, M., Owens, N. J. P., Pollard, R., Rivkin, R. B., Sarmiento, J., Schoemann, V., Smetacek, V., . . . Watson, A. J. (2007). Mesoscale iron enrichment experiments 1993–2005: Synthesis and future directions. *Science, 325*, 612–617.

Boyson, V. F. (1924). *The Falkland Islands.* Clarendon Press.

British association. (1928). *Report of the British Association for the Advancement of Science's* 96th Meeting at Glasgow (pp. 559–560).

Brown, R. N. Rudmose. (1927) *The polar regions.* Methuen.

Brown, S. G. (1962). The movements of fin and blue whales within the Antarctic zone. *Discovery Reports, 33*, 1–11.

Brown, S. G. (1963). A review of Antarctic whaling. *Polar Record, 11*(74), 555–566. https://doi.org/10.1017/S0032247400055650

Brown, S. G., & Lockyer, C. H. (1984). Whales. In *Laws*, R.M. (ed.) *Antarctic Ecology*. Academic Press, 717–781.

Bruce, W. S. (1894). A Few Notes on Seals and Whales seen during Antarctic Voyage, 1892–93. *Report of the British Association for the Advancement of Science, 63*.

Bruce, W. S. (1915). Some Observations on Antarctic cetaceans. *Report on the scientific results of the voyage of the S.Y. Scotia. Scottish National Antarctic Expedition (Zoology), 4*, 491–505, pls. 2.
Budker, P. (1958). Whales and whaling. *Muséum d'histoire naturelle, Paris, France.*
Bull, H. J. (1896). *Cruise of the Antarctic.* Arnold.
Burnett, D. G. (2012). *The sounding of the whale.* Chicago University Press.
Burton, R. (1973). *The life and death of whales.* Andre Deutsch.
Bush, W. M. (1982). *Antarctic and international law.* Oceana Publications.
Calderan, S. V., Black, A., Branch, T. A., Collins, M. A., Kelly, N., Leaper, R., Lurcock, S., Miller, B. S., Moore, M., Olson, P. A., Širović, A., Wood, A. G., & Jackson, J. A. (2020). South Georgia blue whales five decades after the end of whaling. *Endangered Species Research, 43*, 359–373. https://doi.org/10.3354/esr01077
Calman, W. T. (1947). *Stanley Wells Kemp, 5* (pp. 467–476). Obituary Notices of Fellows of the Royal Society.
Calman, W. T. (1951). *Sidney Frederick Harmer, 7* (pp. 358–371). Obituary Notices of Fellows of the Royal Society.
Cartwright, T. J. (1975). *Royal commissions and departmental committees in Britain.* Hodder & Stoughton.
Charnock, H. (1985). George Edward Raven Deacon. *Biographical Memoirs of Fellows of the Royal Society, 31*, 113–142.
Chatterton, E. K. (1925). *Whalers and whaling.* Fisher Book Company and Unwin.
Christensen, L. (1935). *Such is the Antarctic.* Hodder & Stoughton.
Christie, E. H. (1951). *The Antarctic problem.* Allen & Unwin.
Clarke, M. R. (1962). Significance of cephalopod beaks. *Nature, 193*(4815), 560–561. https://doi.org/10.1038/193560a0
Clarke, M. R. (1980). Cephalopoda in the diet of sperm whales of the southern hemisphere and their bearing on sperm whale biology. *Discovery Reports, 37*, 1–324.
Clarke, M. R. (1983). *Cephalopod biomass: Estimation from predation, 44* (pp. 95–107). Memoirs of the National Museum of Victoria.
Clarke, R. (1954). Open boat whaling in the Azores: The history and present methods of a relic industry, *26*, 281–354 Plate XIIIXVIII.
Clarke, R. (1956). Sperm whales of the Azores. *Discovery Reports, 28*, 237–298 Plates I–II.
Clowes, A. J. (1938). Phosphate and silicate in the Southern Ocean, *Discovery Reports, 19*, 1–120 Plates I–XXV.
Cohen, J. E. (1970). Review of "an introduction to mathematical ecology". *American Scientist, 58*, 699 (Pielou, E. C. 1969. *286 p.p.* Wiley Interscience).
Cohen, T. (1987). *Remaking Japan. The American occupation as New Deal.* Free Press.
Coleman-Cooke, J. (1963). *Discovery II in the Antarctic.* Odhams Press.
Constantine, S. (1984). *The making of British colonial development policy 1914–1940.* Frank Cass Publishers.
Crawford, W., & Broadley, H. (1938). *The people's food.* Heinemann.
Cressey, D. (2015). World's whaling slaughter tallied. *Nature, 519*(7542), 140–141. https://doi.org/10.1038/519140a
Cross, J. A. (1982). *Lord Swinton.* Clarendon Press.
Cubaynes, H. C., & Fretwell, P. T. (2022). Whales from space dataset, an annotated satellite image dataset of whales for training machine learning models. *Scientific Data, 9*(1), article no.: 245. https://doi.org/10.1038/s41597-022-01377-4

Cushman, G. T. (2013). *Guano and the opening of the Pacific World.* Cambridge University Press.

Dakin, W. J. (1934). *Whalemen adventures.* Angus and Robertson.

Darnley, R. (1923). A new Antarctic expedition. *Nineteenth Century and After, 93*, 718–727.

Davis, L. E., & Gallman, R. E. (1993). American whaling 1820–1900. In B. L. Basberg, J. E. Ringstad & E. Wexelsen (Eds.), *Whaling and History: Perspectives on the evolution of the industry* (pp. 55–65). Sandefjordmuseene.

Daw, T., & Gray, T. (2005). Fisheries science and sustainability in international policy: A study of failure in the European Union's Common Fisheries Policy. *Marine Policy, 29*(3), 189–197. https://doi.org/10.1016/j.marpol.2004.03.003

Day, D. (2012). *Antarctica: A biography*. Oxford University Press.

Deacon, G. E. R. (1937). The hydrology of the Southern Ocean, *Discovery Reports, 15*, 1–124.

Deacon, M. (2010). Marine science in the United Kingdom before World War II. In A. S. Laughton (Ed.), *Of seas and ships and scientists* (pp. 15–23). Lutterworth Press.

de Baar, H. J. W., Boyd, P. W., Coale, K. H., Landry, M. R., Tsuda, A., Assmy, P., Bakker, D. C. E., Bozec, Y., Barber, R. T., Brzezinski, M. A., Buesseler, K. O., Boyé, M., Croot, P. L., Gervais, F., Gorbunov, M. Y., Harrison, P. J., Hiscock, W. T., Laan, P., Lancelot, C., . . . Wong, C. (2005). Synthesis of iron fertilization experiments: From the Iron Age in the Age of Enlightenment. *Journal of Geophysical Research: Oceans, 110*(C9), Suppl. 16. https://doi.org/10.1029/2004JC002601

de Baar, H. J. W., Buma, A. G. J., Nolting, R. F., Cadée, G. C., Jaques, G. & Tréguer, P. J. (1990). On iron limitation of the Southern Ocean: experimental observations in the Weddell and Scotia Seas. *Marine Ecology Progress Series, 65*, 105 122.

Delury, D. B. (1947). On the estimation of biological populations. *Biometrics, 3*(4), 145–167. https://doi.org/10.2307/3001390

Development Commission. (1920). *Advisory committee on fishery research.* HMSO.

Dibbern, J. S. (2010). Fur seals, whales and tourists: A commercial history of Deception Island, Antarctica. *Polar Record, 46*(3), 210–221. https://doi.org/10.1017/S0032247409008651

Dickinson, A. B. (1986). The first South Georgia population census. *Polar Record, 23*(144), 349–351. https://doi.org/10.1017/S0032247400007191

Dickinson, A. B. (2007). *Seal Fisheries of the Falkland Islands and Dependencies: An historical overview*, illustrated, soft cover. International Maritime Economic History Association (Research in Maritime History 34). ISBN 978-0-9738934-4-1.

Dornan, T., Fielding, S., Saunders, R. A., & Genner, M. A. (2022). Large mesopelagic fish biomass in the Southern Ocean resolved by acoustic properties. *Proceedings of the Royal Society of London, Series B.* https://doi.org/10.1098/rspb.2021.1781

Dorsey, K. (1998). *The dawn of conservation diplomacy. U.S.-Canadian wildlife protection treaties in the progressive era.* Washington University Press.

Dorsey, K. (2013). *Whales and nations.* Washington University Press.

Doughty, C. E., Roman, J., Faurby, S., Wolf, A., Haque, A., Bakker, E. S., Malhi, Y., Dunning, Jr., J. B., & Svenning, J. C. (2016). Global nutrient transport in a world of giants. *Proceedings of the National Academy of Sciences of the United States of America, 113*(4), 868–873. https://doi.org/10.1073/pnas.1502549112

Dower, J. W. (2000). *Embracing defeat. Japan in the wake of World War II.* Norton.

Dudeney, J., & Sheail, J. (2019). *Claiming the ice. Britain and the Antarctic 1900–1950.* Cambridge Scholars Press.

Earl, T., MacLeod, E., Söffker, M., Gasco, N., Massiot-Granier, F., Tixier, P., Darby, C. (2021). Whale depredation in the South Georgia toothfish (*Dissostichus eleginoides*) fishery in the South Atlantic: a comparison of estimation methods. *ICES Journal of Marine Science, 78,* 3817–3833. https://doi.org/10.1093/icesjms/fsab212

Eiji, T. (2003). *The Allied occupation of Japan.* Continuum.

Elliot, G. (1998). *A whaling enterprise: Salvesen in the Antarctic.* Michael Russell Publishers.

Ellis, R. (1992). *Men and whales.* Robert Hale.

Everson, I. (1977). *The living resources of the Southern Ocean. Southern Ocean fisheries survey programme, GLO/SO/77/1.* Food and Agriculture Organization.

Fayle, C. E. (1923). History of the Great War based on official documents. *Seaborne trade, II.* John Murray.

Fayle, C. E. (1924). History of the Great War based on official documents. *The period of unrestricted submarine warfare, III.* John Murray.

Ferguson, N. (2003). *Empire. How Britain made the modern world.* Penguin.

Feron, R. (1969). Technology and production. In *van Stuyvenberg,* J.H. *Margarine* (pp. 83–121). Liverpool University Press.

Fieldhouse, D.K. (1978) *Unilever overseas.* Croom-Helm.

Fitzmaurice, M. (2015). *Whaling and international law.* Cambridge University Press.

Flower, W. H. (1864). Notes on the skeletons of whales in the principal museums of Holland and Belgium, with descriptions of two species apparently new to science. *Proceedings of the Zoological Society of London, 1864,* 384–420.

Flower, W. H. (1898). Whales and British and Colonial whale fisheries, *Essays on museum and other subjects connected with natural history.* Macmillan.

Foxton, P. (1956). The distribution of the standing crop of zooplankton in the Southern Ocean. *Discovery Reports, 28,* 191–236.

Foxton, P. (1961). *Salpa fusiformis* Cuvier and related species. *Discovery Reports, 32,* 1–32 Plates I–II.

Foxton, P. (1966). The distribution and life-history of *Salpa Thompsoni* Foxton with observations on a related species, *Salpa Gerlachei* Foxton. *Discovery Reports* 34, 1–116.

Fraser, F. C. (1936). On the development and distribution of the young stages of krill (*Euphausia superba*). *Discovery Reports, 14,* 1–192. https://doi.org/10.5962/bhl.part.3886

Fraser, F. C. (1945). On a specimen of the southern bottlenosed whale, *Hyperoodon planifrons. Discovery Reports, 23,* 19–36.

Gabriele, C. M., Lockyer, C., Straley, J. M., Jurasz, C. M., & Kato, H. (2009). Sighting history of a naturally marked humpback whale (Megaptera novaeangliae) suggests ear plug growth layer groups are deposited annually. *Marine Mammal Science, 26*(2), 443–450. https://doi.org/10.1111/j.1748-7692.2009.00341.x

Gambell, R. (1968). Seasonal cycles and reproduction in sei whales of the Southern Hemisphere. *Discovery Reports, 35,* 31–134.

Gambell, R. (1972). Sperm whales off Durban. *Discovery Reports, 35,* 199–358.

Gambell, R. (1993). International management of whales and whaling. *Arctic, 48,* 97–197.

Garstang, W. (1900). The impoverishment of the sea. A Critical Summary of the Experimental and Statistical Evidence bearing upon the Alleged Depletion of the Trawling Grounds. *Journal of the Marine Biological Association of the United Kingdom, 6*(1), 1–69. https://doi.org/10.1017/S0025315400072374

Geremia, C., Merkle, J. A., Eacker, D. R., Wallen, R. L., White, P. J., Hebblewhite, M., & Kauffman, M. J. (2019). Migrating bison engineer the green wave. *Proceedings of the National Academy of Sciences of the United States of America, 116*(51), 25707–25713. https://doi.org/10.1073/pnas.1913783116

Gould, R. T. (1928). *Oddities. A book of unexplained facts.* Philip Allan Publishers.

Graham, M. (1935). Modern theory of exploiting a fishery, and application to North Sea trawling. *ICES Journal of Marine Science, 10*(3), 264–274. https://doi.org/10.1093/icesjms/10.3.264

Graham, M. (1954). Dr. E. S. Russell, O.B.E. *Nature, 174*(4432), 675–676. https://doi.org/10.1038/174675a0

Graham, M. (1956). Harvests of the seas. In W. L. Thomas (Ed.), *Man's Role in Changing the Face of the Earth* (pp. 487–503). University Press.

Gran, H. H. (1931). On the conditions for production of plankton in the sea. *Rapports et Proces verbaux des Réunions, Conseil International pour l'exploration de la Mer, 75*, 37–46.

Hambro, C. J. (1940). *I saw it happen in Norway.* Hodder & Stoughton.

Hamilton, J. E. (1914). Report on the Belmullet whaling station. *Report of the British Association for the Advancement of Science, 1914.*

Hamilton, J. E. (1941). A rare porpoise of the South Atlantic, *Phocaena dioptrica* (Lahille, 1912). *Discovery Reports, 21*, 227–234 Plates V–VII.

Hamilton, J. E. (1948). Effect of present-day whaling on the stock of whales. *Nature, 161*(4102), 913–914. https://doi.org/10.1038/161913a0

Hardin, G. (1968). The tragedy of the Commons. The population problem has no technical solution; it requires a fundamental extension in morality. *Science, 162*(3859), 1243–1248. https://doi.org/10.1126/science.162.3859.1243

Hardy, A. C. (1928). The work of the Royal Research Ship *Discovery* in the Dependencies of the Falkland Islands. *Geographical Journal, 72*, 209–234.

Hardy, A. C. (1936). The continuous plankton recorder. *Discovery Reports, 11*, 457–510.

Hardy, A. C. (1936). Observations on the uneven distribution of oceanic plankton. *Discovery Reports, 11*, 511–538.

Hardy, A. C. (1943). Mr. A. C. Gardiner. *Nature, 152 (3861),* 498–499.

Hardy, A. C. (1946). Stanley Wells Kemp 1882–1945. *Journal of the Marine Biological Association of the United Kingdom, 26,* 219–234. https://doi.org/10.1017/S0025315400012108

Hardy, A. C. (1950). *Johan Hjort 1869–1948, 7* (pp. 167–181). Obituary Notices of Fellows of the Royal Society.

Hardy, A. C. (1967). *Great waters.* Collins.

Hardy, A. C. (2004). Stanley Wells Kemp. In *New Oxford Dictionary of National Biography* (pp. 179–180). Oxford University Press.

Hardy, A. C., & Gunther, E. R. (1935). The plankton of the South Georgia whaling grounds and adjacent waters. *Discovery Reports, 11*, 1–456.

Harmer, S. F. (1918). The South Georgia whale fishery. *Nature, 102*(2552), 65–65. https://doi.org/10.1038/102065a0

Harmer, S. F. (1920). The scientific development of the Falkland Islands Dependencies. *Geographical Journal, 56*(1), 61–65. https://doi.org/10.2307/1781297

Harmer, S. F. (1928). Presidential Address, 24th May, 1928. *Proceedings of the Linnean Society of London. Proceedings of the Linnean Society, 140*(1), 51–95. https://doi.org/10.1111/j.1095-8312.1928.tb00080.x

Huxley, T. H. (1885). Inaugural address. The fisheries exhibition, London, 1883. *The Fisheries Literature*. William Clowes, *4*(1). http://aleph0.clarku.edu/huxley/SM5/fish.html

Hyam, R. (1979). The Colonial Office mind 1900–1914. *Journal of Imperial and Commonwealth History*, *8*(1), 30–55. https://doi.org/10.1080/03086537908582545

Isachsen, G. (1929). Modern Norwegian whaling in the Antarctic. *Geographical Review*, *19*(3), 387–403. https://doi.org/10.2307/209147

Jackson, G. (1993). Why did the British not catch rorquals in the nineteenth century? In B. L. Basberg, J. E. Ringstad & E. Wexelsen (Eds.), *Whaling and history. Perspectives on the evolution of the industry*. Kommander Crh (pp. 111–119). $check with JS. Christensen Hvalfangstmuseum.

Jackson, G. (2004). Salve Christian Frederik Salvesen. In *New Oxford Dictionary of National Biography*, *48*, 780–781.

Jackson, G. (2005). *The British whaling trade*. International Maritime Economic History Association.

Jackson, J. A., Kennedy, A., Moore, M., Andriolo, A., Bamford, C. C. G., Calderan, S., Cheeseman, T., Gittins, G., Groch, K., Kelly, N., Leaper, R., Leslie, M. S., Lurcock, S., Miller, B., Richardson, J., Rowntree, V., Smith, P., Stepien, E., Stowasser, G., . . . Carroll, E. L. (2020). Have whales returned to a historical hotspot of industrial whaling? The pattern of southern right whale *Eubalaena australis* recovery at South Georgia. *Endangered Species Research*, *43*, 323–339. https://doi.org/10.3354/esr01072

Jeffries, C. (1956). *The colonial Office*. Allen & Unwin.

Jenkins, J. T. (1921). *A history of the whale fisheries*. Witherby.

Jenkins, J. T. (1923). Modern whaling. In *Bullen, F.T., The cruise of the* Cachelot *round the world after sperm whales* p. 182. John Murray.

Jenkins, J. T. (1927). *The herring and the herring fisheries*. P. S. King.

Jenkins, J. T. (1932). *Whales and modern whaling*. Witherby.

Jenkins, J. T., & Jenkins, J. T. (1933). Whales and modern whaling. *Geographical Journal*, *81*(2), 182–183. https://doi.org/10.2307/1784597

Jennings, C. (2011). Unexploited assets. In. In B. M. Bennett & J. M. Hodges (Eds.), *Science and empire* (pp. 253–274). Palgrave Macmillan.

Johnsen, A. O. (1959). *Den modern hvalfangsts historie: Opprinnelse og utvikling*, *1*. Norges Hvalfangstforbund.

Johnson, L. (2015). *Shetland's whaling tradition. From Willafjord to Enderby Land*. Shetland Ex-Whalers Association.

Jones, A. G. E. (1974). Protecting the whaling fleet during World War II. *British Antarctic Survey Bulletin*, *38*, 37–42.

Jones, R. T. (2022). *Red leviathan. The secret history of Soviet whaling*. Chicago University Press.

Judd, D., & Slinn, P. (1982). *The evolution of the modern Commonwealth*. Macmillan.

Keilhau, W. (1930). Norway and the world war. In J. T. Shotwell (Ed.), *Sweden, Norway, Denmark and Iceland in the world war*. Yale University Press.

Kemp, L., Xu, C., Depledge, J., & Lenton, T. M. (2022). Climate endgame: exploring catastrophic climate change scenarios. *Proceedings of the National Academy of Science*, *119*, e2108146119. https://doi.org/10.1073/pnas.2108146119

Kemp, S. (1926). The discovery expedition. *Nature*, *118*(2974), 628–632. https://doi.org/10.1038/118628a0

Kemp, S. (1928). The discovery expedition. *Nature, 121*(3055), 795–799. https://doi.org/10.1038/121795a0

Kemp, S. (1929). Progress of the discovery investigations. *Nature, 124*(3126), 483–486. https://doi.org/10.1038/124483a0

Kemp, S. (1931). The voyage of the RRS Discovery II. *Geographical Journal, 89*, 168–185.

Kemp, S., & Bennett, A. G. (1932). On the distribution and movements of whales on the South Georgia and South Shetland whaling grounds. *Discovery Reports, 6*, 165–190 Plates VII–XLII.

Kemp, S., Hardy, A. H., & Mackintosh, N. A. (1929). Objects, equipment and methods. *Discovery Reports, 1*, 143–232 Plates VII–XVIII.

Larsen, C. A. (1894). The voyage of the *Jason* to the Antarctic regions. *Geographical Journal, 6*, 333–344.

Larsen, D. (2021). *Plotting for peace. American peacemaker, British codebreakers, and Britain at war 1914–1917.* Cambridge University Press.

Laurie, A. H. (1933). Some aspects of respiration in blue and fin whales. *Discovery Reports, 7*, 365–406 Plate XV.

Laurie, A. H. (1937). The age of female blue whales and the effect of whaling on the stock. *Discovery Reports, 15*, 223–284.

Lavery, T., Roudnew, B., Gill, P., Seymour, J., Seuront, L., Johnson, J., Mitchell, J. G. & Smetacek, V. (2010). Iron defecation by sperm whales stimulates carbon export in the Southern Ocean. *Proceedings of the Royal Society B, 277*, 3527–3531. https://doi.org/10.1098/rspb.2010.0863

Laws, R. M. (1959). The foetal growth rate of whales with special reference to the fin whale *Baleanoptera physalis. Discovery Reports, 29*, 282–308.

Laws, R. M. (1961). Reproduction, growth and age of southern fin whales. *Discovery Reports, 32*, 327–486.

Laws, R. M. (1977). Seals and whales of the Southern Ocean. *Philosophical Transactions of the Royal Society of London, 279*, 81–96.

Laws, R. M. (2017). Large animals and wide horizons. *Janus, Cambridge, UK.*

Lee, A. J. (1992). *The Directorate of Fisheries Research: Its origins and development.* MAFF.

Lillie, D. G. (1910). Observations on the Anatomy and General Biology of some Members of the Larger Cetacea. *Proceedings of the Zoological Society of London, 80*(3), 769–792. https://doi.org/10.1111/j.1096-3642.1910.tb01916.x

Lillie, D. G. (1915). Cetacea. *British Antarctic* ('Terra Nova') Expedition. 1910, Zool. i. Natural History, *3*, p. 85. British Museum.

Liversidge, D. (1963). *The whale killers.* Jarrolds.

Loeb, V., Siegel, V., Holm-Hansen, O., Hewitt, R., Fraser, W., Trivelpiece, W., & Trivelpiece, S. (1997). Effects of sea-ice extent and krill or salp dominance on the Antarctic food web. *Nature, 387*(6636), 897–900. https://doi.org/10.1038/43174

Lowe, R. (1997). Plumbing new depths: Contemporary historians and the Public Record Office. *Twentieth Century British History, 8*(2), 239–265. https://doi.org/10.1093/tcbh/8.2.239

Mackintosh, N. A. (1934). Distribution of the macroplankton in the Atlantic sector of the Antarctic. *Discovery Reports*, 67–160.

Mackintosh, N. A. (1937). The seasonal circulation of the Antarctic macroplankton. *Discovery Reports, 37*, 365–412.

Mackintosh, N. A. (1942). The southern stocks of whalebone whales. *Discovery Reports, 22*, 197–300.

Mackintosh, N. A. (1944). Movements of the Discovery Committee's Ships in the Falkland Islands Dependencies, 1926–38. *Polar Record, 4*(27), 141–142. https://doi.org/10.1017/S0032247400041632

Mackintosh, N. A. (1945). Dr. Stanley W. Kemp, F.R.S. *Nature, 156*(3950), 41–42. https://doi.org/10.1038/156041a0

Mackintosh, N. A. (1946). The natural history of whalebone whales. *Biological Reviews, 21*(2), 60–74. https://doi.org/10.1111/j.1469-185X.1946.tb00453.x

Mackintosh, N. A. (1965). *The stocks of whales*. Fishing News Books.

Mackintosh, N. A. (1966). The distribution of southern blue and fin whales. In K. S. Norris (Ed.), *Whales dolphins and porpoises*. University of California Press.

Mackintosh, N. A. (1972). Life cycle of Antarctic krill in relation to ice and water conditions. *Discovery Reports, 36*, 1–94.

Mackintosh, N. A. (1973). Distribution of post-larval krill in the Antarctic. *Discovery Reports, 36*, 95–156 Plates I – XVII.

Mackintosh, N. A. (1974). Sizes of krill eaten by whales in the Antarctic. *Discovery Reports, 36*, 157–178.

Mackintosh, N. A., & Brown, S. G. (1956). Preliminary estimates of the southern populations of the larger baleen whales. *Norsk Hvalfangsttid, 45*, 469–480.

Mackintosh, N. A., & Wheeler, J. F. G. (1929). Southern blue and fin whales. *Discovery Reports, 1*, 257–540.

MacLeod, R. (2018). The Great War and the role of natural resources. In R. P. Tucker (Ed.,), *Environmental histories of the First World War* p. 103. University Press.

Markowski, S. (1955). Cestodes of whales and dolphins from the Discovery Collections. *Discovery Reports, 27*, 377–395 Plates XX–XXI.

Marr, J. (1962). The natural history and geography of the Antarctic krill (*Euphausia superba* Dana). *Discovery Reports, 32*, 33–464 Plate III.

Marsden, R. (2001). Expedition to investigation. The work of the Discovery Committee. In M. Deacon, T. Rice & C. Summerhayes (Eds.), *Understanding the oceans. A century of ocean exploration*. Routledge.

Marsden, R. (2002). Investigations of the Humboldt Current: The voyage of the *William Scoresby*. In K. R. Benson & P. E. Rohbeck (Eds.), *Oceanographic history. The Pacific and beyond* (pp. 186–195). University of Washington Press.

Marsden, R. (2005). Discovery Investigations: An early attempt at ecologically sustainable development? *Archives of Natural History, 32*(2), 161–176. https://doi.org/10.3366/anh.2005.32.2.161

Martin, J. H., Gordon, R. M., & Fitzwater, S. E. (1990). Iron in Antarctic waters. *Nature, 345*(6271), 156–158. https://doi.org/10.1038/345156a0

Matthews, L. H. (1931). *South Georgia. The British Empire's sub-Antarctic outpost*. Simkin Marshall.

Matthews, L. H. (1932). Lobster krill: Anomuran crustacea that are the food of whales. *Discovery Reports, 5*, 469–484.

Matthews, L. H. (1938a). The humpback whale, *Megaptera nodosa*. *Discovery Reports, 17*, 9–92 Plate II.

Matthews, L. H. (1938b). The sperm whale, *Physeter catodon. Discovery Reports, 17*, 95–168 Plates III–XI.

Matthews, L. H. (1938c). Notes on the southern right whale, *Eubalaena australis. Discovery Reports, 17*, 169–182 Plates XII–XVII.

Matthews, L. H. (1938d). The sei whale, *Balaenoptera borealis. Discovery Reports, 17*, 185–290 Plates XVIII–XIX.

Matthews, L. H. (1978). *The natural history of the whale.* Weidenfeld & Nicolson.

Maurice, H. G. (1927). The fisheries department. In F. L. C. Floud (Ed.), *The Ministry of Agriculture and fisheries* (pp. 273–280). Putnam.

Maurice, H. G. (1939). John Oliver Borley. *Journal of Marine Sciences, 14*, 5–6.

May, R. M. (1978). Whaling: past, present and future. *Nature, 276*, 319–322.

McEvoy, A. F. (1990). *The fisherman's problem. Ecology and law in the California fisheries 1850–1980.* Cambridge University Press.

McHugh, J. L. (1974). The role and history of the International Whaling Commission. In W. B. E. Schevill (Ed.), *The whale problem* (pp. 307–335). Harvard University Press.

McKercher, B. J. C. (2006). The foreign office 1930–39: Strategy, permanent interests and National security. In G. Johnson (Ed.), *The Foreign Office and British diplomacy* (pp. 87–109). Routledge.

McKercher, B. J. C. (2013). Strategy and foreign policy. In C. Baxter, M. L. Dockrill & K. Hamilton (Eds.), *Britain in global politics, 1* (pp. 153–174). Palgrave.

Mill, H. R. (1951). *An autobiography.* Longmans.

Mills, E. L. (1989). *Biological oceanography: An early history.* Cornell University Press.

Ministry of Agriculture and fisheries. (1925). *Report on sea fisheries for the years 1919–1923.* HMSO.

Ministry of Munitions. (1921). *The control of materials. Part IV. Materials for explosives manufacture.* HMSO.

Mørch, J. A. (1908). Improvements in whaling methods. *Scientific American, 99*(5), 75. https://doi.org/10.1038/scientificamerican08011908-75

Mørch, J. A. (1909). Manufacture of whale products. *Scientific American, 99*, 15–16.

Mørch, J. A. (1911). On the natural history of whalebone whales. *Proceedings of the Zoological Society of London, 47*, 661–676.

Mossman, R. G. (1916). The physical conditions of the Weddell Sea. *Geographical Journal, 48*(6), 479–500. https://doi.org/10.2307/1779818

Murdoch, W. G. Burn (1894). *From Edinburgh to the Antarctic.* Longmans.

Murphy, E. J., Watkins, J. L., Trathan, P. N., Reid, K., Meredith, M. P., Thorpe, S. E., Johnston, N. M., Clarke, A., Tarling, G. A., Collins, M. A., Forcada, J., Shreeve, R. S., Atkinson, A., Korb, R., Whitehouse, M. J., Ward, P., Rodhouse, P. G., Enderlein, P., Hirst, A. G., . . . Fleming, A. H. (2007). Spatial and temporal operation of the Scotia Sea ecosystem: A review of large-scale links in a krill centred food web. *Philosophical Transactions of the Royal Society of London. Series B, Biological Sciences, 362*(1477), 113–148. https://doi.org/10.1098/rstb.2006.1957

Murray, J., & Hjort, J. (1912). *The depths of the ocean.* Macmillan.

Niemi, E. (1993). Modern whaling on the north Arctic coast. In B. L. Basberg, J. E. Ringstad & E. Wexelsen (Eds.), *Whaling and History: Perspectives on the evolution of the industry* (pp. 67–79). Sandefjordmuseene.

Nordenskjöld, O., & Andersson, G. (1905). *Antarctica or two years amongst the ice of the South Pole.* Hurst and Blackett.

Olson, R. (2009). *Don't be such a scientist. Talking substance in an age of style.* Island Press.

Ommanney, F. D. (1932a). The vascular networks (retia mirabilia) of the fin whale (*Balaenoptera physalus*). *Discovery Reports*, *5*, 329–363.

Ommanney, F. D. (1932b). The Uro-Genital system of the fin whale (*Balaenoptera physalus*). *Discovery Reports*, *5*, 365–466 Plates II–III.

Ommanney, F. D. (1938). *South latitude.* Longmans.

Ommanney, F. D. (1971). *Lost leviathan.* Hutchinson.

Pallin, L. J., Baker, C. S., Steel, D., Kellar, N. M., Robbins, J., Johnston, D. W., Nowacek, D. P., Read, A. J., & Friedlaender, A. S. (2018). High pregnancy rates in humpback whales (*Megaptera novaeangliae*) around the western Antarctic Peninsula, evidence of a rapidly growing population. *Royal Society Open Science*, 5(5), 180017. https://doi.org/10.1098/rsos.180017

Palmer, S. (1995). A G Bennett – Naturalist, whaling officer, customs officer and postmaster. *Falkland Islands journal.*

Parliamentary Papers. (1920). *Report on the interdepartmental committee on Research and Development in the dependencies of the Falkland Islands. HM Stationery Office, London, UK.*

Parliamentary Papers. (1933–4). HM Stationery Office Convention for the regulation of whaling. Cmd, 4541.

Payton, C. (2013). The Enderby family. In C. Ellmerse & C. Payton (Eds.), *London and the whaling trade. Proceedings of the a Docklands History Group Symposium* (pp. 185–210). Museum of London.

Pershing, A. J., Christensen, L. B., Record, N. R., Sherwood, G. D., & Stetson, P. B. (2010). The impact of whaling on the ocean carbon cycle: Why bigger was better. *PLOS ONE* Aug 26;5(8):e12444, 5(8), e12444. https://doi.org/10.1371/journal.pone.0012444, PubMed: 20865156, PubMed Central: PMC2928761

Pietsch, T. (2011). Between the nation and the world. In B. M. Bennett & J. M. Hodge (Eds.), *Science and empire.* Palgrave Macmillan.

Pinkerton, M. H., Décima, M., Kitchener, J. A., Takahashi, K. T., Robinson, K. V., Stewart, R., & Hosie, G. W. (2020). Zooplankton in the Southern Ocean from the continuous plankton recorder: Distributions and long-term change. *Deep Sea Research Part I*, *162*, 103303. https://doi.org/10.1016/j.dsr.2020.103303

Pitcher, T. J., & Hart, P. J. B. (1990). *Fisheries ecology.* Chapman & Hall.

Pitcher, T. J., & Pauly, D. (1998). Rebuilding ecosystems, not sustainability, as the proper goal of fishery management. In *Fish and Fisheries Series* T. J. Pitcher, D. Pauly & P. J. B. Hart (Eds.) *Reinventing Fisheries Management.* Springer, *23*. https://doi.org/10.1007/978–94–011–4433–9_24

Poupard, M., Ferrari, M., Best, P., & Glotin, H. (2022). Passive acoustic monitoring of sperm whales and anthropogenic noise using stereophonic recordings in the Mediterranean Sea, North West Pelagos Sanctuary. *Scientific Reports*, 12(1), 2007. https://doi.org/10.1038/s41598-022-05917-1

Pugh, R. B. (1959). The colonial office 1801–1925. In E. A. Benians, J. R. M. Butler, P. N. S. Mansergh & E. A. Walker (Eds.), *The Cambridge history of the British empire*, *3*. Cambridge University Press.

Purves, P. E. (1955). The wax plug in the external auditory meatus of the Mysticeti. *Discovery Reports*, *27*, 293–302 Plate XIVXVIII.

Quartermain, L. B. (1967). *South to the pole. The early history of the Ross Sea sector, Antarctica.* Oxford University Press.

Quinn, T. J. (2003). Ruminations on the development and future of population dynamics models in fisheries. *Natural Resource Modeling, 16*(4), 341–392. https://doi.org/10.1111/j.1939-7445.2003.tb00119.x

Rabassa, J., & Borla, M. L. (Eds.). (2007). *Antarctic Peninsula and Tierra Del Fuego: 100 years of Swedish-Argentine scientific cooperation at the end of the world.* Taylor & Francis.

Racovitza, E. G. (1903). Cétacés. Research voyage of S.Y. Belgica, en 1897, *1899.*

Rankin, N. (1951). *Antarctic Isle, wild life in South Georgia.* Collins.

Rayner, G. W. (1939). Preliminary results of the marking of whales by the Discovery Committee. *Nature, 144*(3659), 999–1002. https://doi.org/10.1038/144999a0

Rayner, G. W. (1940). Whale marking, progress and results to December 1939. *Discovery Reports, 14,* 245–284 Plate XLIIILXVIII.

Rayner, G. W. (1948). Whale marking II. Distribution of blue, fin and humpback whales marked from 1932 to 1938. *Discovery Reports, 25,* 31–38 Plate VXXII.

Read, F. L., Hohn, A. A., & Lockyer, C. H. (2018). A review of age estimation methods in marine mammals with special reference to monodontids. *NAMMCO Scientific Publications, 10,* 1–67. https://doi.org/10.7557/3.4474

Redfield, A. C. (1934). On the proportions of organic derivatives in seawater and their relation to the composition of plankton. In R. J. Daniel (Ed.), *James Johnstone memorial volume* (pp. 177–192). Liverpool University Press.

Reynolds, D. (2019). *Island stories.* Collins.

Richards, T. (1993). *The Imperial archive.* Verso.

Richardson, A. J., Bakun, A., Hays, G. C., & Gibbons, M. J. (2009). The jellyfish joyride: Causes, consequences and management responses to a more gelatinous future. *Trends in Ecology and Evolution, 24*(6), 312–322. https://doi.org/10.1016/j.tree.2009.01.010

Ricker, W. E. (1954). Stock and recruitment. *Journal of the Fisheries Research Board of Canada, 11*(5), 559–623. https://doi.org/10.1139/f54-039

Riffenburgh, B. (2016). *C. A. Larsen. Explorer, whaler and family man.* Fram Museum.

Riste, O. (1966). *The neutral ally. Norway's relations with belligerent powers in the First World War.* Universtetforlaget.

Roberts, B. (1947). The exploration of Antarctica. *Nature, 159*(4038), 388–392. https://doi.org/10.1038/159388a0

Roberts, B. B. (1966). Rupert de Burgh Trouton. *Polar Record, 13,* 357–358.

Roberts, B. (1975). Obituary. Neil Alison Mackintosh. *Polar Record, 17,* 422–428.

Roberts, P. (2011). *The European Antarctic.* Palgrave.

Roberts, P., & Howkins, A. (2023) [Introduction]. In A. Howkins & P. Roberts (Eds.). *The Cambridge history of the polar regions* (pp. 17–18). Cambridge University Press.

Rocha, R. C., Jr., Clapham, P. J., & Ivashchenkp, Y. V. (2014). Emptying the oceans: A summary of industrial whales catches in the 20th century. *Marine Fisheries Review, 76,* 37–48.

Rodhouse, P. G., & Clarke, M. R. (1985). Growth and distribution of young *Mesonychteuthis hamiltoni* Robson (mollusca: Cephalopoda): An Antarctic squid. *Vie et Milieu, 35,* 223–230.

Rodhouse, P. G., & Nigmatullin, Ch. M. (1996). Role as consumers. In *Philosophical Transactions of the Royal Society of London. Series B* M. R. Clarke (Ed.), *351*(1343), 1003–1022. https://doi.org/10.1098/rstb.1996.0090

White, R. (1991) *'It's your misfortune and none of my own'. A history of the American West.* University of Oklahoma Press.

Whitmore, F. C. (1975). *Remington Kellogg.* National Academy of Sciences, Biographical Memoir, 157–189.

Wilkins, N. P. (2020). *A species of delusion The Inspectors of Irish Fisheries.* Institute of Public Administration.

Williams, I. P., & Dudeney, J. (2018). *William Speirs Bruce.* Amberley Publishing.

Wilson, C. (1970). *The history of Unilever.* Cassell.

Woolton, F. J. (1959). *Memoirs of the Rt. Hon. The Earl of Woolton.* Cassell.

Woodward, E. L. (1949). British foreign policy in action. *Listener, 41*, 349–350.

Young, J. W. (1997). *Britain and the world in the twentieth century.* Arnold.

Young, O. R., & Osherenko, G. (Eds.). (1993). *Polar politics. Creating international environmental regimes.* Cornell University Press.

Yonge, C. M. (1949). *The sea shore.* Collins.

Zerbini, A. N., Adams, G., Best, J., Clapham, P. J., Jackson, J. A., & Punt, A. E. (2019). Assessing the recovery of an Antarctic predator from historical exploitation. *Royal Society Open Science, 6*(10), 190368. http://doi.org/10.1098/rsos.190368

# *Brief author biographies*

**John Sheail** (DLit London) is an historical geographer and, prior to retirement, an environmental historian with the Centre for Ecology and Hydrology (Natural Environment Research Council) and External Professor of Geography at Loughborough University. His publications include a dozen books and over 200 research papers.

**Paul Rodhouse** is a marine biologist with a background in marine living resources, fisheries oceanography and ecology. His BSc was in biological sciences at Westfield College, London University and MSc and PhD in oceanography at the University of Southampton. After receiving his PhD he worked at the Shellfish Research Laboratory, University College Galway, then in the Ecology and Evolution Department at Stony Brook University, New York. The rest of his career was spent at the British Antarctic Survey where he is now an emeritus fellow. In Galway he ran experiments at Ballynakill where Sidney Kemp started his career. Later, at BAS, he was the principal scientist responsible for commissioned research at the South Georgia government's applied fisheries research laboratory at King Edward Point, which lies next to Discovery House. His first year with BAS was spent on detached duty at the MBA, Plymouth, working on Discovery material with Malcolm Clarke, author of the final Discovery Report.

**John Dudeney** (PhD and BSc London) has over 56 years of Antarctic experience, first a physical scientist and administrator at the British Antarctic Survey, retiring as Deputy Director in 2006, then as a researcher in Antarctic political history and lecturer on Antarctic tour ships. He has travelled many times to Antarctica including two consecutive winters in the late 1960s. John has around 100 publications and was honoured to receive the Polar Medal in 1976, a second clasp in 1995 and an OBE in 2004. In 2020 a small range of Antarctic mountains were named after him – the Dudeney Nunataks. He is a fellow of the Royal Astronomical Society, Fellow of the Royal Geographical Society and a member of the American Geophysical Union.

# *Index*


**A**

Aagaard, Bjarne 177
Acheson, Andrew B. 210
Acheson, Dean 236
Addison, Christopher 99
Allardyce, William L. 23–5, 27, 31–2, 35, 38–9, 42–3, 46–7, 51–2, 53, 55–63, 78
Allen, Edgar J. 158, 165
Allen, Harold 65, 84, 90, 102, 105–6, 169
Allen, Henry 104, 155–6, 184
Allen, Kenneth Radway 3, 255
Amended Ordinance 1913 61
Amery, Leo S. 102, 105, 127–8, 130, 155, 163, 169, 174, 179
Anderson, John 52, 62, 66
Andresen, Adolph A. 32–3, 40
Antonello, Alessandro 17
Aristotle 4
Atcheson, George Jr 233–5, 238
Attlee, Clement 230
Australian whaling industry 179, 236

**B**

Baines, E.C. 211
Baird, Spencer Fullerton 4
Baldwin, Stanley 176
Balfour, John 238
Banks, Joseph 3
Baranov, Fedor 263–4
Barlas, William 91, 111, 208
Barlow, Alan 223
Barlow, John A.N. 177
Barrett-Hamilton, Gerald E.H. 72–3, 77, 78, 101, 166
Bartholomew, George A. 223
Basberg, Bjørn 7, 106, 120
Batterbee, Harry F. 25, 32, 38, 137, 189
Beale, Thomas 6
Bearpark, Arthur F. 73–4
Beattie, Alexander E. 170
Beck, William H. 198
Bennett, Arthur George 91, 93, 108–9, 263, 282
Bennett, Frederick Debell 6
Bergersen, Birger 146, 200, 202, 214, 216, 228
Bevin, Ernest 148
Bevir, Anthony 180–1
Binnie, Edward 8, 33, 53–4, 60–1, 91, 93, 95, 104, 107–8, 110, 111, 178
blue whales 11, 20, 49, 53–4, 56*f*, 57–8, 68, 82, 91–3, 111, 116, 125, 132, 141, 145, 165, 197–8, 200, 208, 216, 221, 252–3, 255–7, 279
Bogen, Peder 30
Borge, Hans 110–11, 121
Borley, John O. 100, 137, 149, 153, 169, 178, 186, 188–9, 190, 247
Brandt, Arthur H. 22
British Australian New Zealand Antarctic Research Expedition (BANZARE) 179
British Museum 43, 48, 68–9, 77, 91, 95, 99, 101, 104, 107, 171, 259, 267

Brown, Sidney 282
Bruce, William Speirs 4, 18, 28, 34, 99–100, 246
Bryde, Ingvald 30
Buckland, Frank 4–5
Bugge, Finn 133
Bull, Henrik Johan 14, 34
Bullen, Frank 6
Burnett, Graham 8, 23, 102, 194
Burn Murdoch, William 34
Burrows, Bernard A.B. 146, 148, 150

**C**

Calman, William T. 179–80
Carpenter, William B. 3
Casey, Richard 185–6
Chamberlain, Neville 180–1, 203
Chapman, Douglas 3, 265
Chapman, Sydney 186, 188
Charcot, Jean-Baptiste 17, 185
Chatterton, Keble 182
Chifley, Joseph 239
Christensen, Christen 14, 21, 55, 107
Christensen, Lars 121–2, 135, 143
Christophersen, Peder 18, 20, 26–7
Clarke, Malcolm Roy 282–3
Clay, Lucius D. 230
climate change 271, 275, 276–8
Clive, Robert 149, 195
Clowes, A.J. 160, 249, 268
Clutterbuck, Alexander 138
Colban, Erik 196, 201–2, 205
Collier, Laurence 146, 148, 194, 196, 202, 206–7, 212–13
Collins, Arthur E. 38
Condell, Claude F. 90–1, 96–7
Constantine, Stephen 23
Cook, James 3, 166
Cowell, Henry 69
Cowell, Hubert R. 38, 46, 48
Cox, Bertram 25, 28, 30, 37, 39
Craigie, Robert 193
Cronin, William 8–9
Cunliffe-Lister, Philip 174–7

**D**

Dahl, Thor 30, 40, 86
Dakin, William John 34
Dannevig, Tarald 87
D'Arcy Cooper, Francis 122, 136–7, 140
Darnley, Rowland 43–4, 46–7, 55, 58–62, 65–6, 68, 71–2, 74–7, 80, 82–3, 85–6, 88–92, 97, 99, 102–3, 105, 119, 120–2, 126–32, 134, 154–7, 159–62, 169, 172–4, 177–8, 185–6, 188, 247
Darwin, Charles 3
Davidson, C. Girard 225–6
Davies, Clement 147–8, 193
Davies, William 66
Davis, Charles T. 25
Day, David 8
Deacon, George E.R. 168, 223, 268, 283
Deacon, Margaret 163
Deception Island 42*f*, 58*f*, 61, 88, 105, 111, 138–9, 164*f*
De Gerlache, Adrien 17
Den Norske Hvalfangerforening (Norwegian Whalers Association) 87, 101, 126, 128, 134, 155, 164, 167
Department of Scientific and Industrial Research (DSIR) 100, 221
Dependencies whaling 74, 82–3, 105–16, 127–30
*see also* Falkland Islands
*Det Stille Aret* 125, 135
Dilwyn, John D. 284
*Discovery* 153–4, 154*f*, 157, 160–3, 179, 248
Discovery House Marine Station 160, 161*f*
*Discovery II* 162–3, 164*f*, 175, 176, 181, 252, 256, 258, 269
Discovery Investigations (DI) 68, 153–219, 245–78
Dixon, Charles W. 183
Dobson, Alban 201–2, 204, 206–9, 211, 213–16, 220, 224–6, 228
Dohrn, Anton 4
Dorman-Smith, Reginald H. 207
Dormer, Cecil 144, 146–9, 202
Dorsey, Kirkpatrick 8–9, 48, 193, 220

Dunn, P.H. 224, 227
Du Toit, Francois J. 193

## E

ecosystem-based fishery management 272
Edward VII 5
Elliot, Gerald 123
Elliot, Walter E. 190
Ellis, Richard 143
Esmark, Hans 11

## F

Fagan, Charles 101
Falkland Islands 6, 7–8, 23, 26–40, 41*f*, 42*f*, 63–4, 66, 74, 83–4, 86, 139
  census 48–9
  research and development 95–104
  *see also* Dependencies whaling; Discovery Investigations (DI); South Georgia
Falkland Islands Ordinances 176
Ferguson, Niall 17
Fiddes, George, V. 44, 65, 74, 90, 97, 99, 102
fin whales 5, 10, 49, 72, 79, 85*f*, 91, 101, 104, 141, 168, 192, 221, 247, 250–9, 261–2, 275
First World War 7, 22–3, 68, 86, 212, 216, 279–80
Fitzmaurice, Malgosia 47, 182
Flannery, Fortescue 153
flensing 11, 12*f*, 53, 183, 241*f*
floating factories 21, 30–3, 47, 49, 53–4, 56–9, 65, 73–4, 84, 86–8, 93–4, 101–2, 104, 108–12, 115*f*, 119, 122, 125*f*, 142, 187
Flower, William 83
Food and Agriculture Organisation (FAO) 224, 227–8
Forbes, Edward 3
Forster, Georg 3, 166
Forster, Johann 3, 166
Foxton, Peter 271
Foyn, Svend 11–12, 14, 22, 120
Fraser, Francis Charles 249, 283
Fraser, Peter 239
Freestone, Brian 130
French, Henry 206

## G

Gambell, Ray 283
Gardiner, Alan 270, 276
Garstang, Walter 172
German whaling industry 205, 207, 218, 230–2, 235
Gosse, Philip Henry 4
*Governøren* 92*f*
Graham, Michael 3
Graves, Hubert A. 236–8, 240
Green, John F.N. 98
Grindle, Gilbert E.A. 35–6, 39, 44, 46, 48, 53, 58, 63–4, 72–3, 87–8, 96, 98–9, 111, 121, 128, 130–2, 155–7, 159–60, 163, 171, 173–4
Grytviken whaling station 12*f*, 18, 19*f*, 20–4, 20*f*, 27, 29, 35, 48, 62, 90, 106, 112, 125, 139, 207–8, 221, 241–2*f*, 249, 251
Gunther, E. Rolfe 159, 283

## H

Halford, Aubrey S. 197
Hambro, Carl J. 141
Hamilton, James Erik 103, 108–10, 112–13, 120, 155, 159, 178, 208, 221, 248, 283
Harcourt, Lewis 52, 62, 74–5
Harcourt, William 59–60, 65–7, 83
Hardin, Garrett 8
Harding, Edward J. 43–4
Hardy, Alister 158–9, 162, 166–7, 178, 180, 183, 269, 276, 283–4
Harford, Frederic D. 24, 26–7
Harmer, Sidney F. 48, 69–71, 77–8, 95–6, 99, 101, 104, 108–9, 112, 140, 153, 157, 165, 169, 176, 182–3, 189, 246–7
Hart, Ian 7–8, 43, 47, 55, 106, 143–4
Hart, Thomas John 276, 284
Haslam, Jonathan 97
hauling slipway 119, 120*f*, 242*f*
Hawes, Charles Boardman 12

Heaton, Henniker 111–13
Heazle, Michael 229
Hektor whale factory 58*f*
Herdman, H.F.P. 258, 284
Heyburn, Ian 23
Hilldring, John H. 236
Hjort, Johan 4, 11, 80–3, 87, 101, 126–8, 134, 136, 166–8, 201, 205, 246, 264–5
Hodson, Arnold 111–12, 170–1, 173–4
Hohman, Elmo H. 10
Holt, Sidney 3, 264–5
Hooker, Joseph Dalton 3
Hopwood, Francis 46
Hore-Belisha, Leslie 174
Howkins, Adrian 154, 184
Hudson, Robert S. 202, 209, 213
Hull, Cordell 194
humpback whales 5, 10, 14, 52, 54, 56*f*, 71–2, 76, 78–9, 85*f*, 91, 109–10, 131, 199–202, 206, 210
  declining numbers 64, 68, 95, 109
  population trends 70*f*
Hunt, Kenneth 23
Hurst, Cecil 128, 179, 184, 185–7
Husby, Emil 28, 59
Huxley, Thomas Henry 3, 5

**I**

ice-front whaling 122–5
Inskip, Thomas 176
Institute of Oceanographic Sciences (IOS) 258
International Council for the Exploration of the Sea (ICES) 81, 205
International Emergency Food Organisation (IEFO) 227
international regulation 182–219
International Whaling Agreement 1937 194–7
International Whaling Commission (IWC) 220, 223, 226–9, 232, 240, 245–6, 257–8, 265, 271, 281
International Whaling Protocol 1938 197–201
Irgens, Johannes 37
Irvin, George 106
Irvin, John H. 43, 65
Irvin, Richard 42–3
Isachsen, Gunnar 108

**J**

Jackson, Gordon 7, 11, 14, 22–3, 135–6, 145, 149, 229
Jacobsen, Fridthjof 106–7
Jahre, Anders A. 123–4, 143
Japanese whaling industry 220, 232–40, 233–4, 280
Jason Harbour whaling station 30, 31*f*, 49, 126
Jenkins, James Travis 106, 182, 187
Jennings, Christian 178
Johnsen, Arne O. 7, 45, 86–7, 95, 145, 187, 199, 212, 216
Johnson, Charles O. 43, 78–9, 246
Johnson, Herschel V. 194, 195, 198–9, 206
Jones, George A. 139

**K**

Kellogg, A. Remington 194, 198–9, 200, 214–16, 220, 228
Kemp, Stanley W. 158–9, 163, 166, 169–72, 175–6, 178, 180, 185, 189, 222, 247–8, 261–6, 284
King Edward Point 20*f*, 62, 110, 160, 161*f*, 166, 171, 242*f*, 273, 273*f*
Kinnes, Robert 34
Kipling, Rudyard 6
Kirkpatrick, Ivone A. 202–3
Klatt, William 230–1
Kodaki, Akiro 200–2, 207, 240
Koht, Halvdan 194, 197
krill 172, 254, 259–60, 269–70, 272, 276
Krogh-Hansen, Hans 57, 59, 63, 87

**L**

Lange, Alexander 21, 31–2, 37, 42, 43–4
Larsen, Carl Anton 14, 17–18, 20–1, 24, 27, 29, 34, 55, 101, 113, 115, 279
Larsen, Lauritz E. 26–7, 31*f*, 35

Thomson, John 201
Tønnessen, Johan N. 7, 13, 45, 86–7, 95, 131, 145, 187, 199, 212, 216, 230
Tornquist, Ernesto 18, 20
Trouton, Rupert 133, 197, 211
Tyrrell, William 185

**U**

Unilever 135–7, 138, 140–1, 143–4, 147–8, 150
unions 145–8, 194, 197, 213, 217, 228
United States
  international involvement 220–40
  whaling industry 10

**V**

Villiers, Francis Hyde 26
Vogt, Benjamin 127

**W**

Wall Street Crash (1929) 134
Warner, George R. 184
Washington Conference 1946 223–7
wastage 49–51, 50*f*, 57, 196
Watt, Norman C. 145, 189
Welles, Sumner 196
Whale and Seal Oil Export and Sale Ordinance 1917 84
Whale Fishery Ordinance 36, 38, 40
whale-oil prices 133–6, 135*f*
whaling leases 24–44, 49, 55, 63–5, 74, 89–90, 105, 110–11, 125–33, 138–9
Whaling Ordinance 1908 33, 36, 38–9, 46, 56
Whaling Ordinance 1924 170, 171
Wheeler, John F.G. 160, 166, 249–52, 254, 285
White, Richard 7
Wilkins, Hubert 169
Williams, Owen 83, 95, 106
Williams, Tom 222
*William Scoresby* 162, 162*f*, 167–8, 172, 181, 248, 252, 255, 256, 258, 269
Willoughby, Francis 4
Wilson, Henry 171
Wilson, Horace 203
Wilson, James Innes 48–9, 51–2, 56, 59, 62–4, 93, 96, 108
Wilson, Samuel 122
Wingfield, Charles 137, 141, 143
Wiseman, Robert 51, 60, 74
Wohlthat, Helmuth 203
Wordie, James 153, 222, 247
Worsley, Frank 160

**Y**

Yarrell, William 4
Young, Douglas 107–8